The physiology of excitable cells

The physiology of excitable cells

SECOND EDITION

DAVID J. AIDLEY

School of Biological Sciences
University of East Anglia, Norwich

CAMBRIDGE UNIVERSITY PRESS

CAMBRIDGE
LONDON · NEW YORK · MELBOURNE

Published by the Syndics of the Cambridge University Press
The Pitt Building, Trumpington Street, Cambridge CBS 1RP
Bentley House, 200 Euston Road, London NW1 2DB
32 East 57th Street, New York, NY 10022, USA
296 Beaconsfield Parade, Middle Park, Melbourne 3206, Australia

First published 1971
Reprinted 1973, 1974, 1975, 1976
Second edition 1978

Printed in United States of America
Typeset by J. W. Arrowsmith Ltd., Bristol, England
Printed and bound by R. R. Donnelley and Sons Company, Crawfordsville, Indiana

Library of Congress Cataloguing in Publication Data

Aidley, D. J.
The physiology of excitable cells.

Includes bibliographical references and index.
1. Cell physiology.   2. Neurophysiology.
I. Title. [DNLM: 1. Cells—Physiology. 2. Neurophysiology. QH631 A288p]
QH631.A36   1978   591.8'76   77-87375

ISBN 0 521 21913 2   hard covers
ISBN 0 521 29308 1   paperback

To Jessica

ERRATA:

Aidley: Physiology of Excitable Cells. 2nd Edition

ISBN   0 521 219132   hard covers

ISBN   0 521 293081   paperback

Errata

pp 75 and 76: Figures 5.4 and 5.5. have been transposed.
               The captions are correct.

Contents

Preface   ix

1   Introduction   1

The biological material, 2   Electricity, 4   Scientific investigation, 5

2   Electrophysiological methods   7

Recording electrodes, 7   Electronic amplification, 8   The cathode ray
oscilloscope, 10   Electrical stimulation, 12

3   The resting cell membrane   13

The structure of the cell membrane, 13   Concentration cells, 16   Ionic
concentrations in the cytoplasm, 18   The Donnan equilibrium system in
muscle, 20   Active transport of ions, 23   The resting potential, 27

4   Electrical properties of the nerve axon   37

Action potentials in single axons, 37   Subthreshold potentials, 42
Impedance changes during activity, 45   The passive electrical properties
of the axon membrane, 47   The local circuit theory, 53   Saltatory
conduction in myelinated nerves, 58   Electrical stimulation parameters,
64   Compound action potentials, 69

5   The ionic theory of nervous conduction   71

The 'sodium theory' of the action potential, 72   Ionic movements during
activity, 74   Voltage clamp experiments, 77   Fixed charges and the
involvement of calcium ions, 92   Experiments on perfused giant axons,
95   The actions of drugs, 99   Gating currents, 101   Counting the
channels, 103

6   The neuromuscular junction   106

The problem of synaptic transmission, 106   The structure of the
neuromuscular junction, 108   The release of acetylcholine by the motor
nerve endings, 110   The end-plate potential, 112   The action of
acetylcholine on the end-plate membrane, 122   The quantal nature of
transmitter release, 132   Facilitation, depression and post-tetanic
potentiation, 140   Presynaptic events producing transmitter release,
142   Acetylcholinesterase, 147

7   Synapses between neurons   149

Synaptic excitation in mammalian spinal motoneurons, 149   Inhibition in
mammalian spinal motoneurons, 159   Molluscan synapses, 168
Electrically transmitting synapses, 170

8   The pharmacology of synapses   177
     The action of drugs on neuromuscular transmission, 178   Other
     cholinergic synapses, 181   Catecholamines, 187   Central inhibitory
     synapses in vertebrates, 193   Gamma-aminobutyric acid, 195   Excitatory
     amino acids, 195   5-hydroxytryptamine, 196   Adenosine triphosphate,
     196

9   The integrated activity of neurons   197
     Electrically excited and non-electrically excited responses, 197   Functional
     and anatomical regions of neurons, 198   Input–output relations of
     neurons, 202   Spontaneous activity, 204

10   The electrical properties of glial cells   205

11   The organization of muscle cells   211
     Isometric contractions, 211   Biochemical aspects of contraction, 214
     The structure of the myofibril, 219   The sliding filament theory, 235
     Biochemical events during the cross-bridge cycle, 249   The excitation-
     contraction coupling process, 250

12   The dynamics of muscular contraction   264
     Contraction against different types of load, 264   The heat production of
     muscle, 267   Analysis of the dynamics of muscular contraction, 272
     Limitations of Hill's analysis, 280   The structural basis of the dynamics of
     contraction, 283   The analysis of twitches, 292   Chemical energetics of
     contraction, 297

13   The comparative physiology of muscle   301
     Excitation processes, 301   The gradation of contraction in skeletal
     muscles, 317   The excitation–contraction coupling process, 318   The
     organization of the contractile apparatus, 319   Mechanical properties, 324

14   The electric organs of fishes   336
     The electroplaques of the electric eel, 336   The electroplaques of some
     other electric fish, 341   The functions and evolution of electric organs,
     346

15   The organization of sensory receptors   348
     The coding of sensory information, 352   The initiation of sensory nerve
     impulses, 354   Further aspects of sensory coding, 362   Transmitter–
     receiver systems, 372   Central control of receptors, 373

16   Some particular sense organs   376
     The acoustico–lateralis system of vertebrates, 376   Mammalian muscle
     spindles, 402   Chemoreception in insects, 410   Electroreceptors, 415

17   The vertebrate eye   421
     The structure of the eye, 422   The duplicity theory, 428   Visual
     pigments *in vitro*, 430   Visual pigments *in situ*, 442   The absolute
     threshold, 450   The electrical activity of the retina, 456   Dark
     adaptation, 473,   Colour vision, 474

References   477

Index   523

Preface

The physiology of nerve, muscle and sensory cells is a subject in which, as in many other fields in the biological sciences, our knowledge is very much greater now than it was thirty or so years ago. The aim of this book is to give an account of some of the experimental evidence on which our present knowledge is based. It is primarily intended for use by students of physiology, zoology or biophysics who are taking courses in the subject, but it may prove useful to those beginning research and to scientists of other disciplines who may be interested in the subject.

I am very grateful to those authors and publishers who have allowed me to reproduce here diagrams which originally appeared in their publications. The sources of each of these will be found in the reference list at the end of the book. In addition, I am especially grateful to Dr H. E. Huxley and Professor R. Miledi for lending me their original photographs that are here reproduced as plates 11.1, 11.2, fig. 11.24 and plate 6.1.

It is a pleasure to thank those of my colleagues who have read and criticized the text at various stages of its preparation. I would thus like to repeat my thanks to Professor Sir Alan Hodgkin, Dr P. L. Miller and Dr G. Shelton, who made many helpful remarks on the content of the first edition as a whole, and to Drs M. C. Brown, G. Duncan, R. T. Tregear and D. C. S. White, who did the same for particular parts of it. In preparing the second edition, I have received much help from Dr Brown, Dr Duncan, Professor E. Rojas and Dr Shelton, in their comments on some of the new material. Needless to say, the responsibility for such shortcomings as remain is entirely my own.

1 Introduction

Suppose a man has a tomato thrown at his head, and that he is able to take suitable evasive action. His reactions would involve changes in the activity of a very large number of cells in his body. First of all, the presence of a red object would be registered by the visual sensory cells in the eye, and these in turn would excite nerve cells leading into the brain via the optic nerve. A great deal of activity would then ensue in different varieties of nerve cell in the brain and, after a very short space of time, nerve impulses would pass from the brain to some of the muscles of the face and, indirectly, to muscles of the neck, legs and arms. The muscle cells there would themselves be excited by the nerve impulses reaching them, and would contract so as to move the body and so prevent the tomato having its desired effect. These movements would then result in excitation of numerous sensory endings in the muscles and joints of the body and in the organs of balance in the inner ear. The resulting impulses in sensory nerves would then cause further activity in the brain and spinal cord, possibly leading to further muscular activity.

A chain of events of this type involves the activity of a group of cell types which we can describe as 'excitable cells': a rather loose category which includes nerve cells, muscle cells, sensory cells and some others. An excitable cell, then, is a cell which readily and rapidly responds to suitable stimuli, and in which the response includes a fairly rapid electrical change at the cell membrane.

The study of excitable cells is, for a number of reasons, a fascinating one. These are the cells which are principally involved in the behavioural activities of animals: these are the cells with which we move and think. Yet just because their functioning must be examined at the cellular and subcellular levels of organization, the complexities that emerge from investigating them are not too great for adequate comprehension; it is frequently possible to pose specific questions as to their properties, and to elicit some of the answers to these questions by suitable experiments. It is perhaps for this reason that the subject has attracted some of the foremost physiologists of this century. As a consequence, the experimental evidence on which our knowledge of the physiology of excitable cells is based is often elegant, clearcut and intellectually exciting, and frequently provides an object lesson in the way a scientific investigation should be carried out. Nevertheless, there are very many investigations still to be done in this field, many questions which have yet to be answered, and undoubtedly very many which have not yet been asked. This book will not have failed in its objects if it manages to convey some of the interest of such investigations.

Most readers of this book will possess a considerable amount of information on basic ideas in the biological and physical sciences. But it may be as well in this introductory chapter to remind them, in a rather dogmatic fashion, of some of the background which is necessary for a more detailed study: to formulate, in fact, a few axioms.

THE BIOLOGICAL MATERIAL

Cells

All large organisms are divided into a number of units called cells, and every cell is the progeny of another cell. This statement constitutes the cell theory. Every cell is bounded by a cell membrane and contains a nucleus in which the genetic material is found. The main part of the living matter of the cell is a highly organized system called the cyto-plasm, which is concerned with the day-to-day activity of the cell. The cell membrane separates this highly organized system inside from the relative chaos that exists outside the cell. In order to maintain and increase its high degree of organization and in order to respond to and alter its environment, the cell requires a continual supply of energy. This energy must be ultimately derived from the environment, usually in the form of chemical energy such as can be extracted by the cell from glucose molecules. Thus, thermodynamically speaking, the cell is an open system maintained in a rather improbable steady state by the continual expenditure of energy. Its life is a continual battle against the second law of thermodynamics (which we may state without gross inac-curacy as 'things tend to get mixed up').

The cells of nervous systems are called *neurons*. Their primary function is the carriage of information in the form of changes in the electrical potential across the cell membrane, especially as unitary events known as nerve impulses or *action potentials*. The idea that the nervous system is composed of discrete cells is known as the neuron theory. This theory, which is merely a particular application of the cell theory, was developed during the nineteenth century, and is now generally accepted. The alternative proposal, that nervous systems are not divided into separate membrane-bounded entities (the reticular theory) was extremely difficult to reconcile with the observations of light microscopists, and seems to be conclusively refuted by the evidence of electron microscopy.

Animals

Every animal has a history: every animal owes its existence to the success of its ancestors in combating the rigours of life, that is to say, in surviving the rigours of natural selection. Hence every animal is adapted to its way of life, and its organs, its tissues and its cells are adapted to performing their functions efficiently. There is an enormous variety in the form and functioning of animals, connected with a similar variety in ways of living. Nevertheless, it seems that in many cases there is only a limited number of ways in which animals can solve particular physiological problems. For instance, there is only a limited number of respiratory pigments, only a limited number of designs for hearing organs, and so on. Hence it is possible for us to detect certain principles in, and make certain generalizations about, the ways in which particular physiological problems are solved in different animals.

An animal is a remarkably stable entity. It is able to survive the impact of a variety of different environments and situations, and its cells and tissues are able to survive a variety of different demands upon their capacities. The main reason for this seems to be that an animal is a complex of self-regulating (homeostatic) systems. These systems are themselves coordinated and regulated so that the physiology and behaviour of the animal form an integrated whole.

Nervous systems

A nervous system is that part of an animal which is concerned with the rapid transfer of information through the body in the form of electrical signals. The activity of a nervous system is initiated partially by the input elements – the sense organs – and partially by endogenous activity arising in certain cells of the system. The output of the system is ultimately expressed via effector organs – muscles, glands, chromatophores, etc.

Primitive nervous systems consist of scattered but usually interconnected nerve cells, forming a nerve net, as in coelenterates. Increase in the complexity of responses is associated with the aggregation of nerve cell bodies to form ganglia, and when the ganglia themselves are collected and connected together, we speak of a *central nervous system*. The *peripheral nervous system* is then mainly composed of nerve fibres originating from the central nervous system. Peripheral nerves contain afferent (sensory) neurons taking information inwards into the central nervous system and efferent (motor) neurons taking information outwards; neurons confined to the central nervous system are known as interneurons. Ganglia which remain or arise outside the central nervous system, and the nerve fibres which lead to and arise from them to innervate the animal's viscera, are frequently described as forming the *autonomic nervous system.*

One of the simplest, but possibly not one of the most primitive, modes of activity of a nervous system is the *reflex*, in which a relatively fixed output pattern is produced in response to a simple input. The stretch reflexes of mammalian limb muscles provide a well-known example (fig. 7.2). Stretching the muscle excites the endings of sensory nerve fibres attached to certain modified fibres (muscle spindles) of the muscle. Nerve impulses pass up the sensory fibres into the spinal cord where they meet motor nerve cells (the junctional regions are called *synapses*) and excite them. The nerve impulses so induced in the motor nerve fibres then pass out of the cord along peripheral nerves to the muscle, where their arrival causes the muscle to contract. Much more complicated interactions occur in the analysis of complex sensory inputs, the coordination of locomotion, the expression of the emotions and instinctive reactions, in learning and other 'higher functions'. We are as yet very far from understanding how these more complicated interactions occur.

ELECTRICITY

Matter is composed of atoms, which consist of positively charged nuclei and negatively charged electrons. Static electricity is the accumulation of electric charge in some region, produced by the separation of electrons from their atoms. Current electricity is the flow of electric charge through a conductor. Current flows between two points connected by a conductor if there is a potential difference between them, just as heat will flow from a hot body to a cooler one placed in contact with it. The unit of potential difference is the *volt*. The current, i.e. the rate of flow of charge, is measured in *amperes*, and the quantity of charge transferred is measured in *coulombs*. Thus one coulomb is transferred by a current of one ampere flowing for one second.

In many cases it is found that the current (I) through a conductor is proportional to the potential difference (V) between its ends. This is *Ohm's law*. Thus if the constant of proportionality, the *resistance* (measure in *ohms*) is R, then

$$V = I \cdot R.$$

The specific resistance of a substance is the resistance of a 1 cm cube of the substance. The resistance of a wire of constant specific resistance is proportional to its length and inversely proportional to its cross-sectional area. The reciprocal of resistance is called *conductance*. The total resistance of a number of resistances arranged in series is the sum of the individual resistances, whereas the total conductance of a number of resistances in parallel is the sum of their individual conductances.

Two plates of conducting material separated by an insulator form a capacitor. If a potential difference V is applied across the capacitor, a quantity of charge Q, proportional to the potential difference, builds up on the plates of the capacitor. Thus

$$Q = V . C$$

where C, the constant of proportionality, is the *capacitance* of the capacitor. When the voltage is changing, charge flows away from one plate and into the other, so that we can speak of the current, I, through a capacitor, given by

$$I = C . \mathrm{d}V/\mathrm{d}t$$

where $\mathrm{d}V/\mathrm{d}t$ is the rate of change of voltage with time. The capacitance of a capacitor is proportional to the area of the plates and the dielectric constant (a measure of the ease with which the molecules of a substance can be polarized) of the insulator between them, and inversely proportional to the distance between the plates. The total capacitance of capacitors in parallel is the sum of the individual capacitances, whereas the reciprocal of the total capacitance of capacitors in series is the sum of the reciprocals of their individual capacitances.

SCIENTIFIC INVESTIGATION

Science is concerned with the investigation and explanation of the phenomena of the natural world. Any particular investigation usually starts with an idea – a hypothesis – about the relations between some of the factors in the system to be studied. The hypothesis must then be tested by suitable observations or experiments. This business of testing the hypothesis is what distinguishes the scientific method from other attempts at the acquisition of knowledge, and hence it follows that a scientific hypothesis must be capable of being tested. We must therefore understand what is meant by 'testing' a hypothesis.

In mathematics and deductive logic it is frequently possible to prove, given a certain set of axioms, that a certain idea about a particular situation is true or not true. For instance, it is possible to prove absolutely conclusively that, in the system of Euclidean geometry, the angles of an equilateral triangle are all equal to one another. But this absolute proof of the truth of an idea is not possible in science. For example, consider the hypothesis 'No dinosaurs are alive today'. This statement would be generally accepted by biologists as being almost certainly true, but, of course, it is just possible that there are some dinosaurs alive which have never been seen. A few years ago the statement 'No coelacanths are alive today' would also have been accepted as almost certainly correct.

However, in many cases, it *is* possible to prove that a hypothesis is false. The hypothesis 'No coelacanths are alive today' has been proved, conclusively, to be false. If we were to find just one living dinosaur, the hypothesis 'No dinosaurs are alive today' would also have been shown to be false. It follows from this argument that *in order to test a hypothesis it is necessary to attempt to disprove it.* When a hypothesis has successfully survived a number of attempts at disproof, it seems more likely that it provides a correct description of the situation to which it applies.

If we can only test a hypothesis by attempting to disprove it, it follows that a scientific hypothesis must be formulated in such a way that it is open to disproof – so that we can think of an experiment or observation in which one of the possible results would disprove the hypothesis. Any idea which we cannot see how to disprove is not a scientific hypothesis.

2 Electrophysiological methods

Excitable cells can be and are studied by the great variety of techniques that are available for the study of cells in general. These include light and electron microscopy, X-ray diffraction measurements, experiments involving radioactive tracers, cell fractionation techniques, biochemical methods, and so on. We shall not deal with these methods here. The techniques which are particular to the study of excitable cells are those involving the measurement of rapid electrical events. Consequently, in order to understand the subject, it is necessary to have some idea of how these measurements are made.

RECORDING ELECTRODES

If we wish to record the potential difference between two points, it is necessary to position electrodes at these points and connect them to a suitable instrument for measuring voltage. It is desirable that these electrodes should not be affected by the passage of small currents through them, i.e. that they should be nonpolarizable. For many purposes fine silver wires are quite adequate. Slightly better electrodes are made from platinum wire or from silver wire that has been coated electrolytically with silver chloride. For very accurate measurements of steady potentials, calomel half-cells (mercury/mercuric chloride electrodes) may have to be used.

If the site we wish to record from is very small in size (such as occurs in extracellular recording from cells in the central nervous system), the electrode must have a very fine tip, and be insulated except at the end. Successful electrodes of this type have been made from tungsten wire which is sharpened by dipping it into a solution of sodium nitrate while current is being passed from the electrode into the solution; insulation is produced by coating all except the tip of the electrode with a suitable varnish.

The manufacture of a suitable electrode is rather more difficult if we wish to record the potential inside a cell. Apart from a few large cells such as squid giant axons, this necessitates the use of an electrode which is fine enough to penetrate the cell membrane without causing it any

7

appreciable damage. The problem was solved with the development of glass capillary microelectrodes. There are several ways of making these electrodes, but they all depend on heating a small section of a hard glass tube and then very rapidly extending it as it cools. The heated section gets thinner and cooler as the pulling proceeds, so that finally the pulling force exceeds the cohesive forces in the glass, and the tube breaks to give two micropipettes. If the machine designed to do this has been correctly adjusted, the outside diameter of the micropipette at its tip will be about 0.5 μm. The micropipette is now filled with a strong electrolyte by various means, such as boiling it in 3 M potassium chloride solution for half an hour. The connection to the recording apparatus is achieved by inserting a silver wire into the wide end of the tube (although more complicated methods are sometimes used), and the electrode is now ready for use. An electrode of this type has a very high resistance, usually in the range 5 to 40 MΩ; in fact the suitability of an electrode is usually tested by measuring its resistance, since the tip is too small to be examined satisfactorily by light microscopy.

ELECTRONIC AMPLIFICATION

The potential differences which are measured in investigations on the activity of excitable cells vary in size from just over 0.1 V down to as little as 20 μV or so. Before being measured by a recording instrument, these voltages usually have to be amplified. This is done by means of suitable electronic circuits involving thermionic valves or transistors, the details of which need not concern us. However there are three aspects of any amplifier used for electrophysiological recording purposes which we must consider: the frequency response, the noise level and the input resistance.

All amplifiers show a higher gain (the gain is the ratio of the output to the input voltages) at some frequencies than at others. A typical amplifier for use with extracellular electrodes might have a constant gain over the range 10 hertz to 50 kilohertz, the gain falling at frequencies outside this range (fig. 2.1). An amplifier of this type is known as an AC amplifier, since it measures voltages produced by alternating current. A DC amplifier is one which can measure steady potential differences, i.e. its frequency response extends down to zero hertz. If we wish to measure steady potentials, or slow potential changes without distortion, then obviously we must use a DC amplifier. Amplifiers used with intracellular electrodes are usually DC amplifiers, so that the steady potential difference across the cell membrane can be measured, as well as the rapid (high-frequency) changes involved in its activity.

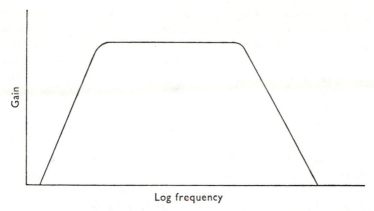

Fig. 2.1. The frequency response of an AC amplifier.

Any amplifier will produce small fluctuations in the output voltage even when there is no input signal. These fluctuations are known as *noise*, and are caused by random electrical activity in the amplifier. The existence of noise sets a lower limit to the signal voltage that can be measured, since it is difficult to distinguish very small signals from the noise. Consequently it is necessary to use an amplifier with a low noise level if we wish to measure signals of very small size.

If we connect a potential difference across the input terminals of an amplifier, a very small current flows between them, which is proportional to the potential difference. The proportionality factor is called

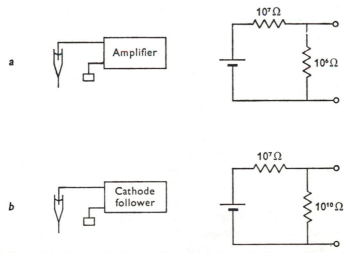

Fig. 2.2. Equivalent circuits of a glass capillary microelectrode whose resistance is 10 MΩ connected to *a*: an 'ordinary' amplifier with an input resistance of 1 MΩ, and *b*: a cathode follower circuit with an input resistance of 10 000 MΩ.

the *input resistance* of the amplifier. It is determined by application of Ohm's law and is measured in ohms; for instance, the input resistance of a cathode ray oscilloscope amplifier is usually about $1\,M\Omega$. Now suppose we connect an intracellular microelectrode whose resistance is, say, $10\,M\Omega$ to an amplifier with an input resistance of $1\,M\Omega$. The equivalent circuit is shown in fig. 2.2a. The two resistances form a potential divider, so that the voltage input to the amplifier is only $10^6/(10^6+10^7)$ i.e. one-eleventh of the signal voltage. Obviously this is of little use for measuring the signal voltage. Hence, when using high resistance electrodes, it is necessary to use an input stage with a very high input resistance. This is done by using a special valve in what is known as the 'cathode follower' circuit configuration. Suppose the input resistance of such a device is $10\,000\,M\Omega$ (fig. 2.2b); then the voltage recorded is $10^{10}(10^{10}+10^7)$, i.e. 99.9 per cent of the signal voltage, which is quite accurate enough for most purposes.

THE CATHODE RAY OSCILLOSCOPE

We now have to consider how the voltage output from an amplifier is to be recorded and measured. If the voltage is steady, or only changing very slowly, we could use an ordinary voltmeter in which the current through a coil of wire placed between the poles of a magnet causes a pointer on which the coil is mounted to move. However, a device of this nature is no use for measuring rapid electrical changes since the inertia of the pointer is too great. What we need, in effect, is a voltmeter with a weightless pointer. This is provided by the beam of electrons in the cathode ray tube of a cathode ray oscilloscope.

The cathode ray tube is an evacuated glass tube containing a number of electrodes and a screen, which is coated with phosphorus compounds so that it luminesces when and where it is bombarded by electrons, at one end (fig. 2.3). The cathode is heated and made negative (by about 2000 V) to the anode; consequently electrons are emitted from the cathode and accelerate towards the anode. Since the anode has a hole in its centre, some of the electrons continue moving at a constant high velocity beyond it; these form the electron beam. The intensity and focusing of the beam can be controlled by other electrodes, not shown in fig. 2.3, placed in the vicinity of the anode and cathode. When the electron beam hits the screen, it produces a spot of light at the point of impact. Between the anode and the screen, the beam passes between two pairs of plate electrodes placed at right angles to each other, known as the X and Y plates. If a potential difference is connected across one of these pairs, the electrons in the beam will move towards the positive plate. Thus when the electrons pass out of the electric field of the pair of

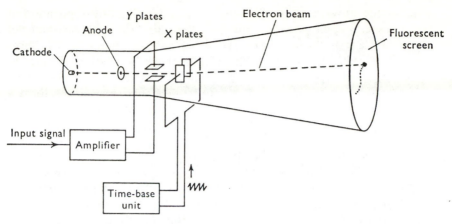

Fig. 2.3. Simplified diagram to show the principal components of a cathode ray oscilloscope.

plates, their direction will have been changed, and so the light spot on the screen will move, by an amount proportional to the potential difference between the plates. The Y plates are connected to the output of an amplifier whose input is the signal voltage to be measured; hence the signal voltage appears as a vertical deflection of the spot on the screen. The X plates are usually connected to a waveform generating circuit (the 'time-base' generator) which produces a sawtooth waveform. This sawtooth waveform thus moves the spot horizontally across the screen at a constant velocity, flying back and starting again at the end of each 'tooth'. As a consequence of these arrangements, the spot on the screen traces out a graph with signal voltage on the y-axis and time on the x-axis. By making the rise-time of the time-base sawtooth sufficiently fast, it is possible to measure the form of very rapid voltage changes.

Many oscilloscopes have tubes with two beams, each with separate Y plates and amplifiers, so that one can measure two signals at once. The time-base unit can frequently be arranged so that a single sawtooth wave, leading to a single sweep of the beam, can be initiated (or 'triggered') by some suitable electrical signal; this facility is essential for much electrophysiological work. In some cases it is possible to connect the X plates to another input amplifier, instead of to the time-base generator, so that a signal related to some quantity other than time is measured on the x-axis.

In order to obtain a permanent record of an oscilloscope trace, it is necessary to photograph it. This can be done in one of three ways. In the first method (as in fig. 4.2, for example), the film remains still and records the movement of the spot across the screen. In the second method the time-base generator is disconnected so that the spot stays in the same place on the horizontal axis, and the film is moved past it (as in fig. 15.20). Finally, the film can be moved vertically while the spot moves horizontally, so that a number of successive traces can be seen (as in fig. 8.5).

ELECTRICAL STIMULATION

An electrical stimulus must be applied via a pair of electrodes. Stimulating electrodes may be of any of the types previously described for recording purposes. The simplest way of providing a stimulating pulse is to connect the electrodes in series with a battery and a switch, but this is not satisfactory if brief pulses are required. In the past, stimulating pulses have been produced by such means as discharge of condensors or by using an induction coil, but nowadays most investigators use electronic stimulators which produce 'square' pulses of constant voltage, beginning and ending abruptly. A good stimulator unit will be able to produce pulses which can be varied in strength, duration and frequency. We need not be concerned here with the design of these instruments (see Young, 1973).

3 The resting cell membrane

If an intracellular microelectrode is inserted into a nerve or muscle cell, it is found that the inside of the cell is electrically negative to the outside by some tens of millivolts. This potential difference is known as the *resting potential*. If we slowly advance a microelectrode so that it penetrates the cell, the change in potential occurs suddenly and completely when the electrode tip is in the region of the cell membrane; thus the cell membrane is the site of the resting potential. In this chapter we shall consider some of the properties of the cell membrane that are associated with the production of the resting potential.

THE STRUCTURE OF THE CELL MEMBRANE

Plasma membranes are usually composed of roughly equal amounts of protein and lipid, plus a small proportion of carbohydrate. Human red cell membranes, for example, contain about 49 per cent protein, 44 per cent lipid and 7 per cent carbohydrate. Intracellular membranes tend to have a higher proportion of protein, whereas the protein content of myelin (p. 59) is only about 23 per cent. Most of the membrane lipids are phospholipids, with some cholesterol and sphingomyelin.

When lipids are spread on the surface of water, they form a monolayer in which the polar ends of the molecules are in contact with the water surface and the nonpolar hydrocarbon chains are oriented more or less at right angles to it. This monolayer can be laterally compressed until the lipid molecules are in contact with each other; at this point the lateral pressure exerted by the monolayer has reached a maximum, as can be seen by the use of a suitable surface balance. Using such a balance, Gorter and Grendel (1925) measured the minimum area of the monolayer produced by the lipids extracted from red blood cells, and compared it with the surface area of the cells. They found that the monolayer area was almost double the surface area of the cells, and concluded that the lipids in the cell membrane are arranged in a layer two molecules thick. Davson and Danielli (1943) later suggested that this bimolecular layer was stabilized by a thin layer of protein molecules on each side of the lipid layer.

The idea that the essential barrier to movement of substances across the cell membrane is a layer of lipid molecules receives some corroboration from the observation that lipid-soluble substances appear to penetrate the cell membrane more readily than many non-lipid-soluble substances. The electrical capacitance of cell membranes is usually about $1 \, \mu F/cm^2$; this is what one would expect if the membrane were a bimolecular layer of lipid 50 Å thick with a dielectric constant of 5.

High resolution electron microscopy of sections of cells (usually fixed with potassium permanganate) shows that the cell membrane appears as two dense lines separated by a clear space, the whole unit being about 75 Å across (see Robertson, 1960). This accords well with Davson and Danielli's model, since one would expect electron dense stains to be taken up by the polar groups of the lipid molecules and the proteins associated with them, but not by the non-polar lipid chains in the middle of the membrane. X-ray diffraction studies of myelin, which is composed of overlapping membranes of Schwann cells, shows a repeat distance of 85.5 Å, which probably corresponds with the 75 Å thick membranes seen in electron micrographs.

It has become clear in recent years that much of the intrinsic protein of the cell membrane penetrates the lipid bilayer, as is shown in figs. 3.1 and 3.2 (Singer and Nicolson, 1972; Bretscher, 1973). Much of the

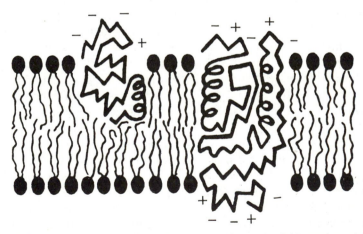

Fig. 3.1. Schematic section of a cell membrane, according to the fluid mosaic model. The phospholipid molecules are arranged as a bilayer with their polar heads (the black blobs) outwards and their nonpolar hydrocarbon 'tails' inwards. Two protein molecules are shown, each partly embedded in the lipid phase. The ionic residues of the protein chain (shown by the + and − signs) protrude from the membrane into the polar aqueous environment, while the nonpolar residues remain within the membrane. (From Singer and Nicolson, 1972.)

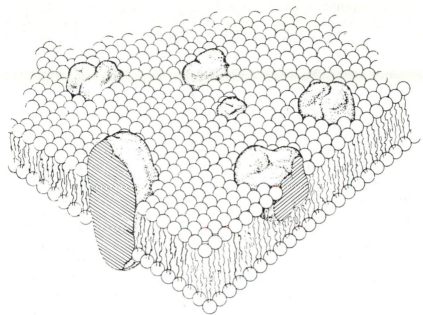

Fig. 3.2. Three dimensional impression of the fluid-mosaic model of the cell membrane, showing protein molecules (stippled), some of them in small aggregations, partly immersed in the lipid bilayer. (From Singer and Nicolson, 1972.)

evidence for this comes from the use of sophisticated chemical techniques which we shall not deal with here, but there is also some evidence from a special technique of electron microscopy known as freeze-fracture. A portion of tissue is frozen and then broken with a sharp knife. The cell membranes then cleave along the middle so as to separate the inner and outer lipid leaflets. A replica of the fractured face is made, 'shadowed' with some electron dense material and then examined in the electron microscope. Small particles are seen projecting from both faces, but especially from the inner one. If these are protein molecules (and it is difficult to see what else they could be) then they are clearly embedded within the lipid bilayer, rather than simply applied to its surface as in the Davson and Danielli model. Singer and Nicolson suggest that the nonpolar parts of such protein molecules are embedded in the nonpolar environment formed by the hydrocarbon chains of the lipid molecules, whereas their polar sections project from the membrane into the polar environment provided by the aqueous media on each side of the membrane (fig. 3.1).

Singer and Nicolson view the membrane as a mosaic in which a variety of protein molecules serve different functions. There is evidence that some of the protein molecules are able to move in the plane of the membrane (rather like icebergs in the surface waters of some polar sea), and hence the structure illustrated in figs. 3.1 and 3.2 has become known as the fluid mosaic model. For our purposes it is important to notice that some of the protein molecules which are embedded in the membrane may well act as specific sites at which particular substances may be transported across the membrane or temporarily bound to it.

CONCENTRATION CELLS

Consider the system shown in fig. 3.3. The two compartments contain different concentrations of an electrolyte XY in aqueous solution, and are separated by a membrane which is permeable to the cation X^+ but impermeable to the anion Y^-. The concentration in compartment 1 is higher than that in compartment 2. Obviously X^+ will tend to move down its concentration gradient, so that a small number of cations will move from compartment 1 to compartment 2, carrying positive charge with them. This movement of charges causes a potential difference to be set up between the two compartments. The higher the potential gets, the harder it is for the X^+ ions to move against the electrical gradient. Hence an equilibrium position is reached at which the electrical gradient (tending to move X^+ from 2 to 1) just balances the chemical or concentration gradient (tending to move X^+ from 1 to 2). Since the potential difference at equilibrium (known as the *equilibrium potential* for X^+) arises from the difference in the concentration of X^+ in the two compartments, the system is known as a *concentration cell*.

What is the value of this potential difference? Suppose a small quantity, δn moles, of X is to be moved across the membrane, up the concentration gradient, from compartment 2 to compartment 1. Then, applying elementary thermodynamics, the work required to do this, δW_c, is given by

$$\delta W_c = \delta n \, . \, RT \log_e \frac{[X]_1}{[X]_2}$$

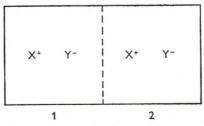

Fig. 3.3. Concentration cell. See text for details.

where R is the gas constant (8.314 joules degree^{-1} mole^{-1}), T is the absolute temperature, and $[X]_1$ and $[X]_2$ are the molar concentrations* of X in compartments 1 and 2 respectively. Now consider the electrical work, δW_e, required to move δn moles of X against the electrical gradient, i.e. from compartment 1 to compartment 2. This is given by

$$\delta W_e = \delta n \,.\, zFE$$

where z is the charge on the ion, F is Faraday's constant ($96\,500$ coulombs/mole) and E is the potential difference in volts between the two compartments (measured as the potential of compartment 2 with respect to compartment 1). Now, at equilibrium, there is no net movement of X, and therefore

$$\delta W_e = \delta W_c$$

or
$$\delta n \,.\, zFE = \delta n \,.\, RT \log_e \frac{[X]_1}{[X]_2}$$

i.e.
$$E = \frac{RT}{zF} \log_e \frac{[X]_1}{[X]_2}. \tag{3.1}$$

This equation, known as the *Nernst equation* after its formulator, is most important for our purposes since it helps us to understand the origins of electric potentials in excitable cells, which are in part dependent upon differences in the ionic composition of the cytoplasm and the external fluid.

Equation (3.1) can be simplified by enumerating the constants to become, at 18 °C,

$$E = \frac{25}{z} \log_e \frac{[X]_1}{[X]_2} \tag{3.2}$$

or
$$E = \frac{58}{z} \log_{10} \frac{[X]_1}{[X]_2}, \tag{3.3}$$

where E is now given in millivolts. For instance, in fig. 3.3, suppose $[X]_1$ were ten times greater than $[X]_2$, then E would be $+58$ mV if X were K^+ and $+29$ mV if X were Ca^{++}; if the membrane were permeable to Y and not to X, then E would be -58 mV if Y were Cl^-, and -29 mV if Y were SO_4^{--}.

We must digress a little here to consider the question, how many X ions have to cross the membrane to set up the potential? The answer

---

* Or, more strictly, the activities of X in the two compartments. The simplification given here is valid as long as the activity coefficient of X is the same in each compartment.

depends upon the valency of the ion, the value of the potential set up and the capacitance of the membrane. Let us consider a membrane with a capacitance (C) of $1\ \mu\text{F}/\text{cm}^2$ and a potential of $70\ \text{mV}$ across it. Then the charge on $1\ \text{cm}^2$ is given by

$$Q = CV,$$

where Q is measured in coulombs, C in farads and V in volts. The number of moles of X moved will be CV/zF, where z is the charge on the ion and F is Faraday's constant. In this case, assuming that X is monovalent,

$$\frac{CV}{zF} = \frac{10^{-6} \times 7 \times 10^{-2}}{96\ 500}$$

$$= 6.8 \times 10^{-13}\ \text{moles/cm}^2$$

which is a very small quantity. For instance (to anticipate a little), supposing we are dealing with a squid giant axon, diameter 1 mm, with a membrane capacitance of $1\ \mu\text{F}/\text{cm}^2$, and a membrane potential of $70\ \text{mV}$ set up primarily by a potassium ion concentration cell: $1\ \text{cm}^2$ of membrane will enclose a volume containing about 3×10^{-5} moles of potassium ions, and thus a loss of 6.8×10^{-13} moles would be undetectable. In this case the difference between the numbers of positive and negative charges inside the axon would be about $0.000\ 002$ per cent.

It is important to realize that the movements of ions from one compartment to another which we have been discussing are *net* movements. At the equilibrium position, although there is now no net ionic movement, X ions still move across the membrane, but the rate of movement (the *flux*) is equal in each direction. Before the equilibrium position is reached, the flux in one direction will be greater than that in the other. Fluxes can only be measured by using isotopic tracer methods. In describing fluxes across cell membranes, movement of ions into the cell is called the *influx*, and movement out, the *efflux*.

IONIC CONCENTRATIONS IN THE CYTOPLASM

Various microchemical methods are available for determining the quantities of ions present in a small mass of tissue. These include various microtitration techniques, flame photometry (in which the quantity of an element is determined from the intensity of light emission at a particular wavelength from a flame into which it is injected) and activation analysis (in which the element is converted into a radioactive isotope by prolonged irradiation in an atomic pile). All estimations made on a mass of tissue must include corrections made for the fluid present in the

extracellular space. The size of the extracellular space is usually estimated by measuring the concentration in the tissue of a substance which
is thought not to penetrate the cell membrane, such as inulin. Table 3.1

TABLE 3.1. *Ionic concentrations in frog muscle fibres and in plasma.*
(Simplified after Conway, 1957.)

	Concentration in muscle fibres (mM)	Concentration in plasma (mM)
K	124	2.25
Na	10.4	109
Cl	1.5	77.5
Ca	4.9	2.1
Mg	14.0	1.25
HCO$_3$	12.4	26.6
Organic anions	*ca.* 74	*ca.* 13

gives a simplified balance sheet of the ionic concentrations in frog
muscle determined in this way. In the giant axons of squids, which may
be up to 1 mm in diameter, the situation is much more favourable, since
the axoplasm can be squeezed out of an axon like toothpaste out of a
tube; table 3.2 shows the concentrations of the principal constituents.

TABLE 3.2. *Ionic concentrations in squid axoplasm and blood.*
(Simplified after Hodgkin, 1958.)

	Concentration in axoplasm (mM)	Concentration in blood (mM)
K	400	20
Na	50	440
Cl	40–150	560
Ca	0.4	10
Mg	10	54
Isethionate	250	—
Other organic anions	*ca.* 110	—

The main features of the ionic distribution between the cytoplasm and
the external medium are similar in both the squid axon and the frog
muscle fibre. In each case the intracellular potassium concentration is
much greater than that of the blood, whereas the reverse is true for
sodium and chloride. Moreover, the cytoplasm contains an appreciable
concentration of organic anions. These features seem to be common to

all excitable cells. As we shall see, these inequalities in ionic distribution are essential to the electrical activity of nerve and muscle cells. They arise by means of two systems which we must now consider: these are (1) a Donnan equilibrium system, in combination with (2) active transport processes which are driven by metabolic energy.

THE DONNAN EQUILIBRIUM SYSTEM IN MUSCLE

In the system shown in fig. 3.4, two potassium chloride solutions are separated by a membrane which is permeable to both potassium and chloride ions. The system is at constant volume so that, although the membrane is permeable to water, no net flow of water can occur between compartments i and o. Let us assume that the two potassium chloride solutions are initially at the same concentration. A quantity of a potassium salt KA is now added to compartment i, and it dissolves to give K^+ and A^- ions; we shall assume that the anion A^- cannot pass through the membrane. The concentration of potassium ions in compartment i, $[K]_i$, is now greater than in compartment o, $[K]_o$, so that potassium ions move from i to o. Since the total number of positive and negative changes in i must be approximately* equal (the same applies to o, of course), chloride ions move from i to o also. Now since $[K]_i \neq [K]_o$ and $[Cl]_i \neq [Cl]_o$, a concentration cell potential must arise between the two compartments in each case, such that, applying the Nernst equation,

$$E_K = \frac{RT}{F} \log_e \frac{[K]_o}{[K]_i} \qquad (3.4)$$

and

$$E_{Cl} = \frac{RT}{-F} \log_e \frac{[Cl]_o}{[Cl]_i} \qquad (3.5)$$

where E_K is known as the potassium equilibrium potential and E_{Cl} is known as the chloride equilibrium potential. But there can only be a single potential difference across the membrane; hence potassium chloride will continue to move from i to o until $E_K = E_{Cl}$. When this stage is reached, the system is in equilibrium and it follows that

$$\frac{RT}{F} \log_e \frac{[K]_o}{[K]_i} = \frac{RT}{-F} \log_e \frac{[Cl]_o}{[Cl]_i}.$$

Therefore

$$[K]_o/[K]_i = [Cl]_i/[Cl]_o$$

or

$$[K]_i \times [Cl]_i = [K]_o \times [Cl]_o. \qquad (3.6)$$

Equation (3.6) is the Donnan rule, which states that the product of the

* See p. 18 for what is meant by 'approximately' in this context.

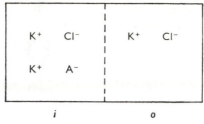

Fig. 3.4. Donnan equilibrium system. See text for details.

concentrations of the diffusible ions in one compartment is equal to the product of the concentrations of the diffusible ions in the other compartment, at equilibrium.

Thus the presence of the indiffusible anions in compartment i results in an inequality of distribution of the diffusible ions, potassium and chloride. However, we cannot apply this simple system to an animal cell, because an animal cell is not a constant volume system, and therefore the osmotic concentration must be the same on each side of the membrane. In the simple system of fig. 3.4 this is not so, as the following analysis shows. Applying the condition for approximate electrical neutrality in each compartment,

$$[K]_o = [Cl]_o$$

and
$$[K]_i > [Cl]_i.$$

Now
$$[K]_i \times [Cl]_i = [K]_o \times [Cl]_o,$$

therefore
$$[K]_i + [Cl]_i > [K]_o + [Cl]_o,$$

therefore
$$[K]_i + [Cl]_i + [A] > [K]_o + [Cl]_o,$$

i.e. the osmotic concentration in i is greater than that in o. Thus if the constant volume constraint were removed from the system, water would move from compartment o to compartment i. This would change the ionic concentrations in the two compartments and so disturb the Donnan equilibrium. Hence potassium chloride would move from i to o, again upsetting the osmotic equilibrium. These processes would continue until the concentration of A in i was infinitesimal and the concentrations of potassium and chloride were equal throughout the system.

Now consider a similar system (fig. 3.5) in which the membrane is impermeable to sodium ions and some sodium chloride is added to compartment o. Potassium and chloride will move until the Donnan equilibrium is established, i.e. until

$$[K]_i \times [Cl]_i = [K]_o \times [Cl]_o.$$

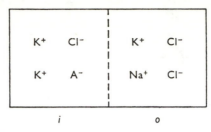

Fig. 3.5. Double Donnan equilibrium system. See text for details.

But now $[K]_o$ is less than $[Cl]_o$, so that if the constant volume constraint is removed, it *is* possible for the system to be in Donnan equilibrium and osmotic equilibrium at the same time. In effect, the osmotic effect of the indiffusible anion in *i* is balanced by that of the indiffusible cation (sodium) in *o*.

We must conclude that a system similar to that shown in fig. 3.5 might account for the inequalities of ionic distribution seen in nerve and muscle cells. If this is so, then it should be possible to show that the $[K][Cl]$ product is equal inside and outside after equilibrium has been reached in various conditions. This experiment has been performed by Boyle and Conway (1941), who developed the theory outlined above. They soaked frog muscles in solutions containing various potassium chloride concentrations for 24 hours, and then determined the intracellular potassium and chloride concentrations. Table 3.3 shows the results

TABLE 3.3. *Intracellular potassium and chloride concentrations in frog muscle after equilibration with various solutions for 24 hours at 2–3 °C.* (From Boyle and Conway, 1941.)

$[K]_o$ (mM)	$[Cl]_o$ (mM)	$[K]_i$ (mM)	$[Cl]_i$ (mM)	$[K]_o[Cl]_o$ 10^{-3}	$[K]_i[Cl]_i$ 10^{-3}	$\dfrac{[K]_o[Cl]_o}{[K]_i[Cl]_i}$
3	79	91	7.2	0.24	0.68	0.36
6	82	92	7.2	0.49	0.66	0.74
12	88	101	9.9	1.05	1.00	1.05
18	94	107	16.1	1.69	1.72	0.98
30	106	120	24.9	3.18	2.99	1.06
60	136	142	60.6	8.16	8.61	0.94
90	166	184	86.0	14.9	15.8	0.94
120	196	212	144.2	23.5	24.2	0.97
150	226	240	143.1	33.9	34.4	0.99
210	286	282	186.7	60.0	52.8	1.14
300	376	353	308	112.8	118.7	1.05
					Mean	1.01

of these determinations; it is evident that they conform to the Donnan equilibrium hypothesis at all external potassium concentrations above about 10 mM. Further evidence for the hypothesis, and consideration of the discrepancies at low external potassium ion concentrations, will be provided later.

ACTIVE TRANSPORT OF IONS

Sodium

The simple Donnan equilibrium hypothesis assumes that the cell membrane is impermeable to sodium ions. However this is not so; resting nerve and muscle cells do take up radioactive sodium ions (for example, the resting influx of sodium into giant axons of the cuttlefish *Sepia* is about 35 pmoles $cm^{-2} sec^{-1}$). If nothing were done about this influx, the system would in time run down, so as to produce equal concentrations of each ion on both sides of the membrane. In fact, the cell prevents this by means of a continuous extrusion of sodium ions (frequently known as the 'sodium pump'). Since such extrusion must occur against an electrochemical gradient, we would expect this process to be an active one, involving the consumption of metabolic energy.

Let us first consider some experiments by Hodgkin and Keynes (1955) on the extrusion of sodium from *Sepia* giant axons. An axon was placed in sea water containing radioactive sodium ions and stimulated repetitively for some time, so that (as we shall see in chapter 5) the interior of the axon became loaded with radioactive sodium. It was then placed in a capillary tube through which nonradioactive sea water could be drawn (fig. 3.6), and the efflux of sodium determined by measuring

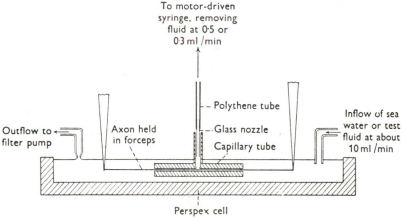

Fig. 3.6. Apparatus used to measure the efflux of radioactive sodium from *Sepia* giant axons. (From Hodgkin and Keynes, 1955.)

the radioactivity of samples of this sea water at intervals. This particular arrangement has two advantages: the efflux from the cut ends of the axon is not measured, and the measured efflux occurs from a known length of axon into a known volume of sea water. It was found that the relation between the logarithm of the efflux and time is a straight line of negative slope (as is shown in the first section of the graph in fig. 3.7), indicating that the efflux of radioactive sodium falls exponentially with time. This is just what we would expect to see if both the internal sodium ion concentration and the rate of extrusion of sodium ions are constant, since under these conditions a constant proportion of the radioactive sodium inside the axon will be removed in successive equal time intervals.

When 2:4-dinitrophenol (DNP) was added to the sea water, the sodium efflux fell markedly, and then recovered when the DNP was washed away (fig. 3.7). Now DNP is an inhibitor of metabolic activity, and probably acts by uncoupling the formation of the energy-rich compound adenosine triphosphate (ATP) from the electron chain in aerobic respiration, hence this experiment implies that the extrusion of sodium is probably dependent on metabolic energy supplied directly or indirectly in the form of ATP. More conclusive evidence for this view is provided by a series of experiments by Caldwell, Hodgkin, Keynes and Shaw (1960), one of which is shown in fig. 3.8. A solution containing radioactive sodium ions was injected into a giant axon of the squid

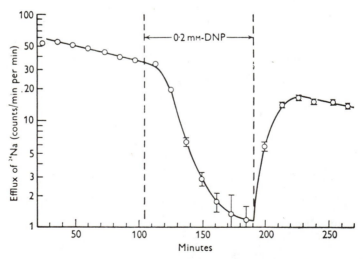

Fig. 3.7. The effect of the metabolic inhibitor 2:4-dinitrophenol (DNP) on the efflux of radioactive sodium from a *Sepia* giant axon. (From Hodgkin and Keynes, 1955.)

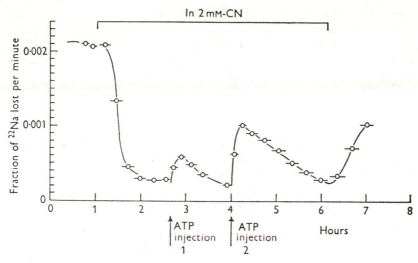

Fig. 3.8. The effect of ATP injection on the sodium efflux from a squid giant axon poisoned with cyanide. More ATP was introduced with the second injection than with the first. (From Caldwell *et al.*, 1960.)

Loligo by means of a very fine glass tube inserted longitudinally into the axon, and the efflux was then measured. Soon after the start of the experiment the axon was poisoned with cyanide. Cyanide ions interfere with the energy-producing processes in aerobic respiration (by inhibiting the action of cytochrome oxidase), and so the efflux of sodium fell to a low level. Injections of ATP, arginine phosphate or phosphenol pyruvate into the axon produced transient increases in the rate of sodium extrusion, so confirming the view that the sodium pump is driven by the energy derived from the breakdown of energy-rich phosphate compounds.

Hodgkin and Keynes also showed that the sodium efflux from *Sepia* axons is dependent upon the external potassium ion concentration; potassium-free sea water reduced the efflux to about a third of its normal value. This implies that the sodium extrusion process is coupled, albeit somewhat loosely, to the uptake of potassium.

In terms of the fluid mosaic model of the cell membrane (fig. 3.2), we may suppose that sodium ions are pumped at particular sites each consisting of a particular protein molecule or group of molecules. How many of these sites are there in any particular area of membrane? The problem has been approached by measuring the binding of the drug ouabain to nerve axons. Ouabain is a glycoside found in certain plants and used as an arrow poison in parts of Africa. It acts by inhibiting the sodium pump of cell membranes. By measuring the uptake of radioactive ouabain it is possible to estimate the number of sodium-pumping sites in a system, assuming that each site binds one molecule of ouabain. In squid giant axons, Baker and Willis (1972) estimate that there are 1000 to 5000 sites per μm^2 of cell membrane, and Ritchie and Straub (1975) estimate that for garfish olfactory nerve there are 350 sites per μm^2.

Chloride

In squid giant axons, Keynes (1963) has shown that, besides the potassium-linked sodium pump, there is another active transport system, concerned with the inward movement of chloride ions. If chloride were passively distributed across the axonal membrane, then the internal chloride concentration would be given by

$$E = \frac{RT}{-F} \log_e \frac{[Cl]_o}{[Cl]_i}$$

(where E is the resting potential), or, in isolated axons,

$$-60 = -58 \log_{10} \frac{560}{[Cl]_i},$$

i.e. $[Cl]_i = 55$ mM.

In fact, however, the average internal chloride concentration in Keynes's experiments was 108 mM, or about twice what one would expect. There are two possible explanations for this situation: either about half of the internal chloride was bound in non-ionic form, or there was an active inward transport of chloride. Measurements on extruded axoplasm with a chloride-sensitive electrode (a silver wire coated with silver chloride) showed that the activity coefficient of chloride in the axoplasm was much the same as that in free solution, which indicates that very little of the chloride can be bound. The chloride influx was halved by treating the axon with DNP, thus confirming the conclusion that there is some active uptake of chloride ions. The functional importance of this 'chloride pump' is obscure.

Fig. 3.9 summarizes the conclusions reached in this and the previous sections.

Calcium

The concentration of calcium in axoplasm is low, at about 400 μM (table 3.2). If a small quantity of radioactive calcium ions is injected into a squid axon, the resulting patch of radioactivity does not spread out by diffusion or move in a longitudinal electric field (in contrast to potassium, for example – see fig. 3.12). This suggests that nearly all the

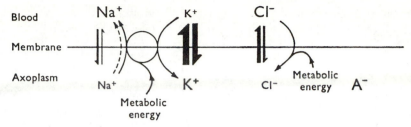

Fig. 3.9. A schematic representation of the movements of the major monovalent ions across the squid giant axon membrane in the resting condition. Passive fluxes are represented by straight arrows whose thickness suggests the magnitude of the flux. Curved arrows indicate active transport. Concentrations are suggested by the size of the chemical symbols. A⁻ represents indiffusible organic anions. In frog muscle cells the chloride fluxes are relatively larger, and chloride movement seems to be largely passive.

calcium is bound in some way, and that only a small proportion is in free solution in the ionic form (Baker and Crawford, 1972). By using the protein aequorin, which emits light in the presence of calcium ions, it is possible to show that the concentration of ionized calcium is as low as $0.1 \mu\text{M}$. There is thus a very great concentration gradient of calcium ions across the membrane, and it is therefore not surprising that there is a continual influx of calcium ions. This is balanced by an outward calcium pump, whose energy supply probably comes from the breakdown of ATP (Baker, 1976). There is a similar outwardly-directed pump for magnesium ions (Baker and Crawford, 1972).

Baker points out that the very low intracellular concentration of ionized calcium allows calcium ions to be used as a potent physiological trigger. Thus release of one per cent of the bound calcium in squid axoplasm would increase the free ionic concentration forty-fold, from 0.1 to $4 \mu\text{M}$. We shall see in later chapters that such trigger actions are fundamental to the release of synaptic transmitter substances and the initiation of muscular contraction, and perhaps to the excitation of photoreceptor cells.

THE RESTING POTENTIAL

The potassium electrode hypothesis

According to the Donnan equilibrium theory developed by Boyle and Conway, we would expect the resting potential to be given by the Nernst equations for potassium and chloride ions, i.e. by (3.4) and (3.5). Furthermore, since the internal chloride concentration is low, a relatively small movement of chloride ions will produce a large change in E_{Cl}, whereas a similar movement of potassium ions would produce only a small change in E_K. Hence we would expect the membrane potential in the resting condition to be determined mainly by the potassium ion gradient, so that, at 18 °C, the resting potential would be given by

$$E = 58 \log_{10} \frac{[K]_o}{[K]_i}. \tag{3.7}$$

The best method of readily testing this hypothesis is to vary the external potassium concentration and observe the change in membrane potential; (3.7) predicts that the relationship should be a straight line with a slope of 58 mV per unit increase in $\log_{10}[K]_o$. This statement is subject to the proviso that the internal potassium ion concentration does not change and that any potential due to chloride ions does not interfere. These conditions can be ensured either by eliminating chloride from the system (by using sulphate solutions, for example) or by ensuring that the $[K]_o[Cl]_o$ product remains constant. Fig. 3.10 shows the membrane potential of isolated frog muscle fibres in sulphate solutions containing different potassium ion concentrations (Hodgkin and Horowicz, 1959).

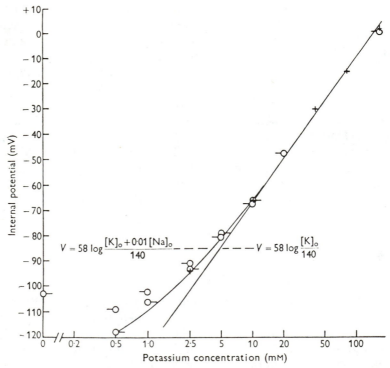

Fig. 3.10. The effect of the external potassium ion concentration on the membrane potential of isolated frog muscle fibres. The external solutions were chloride-free, the principle anion being sulphate. (From Hodgkin and Horowicz, 1959.)

An alternative way of testing (3.7) is by measuring $[K]_i$ as well as $[K]_o$ under different conditions; this has been done by Conway (1957), with the results shown in fig. 3.11. In both these experiments it is evident that (3.7) fits the results very well at potassium concentrations above about 10 mM, but below this value the membrane potential is rather less than one would expect.

What is the reason for this departure from the simple potassium electrode hypothesis at low external potassium ion concentrations? We have seen that sodium ions are distributed across the cell membrane so as to produce the potentiality of a concentration cell with the inside positive, and since the membrane is not completely impermeable to sodium ions, we might expect this concentration cell to play some part in

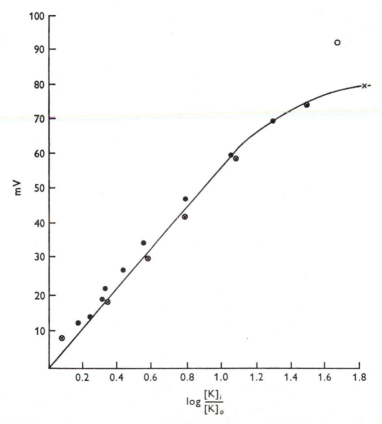

Fig. 3.11. The relation between the potassium ion concentration gradient and membrane potential in frog muscle fibres. Muscles were equilibrated for some hours in solutions containing different concentrations of potassium chloride; the internal potassium concentration was determined by chemical analysis. (From Conway, 1957.)

determining the membrane potential. Just how this effect will make itself felt will depend upon the properties of the membrane. A useful approach, known as the 'constant field theory' has been developed by Goldman (1943). This theory assumes that ions move in the membrane under the influence of electrical fields and concentration gradients just as they do in free solution, that the concentrations of ions in the membrane at its edges are proportional to those in the aqueous solutions in contact with it, and that the electrical potential gradient across the membrane is constant. From these assumptions it is possible to show (see Hodgkin and Katz, 1949) that, when there is no current flowing through the membrane, the membrane potential is given by

$$E = \frac{RT}{F} \log_e \frac{P_K[K]_o + P_{Na}[Na]_o + P_{Cl}[Cl]_i}{P_K[K]_i + P_{Na}[Na]_i + P_{Cl}[Cl]_o}, \qquad (3.8)$$

where P_K, P_{Na} and P_{Cl} are permeability coefficients. (These permeability coefficients are measured in cm/sec and defined as $u\beta RT/aF$, where u is the mobility of the ion in the membrane, β is the partition coefficient between the membrane and aqueous solution, a is the thickness of the membrane, and R, T and F have their usual significance.) If chloride ions are omitted from the system, or if they are assumed to be in equilibrium so that the equilibrium potential for chloride is equal to the membrane potential, (3.8) becomes

$$E = \frac{RT}{F} \log_e \frac{[K]_o + \alpha[Na]_o}{[K]_i + \alpha[Na]_i}, \qquad (3.9)$$

where α is equal to P_{Na}/P_K. Equation (3.9) is plotted in fig. 3.10, assuming α to be 0.01; it is clear that it provides a good fit for the experimental results. Results similar to those shown in fig. 3.10 have also been obtained from observations on the membrane potentials of nerve axons (Hodgkin and Katz, 1949; Huxley and Stämpfli, 1951). We can therefore conclude that sodium ions play some small part in determining the membrane potential of nerve and muscle cells, their effect being greater when the external potassium ion concentration is low.

One further piece of information is required before we can accept the hypothesis that the resting potential is determined mainly by the Nernst equation for potassium ions. It is necessary to show that the internal concentration of free potassium ions (which is involved in (3.7) to (3.9)) is the same, or almost the same, as the total internal potassium concentration (which is what can be measured by chemical methods of analysis). For instance, if half the internal potassium were found in non-ionic form, the predicted values for membrane potentials would have to be reduced by nearly 20 mV. This question was investigated by Hodgkin and Keynes (1953) on *Sepia* giant axons. They placed an axon in oil, but with a short length of it passing through a drop of sea water containing radioactive potassium ions. Thus, after a time, this short length of axon contained radioactive potassium. It was then placed in a longitudinal electric field, so that the potassium ions would move along the axon towards the cathode if they were free to do so, and the longitudinal distribution of radioactivity was measured at intervals by means of a Geiger counter masked by a piece of brass with a thin slit in it. The result of one of these experiments is shown in fig. 3.12, from which it is clear that the internal radioactive potassium ions are free to move in an electric field. From the rate of movement, it was possible to calculate the ionic mobility and diffusion coefficient of radioactive potassium ions in the axon: the values were found to be close to those in a 0.5 M solution of potassium chloride. Thus the radioactive potassium is present inside the axon in a free, ionic form. Since Keynes and Lewis (1951) had previously shown that radioactive potassium exchanges with at least 97 per cent of the potassium present in crab axons, it seemed very reasonable to conclude that almost all the potassium in the axoplasm is effectively in free solution and so can contribute to the production of the resting potential.

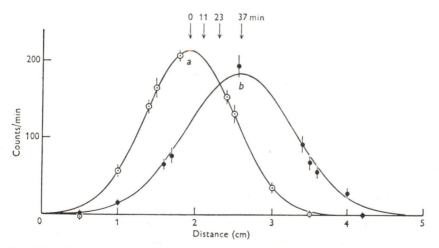

Fig. 3.12. Movement of radioactive potassium in a longitudinal electric field in a *Sepia* giant axon. The two curves show the distribution of radioactivity immediately before (*a*) and 37 minutes after (*b*) application of the longitudinal current. Arrows at 11 and 23 minutes show the positions of peak radioactivity at these times. (From Hodgkin and Keynes, 1953.)

Non-equilibrium conditions

We must now consider the contribution of chloride ions to the resting potential. This, together with some related problems, has been extensively investigated by Hodgkin and Horowicz (1959) on isolated frog muscle fibres. The great advantage of using single fibres is that, with a suitable apparatus, it is possible to change the ionic concentrations at the cell surface within a fraction of a second, so eliminating the inevitable diffusion delays involved in work with whole muscles. When the external solution was changed for one with different potassium and chloride concentrations but with the same $[K]_o[Cl]_o$ product, the new membrane potential (given by (3.4) and (3.5)) was reached within two or three seconds, and thereafter remained constant. However, if the chloride concentration was changed without altering the potassium concentration, the membrane potential jumped rapidly to a new value, but this transient effect gradually decayed over the next few minutes and the potential returned to very nearly its original value (fig. 3.13). The explanation of this effect is based on the Donnan equilibrium hypothesis. The chloride equilibrium potential is given by (3.4), or, at 18 °C,

$$E_{Cl} = -58 \log_{10} \frac{[Cl]_o}{[Cl]_i}. \tag{3.10}$$

At the start of the experiment shown in fig. 3.13 (when $[Cl]_o$ is 120 mM)

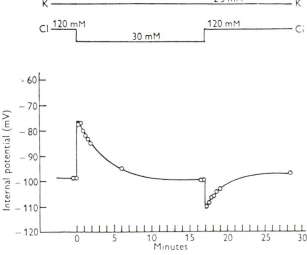

Fig. 3.13. The effect of a sudden reduction in the external chloride concentration on the membrane potential of an isolated frog muscle fibre. (From Hodgkin and Horowicz, 1959.)

we assume that chloride is in equilibrium, i.e. that E_{Cl} is equal to the membrane potential of -98.5 mV; this gives a value of 2.4 mM for $[Cl]_i$. When $[Cl]_o$ is reduced to 30 mM, E_{Cl} will change by 35 mV to -63.5 mV. The membrane potential changes to -77 mV, which is intermediate between E_K and E_{Cl}, indicating that the membrane is permeable to both potassium and chloride ions. Then, in order to restore the Donnan equilibrium, potassium chloride moves out of the cell until E_{Cl} is equal to E_K. This point is reached when $[Cl]_i$ has fallen to about 0.6 mM and, since there is also some movement of water out of the cell in order to maintain osmotic equilibrium, $[K]_i$ is practically unchanged. Hence the new steady potential (reached after 10–15 min in fig. 3.13) is the same as it originally was. When $[Cl]_o$ is returned to 120 mM, there is a similar transient membrane potential change in the opposite direction, and equilibrium is then restored by movement of potassium chloride and water into the fibre. The fact that these experiments can be so readily interpreted in this way provides further evidence in favour of Boyle and Conway's Donnan equilibrium hypothesis.

Fig. 3.14c shows the results of a similar experiment in which $[K]_o$ is changed from 2.5 to 10 mM. This will change E_K by 35 mV, and so the membrane potential jumps to a new value intermediate between E_K and E_{Cl}. Equilibrium is then reached by entry of potassium chloride (and water) into the fibre, so that the membrane potential gradually moves from -73 to -65 mV. On returning $[K]_o$ to 2.5 mM, there is a small instantaneous repolarization, and then equilibrium (and the normal resting potential) is slowly restored by loss of potassium chloride from

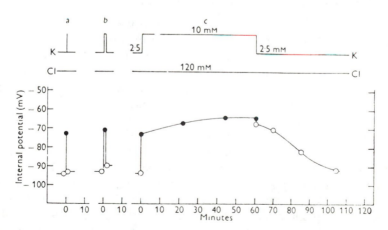

Fig. 3.14. The effect of changes in the external potassium ion concentration on the membrane potential of an isolated frog muscle fibre. (From Hodgkin and Horowicz, 1959.)

the fibre. Notice that, in this case, the fourfold increase in $[K]_o$ causes an instantaneous depolarization of 21 mV, whereas the later fourfold decrease causes an instantaneous repolarization of only 3 mV. This must indicate that there is some rectification process in the potassium channel, so that potassium ions can move inwards very easily but outwards with much more difficulty. The larger repolarizations in figs. 3.14a and b are due to the fact that the internal chloride concentration has not had time to change much from its normal level, so that E_{Cl} is still near the normal resting potential. Hodgkin and Horowicz calculated that P_K is about 8×10^{-6} cm/sec for inward current but may be as little as 0.05×10^{-6} cm/sec for outward current, whereas P_{Cl} remains at about 4×10^{-6} cm/sec irrespective of the direction of the current flow.

A digression on rectification

When the resistance of a conductor is not independent of the voltage across it, the conductor is said to show rectification. Rectification in the potassium channel of muscle fibre membranes, as described in the previous paragraph, had previously been observed by Katz (1949), and has been called *anomalous rectification*. The reason for describing this as 'anomalous' is that the constant field theory does predict some degree of rectification, but in the opposite direction; this is known as *constant field rectification*. Finally, as we shall see in chapter 5, depolarization of electrically excitable membranes results in a large increase in the permeability of the membrane to potassium ions; this effect is sometimes known as *delayed rectification*.

The electrogenic nature of the sodium pump

Consider a membrane ion pump which is concerned solely with the active transport of one species of ion in one direction. There would then be a current flow across the membrane and a change in membrane potential caused by the depletion of charge on one side of the membrane. There would be similar consequences from a pump which coupled inflow of one ion to outflow of another but in which the numbers of ions transported in the two directions were not equal. A pump in which such a net transfer of charge does take place is described as being electrogenic. In mammalian red blood corpuscles, for example, there is very clear evidence that two potassium ions are taken up for every three sodium ions extruded (Post and Jolly, 1957); there must thus be a net flow of charge equal to one third of the sodium ion flow.

The sodium pump of nerve axons was once thought to be electrically neutral, because of a supposed one-for-one exchange of sodium for potassium ions. More recently, however, experiments on a variety of cells have indicated that this is not so and that the pump is electrogenic

(Ritchie, 1971). Some of the best evidence for this conclusion comes from experiments by Thomas (1969) on snail neurones; let us examine them.

The essence of Thomas's method was to inject sodium ions into the cell body of a neuron and observe the subsequent changes in membrane potential. In order to perform the injection two microelectrodes filled with a solution of sodium acetate were inserted into the cell and current passed between them; sodium ions would thus be carried out of the cathodal electrode into the cytoplasm, so raising the internal sodium concentration. A third microelectrode was used to record the membrane potential.

Thomas found that injection of sodium ions into the cell is followed by a hyperpolarization of up to 15 mV, after which the membrane potential returned to normal over a period of about 10 minutes, as is shown in the first parts of the traces in fig. 3.15. Fig. 3.15*a* shows that ouabain greatly reduces this hyperpolarization, suggesting that it is connected with an active sodium extrusion system. In fig. 3.15*b*, the second injection of sodium ions occurs while the cell is in a potassium-free environment, when there is no hyperpolarization, suggesting that sodium extrusion is coupled to potassium uptake. (Incidentally this experiment also eliminates the hypothesis that the hyperpolarization is due to a local reduction in potassium ion concentration at the outer surface of the cell, caused by such a coupled uptake of potassium.) Injections of potassium or lithium ions produced no membrane hyper-polarization.

These observations suggest that, while sodium extrusion is coupled to potassium uptake, the number of sodium ions moved is greater than the number of potassium ions. There is thus a net flow of positive charge outward during the action of the pump, seen as a hyperpolarization of the membrane. It is possible to estimate the ratio of sodium ion to potassium ion movements? Thomas attacked this problem by using an ingenious method known as the voltage clamp technique. We shall examine the principle of this method in chapter 5; suffice it to say here that it involves a negative feedback control system whereby the membrane potential is maintained at a constant level ('clamped') by passing current through it via another electrode. Membrane currents can thus be measured at constant membrane potentials. Thomas found that sodium injection was followed by an outwardly directed current which took about 10 minutes to fall to zero.

By integrating this current with respect to times it was possible to estimate the amount of charge transferred. Thomas found that this was always much less than the quantity of charge injected as sodium ions. And yet he also found that all of the injected sodium was extruded

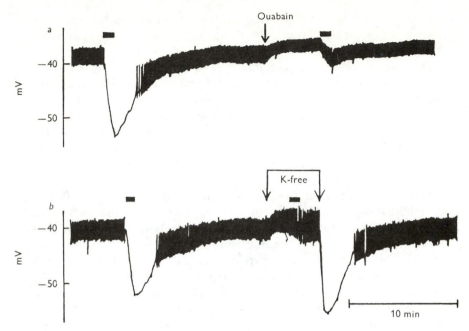

Fig. 3.15. Responses of snail neurons to injections of sodium ions, demonstrating the electrogenic nature of the sodium pump. The records show pen recordings of the membrane potential; black bars indicate the duration of the injecting currents. The first response in each trace shows the hyperpolarization which normally follows sodium injection. After treatment with ouabain (*a*) this is greatly reduced. In a potassium free external solution (*b*), the hyperpolarization does not occur until external potassium is replaced, indicating that sodium extrusion is coupled to potassium uptake. (The thickness of the trace at less negative membrane potentials is caused by spontaneous activity in the neuron, the individual action potentials being too rapid to be registered fully by the pen recorder.) (From Thomas, 1969.)

during the period of membrane current flow; he did this by using an intracellular sodium-sensitive microelectrode (one which produces an electrical signal proportional to the sodium ion concentration in its environment). This means that the sodium outflow must be partly balanced by an inflow of some other cation, for which the obvious candidate is potassium. After making certain corrections, Thomas concluded that the net charge transfer in the pump was about 27 per cent of the sodium ion flow. This figure is fairly near to the 33 per cent that would be expected from a system like that in red blood cells, where three sodium ions move for every two potassium ions. Measurements on a variety of different tissues have produced different ratios, however (see Thomas, 1972), and therefore it would seem unwise to be too dogmatic about the precise numbers; it may be that these are to some extent changeable in different circumstances.

4 Electrical properties of the nerve axon

The most striking anatomical feature of nerve cells is that part of the cell is produced into an enormously elongated cylindrical process, the axon. It is this part of the cell which we shall be concerned with in this chapter and the next. The essential function of the axon is the propagation of nerve impulses.

ACTION POTENTIALS IN SINGLE AXONS

Let us consider a simple experiment on the giant fibres in the nerve cord of the earthworm. These fibres are anatomically not axons, since they are multicellular units divided by transverse septa in each segment, but physiologically each fibre acts as a single axon. These are three giant fibres, one median and two lateral, which run the length of the worm;

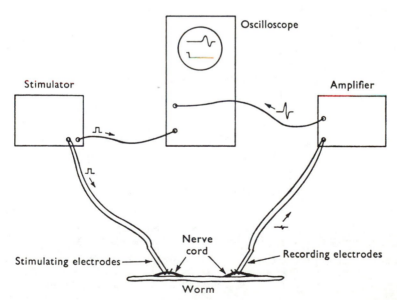

Fig. 4.1. Arrangement for recording action potentials from the giant fibres in the nerve cord of the earthworm.

the laterals are interconnected at intervals. The experimental arrangement for eliciting and recording impulses in the giant fibres is shown in
fig. 4.1. The stimulator produces a square voltage pulse which is applied
to the nerve cord at the stimulating electrodes. The recording electrodes
pick up the electrical changes in the nerve cord and feed them into the
amplifier. Here they are amplified about 1000 times, and then passed to
the oscilloscope where they are displayed on the screen of the cathode
ray tube. The output of the stimulator is also fed into the oscilloscope so
that it is displayed on the second trace on the screen. The timing of the
oscilloscope sweep is arranged so that both traces start at the moment
that the pulse from the stimulator arrives.

We begin with a stimulating pulse of low intensity (fig. 4.2a). The
lower trace shows the stimulating pulse, but nothing appears on the
upper trace except a deflection which is coincident with the stimulus and
(as we can show by varying the stimulus intensity) proportional to its
size. This phenomenon is known as the *stimulus artefact*; it is not a

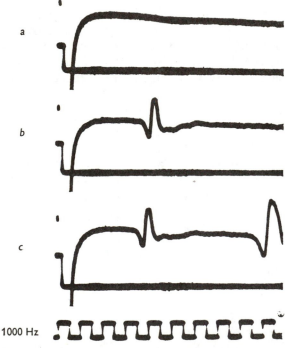

1000 Hz

Fig. 4.2. Oscilloscope records from the experiment shown in fig. 4.1. In each case the
upper trace is a record of the potential changes at the recording electrodes and the lower
trace (at a much lower amplification) monitors the stimulus pulse. Further details in the
text.

property of the worm, and is merely due to the recording electrodes picking up the electric field set up by the stimulus pulse. As we increase the intensity of the stimulus, a new deflection appears on the upper trace (fig. 4.2*b*). This is the nerve impulse, or *action potential*, in the median fibre. Further increase in the stimulus intensity does not change the size of this deflection, and if we turn the stimulus intensity down again it suddenly disappears, without any preparatory decrease in size. In other words, the size of the action potential is independent of the size of the stimulus; it is either there or not there. This phenomenon is known as the '*all-or-nothing*' *law*. The stimulus intensity which is just sufficient to produce an action potential is called the *threshold stimulus intensity*. If we further raise the stimulus intensity, a second deflection may appear at a later time on the trace (fig. 4.2*c*); this is caused by the action potentials of the two lateral fibres.

Now let us perform another experiment. Keeping the stimulus intensity constant, above the threshold for the median fibre, the recording electrodes are moved nearer to the stimulating electrodes. We find that the action potential occurs earlier on the trace. If we measure the time between the stimulus and the action potential at a number of different distances between the two pairs of electrodes, we can plot a graph showing the position of the action potential at various times after the stimulus (fig. 4.3). The points lie very nearly on a straight line which passes near the origin. This means that the action potential arises at the stimulating electrodes and is then conducted along the fibre at a constant velocity given by the slope of this straight line.

In this experiment on conduction in earthworm giant fibres, we have been measuring potential differences between points at different places outside the axon. Obviously it would be informative to measure potential changes across the membrane, using an electrode inside the cell. This was first done in 1939 by Hodgkin and Huxley at Plymouth and by Curtis and Cole at Woods Hole, Massachusetts, using the giant axons innervating the mantle muscle of squids. As was mentioned in the previous chapter, these axons are of unusually large diameter (0.1 to 1 mm), and have been much used by nerve physiologists since their description by J. Z. Young in 1936.

The intracellular electrode used in these experiments was a glass capillary tube, about 100 μm in diameter, filled with sea water (Hodgkin and Huxley, 1945) or a potassium chloride solution isotonic with sea water (Curtis and Cole, 1942). It was inserted through the cut end of axon and pushed along so that its tip was level with an undamaged part of the axon. The potential difference between the electrode and the external sea water was then measured. The axon could be electrically

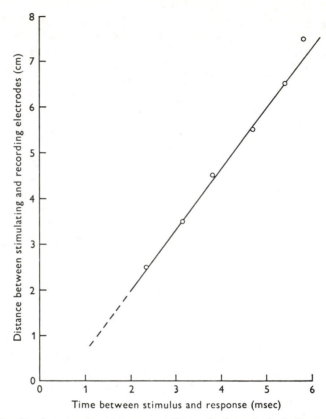

Fig. 4.3. Results of an experiment to measure the conduction velocity in the median giant fibre of an earthworm. In this case it was about 12.5 m/sec.

stimulated via a pair of external electrodes. Fig. 4.4 shows a typical record of an action potential obtained by this method. The trace begins with the inside of the axon negative to the outside (this is the resting potential), but during the action potential the membrane potential changes so that the inside becomes positive. The discovery of this 'overshoot' was unexpected at the time; its significance will be considered in the next chapter.

In many cases a careful examination of the time course of an action potential shows that the membrane potential does not immediately settle down at the resting level after its completion. As is shown in fig. 4.5, the action potential may be followed by (i) a *positive phase*, evident in fig. 4.4, in which the membrane potential 'underswings' towards a more negative value than the normal resting potential, (ii) a *negative after-potential*, in which the membrane potential is less negative than normal, and finally (iii) a *positive after-potential*, in which the membrane

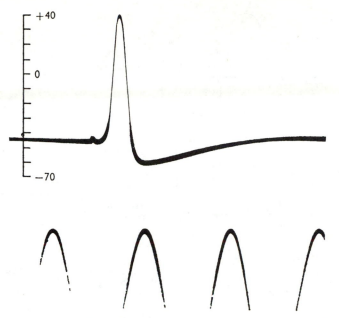

Fig. 4.4. An action potential recorded intracellularly from a squid giant axon. The vertical scale shows the potential (mV) of the internal electrode with respect to the external sea water. The action potential is preceded by a small stimulus artefact. The sine wave at the bottom is a time marker, frequency 500 Hz. (From Hodgkin and Huxley, 1945.)

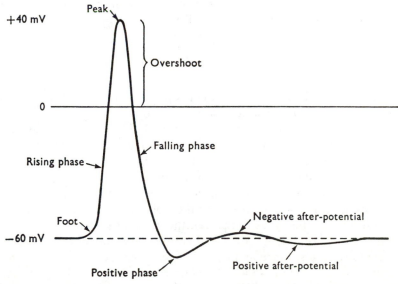

Fig. 4.5. Diagram to show the nomenclature applied to an action potential and the after-potentials which follow it.

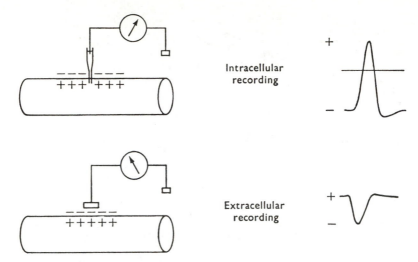

Fig. 4.6. Diagram to show the difference in sign of action potentials recorded by intracellular and by extracellular electrodes. Extracellular records are in fact frequently shown with negative potentials upwards.

potential is again more negative than usual. The reason for the apparently paradoxical nomenclature of these after-potentials is that they were first observed using extracellular electrodes, which of course measure the potential on the outside of the axon membrane instead of the inside, so that the directions of all potential changes are reversed; this point is illustrated in fig. 4.6.

SUBTHRESHOLD POTENTIALS

We will now consider some experiments performed by Hodgkin (1938) on isolated axons from the legs of the shore crab, *Carcinus*, with the object of examining the responses to subthreshold electrical stimuli. The axon was stimulated with brief (60 μsec) electrical shocks, and the response recorded in the vicinity of the stimulating electrodes. The results of a typical experiment are shown in fig. 4.7.

The responses to inward current are shown as the deflections below the baseline in fig. 4.7. Notice that the voltage does not immediately return to the baseline after the end of the stimulating pulse, but decays gradually. The reason for this is that the nerve membrane behaves electrically as a resistance and capacitance in parallel. We can see how this system works by reference to a simple electrical model (fig. 4.8). A battery E is connected to a resistance R and capacitance C via a switch S, so that a voltage V appears across the resistance–capacitance

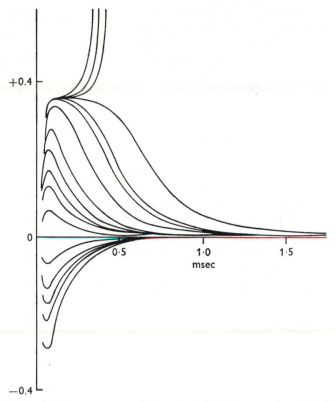

Fig. 4.7. Subthreshold responses recorded extracellularly from a crab axon in the vicinity of the stimulating electrodes. The ordinate is a voltage scale on which the height of the action potential is taken as one unit. (From Hodgkin, 1938, by permission of the Royal Society.)

network when the switch is closed. As a consequence there is a positive charge on one plate of the capacitance and a negative charge on the other. When the switch is opened, this charge flows through the resistance R as shown by the arrow, so that there is a potential difference

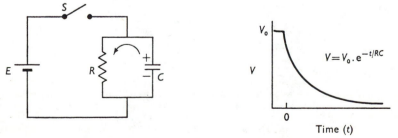

Fig. 4.8. Resistance and capacitance in parallel.

across R which falls exponentially as the capacitance discharges. Thus, returning to fig. 4.7, the voltage decay following passage of inward current through the membrane (anodal stimulation) is caused by the discharge of the membrane capacitance through the membrane resistance. This type of electrical change (i.e. one that can be attributed to current flow through the resistance and capacitance of the resting membrane) is known as an *electrotonic potential.*

The responses above the baseline in fig. 4.7 were obtained with the cathode of the stimulating electrodes next to the recording electrode (i.e. with cathodal stimulation, producing a flow of current outwards through the membrane). When the stimulus intensity was very small, the responses were mirror images of the responses to anodal stimulation, and can similarly be described as electrotonic potentials. However when the stimulus intensity was raised above half the threshold level, the resulting potential did not return to the baseline as rapidly as in the response to anodal stimulation of the same intensity. Hodgkin suggested that these responses were composed of a *local response* added to the electrotonic potential; the time course of the local response could therefore be obtained by subtracting the expected electrotonic response from the whole response. Finally, if the stimulus was large enough to produce a response which crossed a particular level of membrane potential, a

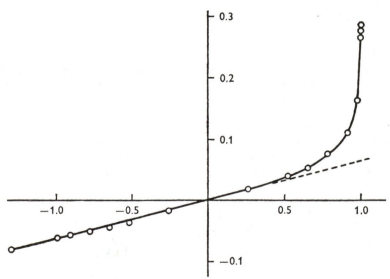

Fig. 4.9. The relation between stimulus and response in a crab axon, derived from fig. 4.7. The abscissa shows the stimulus intensity, measured as a fraction of the threshold stimulus. The ordinate shows the recorded potential 0.29 msec after the stimulus, measured as a fraction of the action potential height. (From Hodgkin, 1938, by permission of the Royal Society.)

propagated action potential was produced. This critical level of membrane potential is called the *threshold membrane potential*.

The results of this experiment are brought together in graphical form in fig. 4.9. For anodal stimuli, and for cathodal stimuli of intensities below about half threshold, the relation between stimulus and response is a straight line, so that the behaviour of the system (assuming the membrane capacitance to be constant) is in accordance with Ohm's law. Above this value this is not so, and thus we must assume that the resistance or capacitance of the membrane changes during local responses and action potentials.

IMPEDANCE CHANGES DURING ACTIVITY

If we pass current (I) through a resistor, the voltage (V) across the resistor is proportional to the current (Ohm's law again), and the constant of proportionality is called the resistance (R), so that

$$V = IR.$$

Similarly, if we pass alternating current through a network containing a resistance and capacitance in parallel, the voltage appearing across the network is proportional to the current, and the constant of proportionality is called the *impedance* (Z), so that

$$V = IZ.$$

The impedance is a complex quantity containing both resistive and capacitative elements, and dependent upon the frequency of the alternating current.

One method of determining the value of an unknown resistance R_x is by means of a simple Wheatstone bridge circuit (fig. 4.10*a*). The two

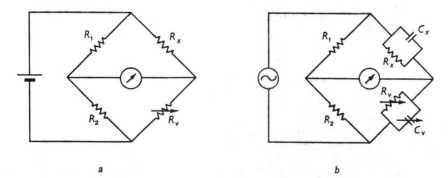

a   b

Fig. 4.10. Wheatstone bridges. *a*: Simple DC bridge for the measurement of resistance. *b*: AC bridge used to measure resistance and capacitance in parallel.

fixed resistances R_1 and R_2 are known, and R_v is a known variable resistance. R_v is altered until there is no voltage between C and D. In this condition (when the bridge is said to be 'balanced') it is evident that

$$R_1/R_2 = R_x/R_v$$

from which R_x can be calculated.

A similar method can be used to measure impedance. In fig. 4.10b the unknown circuit contains a resistance R_x in parallel with a capacitance C_x. If the two fixed resistances are equal, then the bridge is balanced when the variable resistance R_v is equal to R_x and the variable capacitance C_v is equal to C_x. This is the principle of the method used by Cole and Curtis (1939) to measure the impedance changes in the axon membrane during the passage of an action potential. A squid giant axon was placed in a trough so that it passed between two plate electrodes which were connected to a bridge circuit similar to that shown in fig. 4.10b. A high frequency alternating current (2 to 1000 kHz) was applied to the bridge, and the output was converted to a 175 kHz signal of the same amplitude and then displayed on an oscilloscope. Thus the oscilloscope trace was a thin line when the bridge was in balance, and became a broadened band (the 175 kHz AC signal) when it was out of balance. Knowing the values of the variable resistance and capacitance necessary to balance the bridge, and also the geometry of the axon–electrode system, it was possible to calculate the resistance and capacitance of the axon membrane.

Fig. 4.11a shows the effect of stimulation of the axon, with the bridge initially balanced in the resting condition. A short time after the start of the action potential, the oscilloscope trace broadens, indicating that the bridge is out of balance and therefore that the impedance of the system has changed. By starting with the bridge out of balance (as in figs. 4.11b and c), it is possible to find points at which the bridge is in balance at

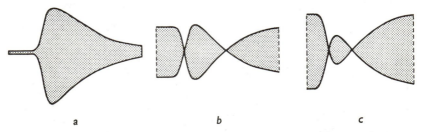

a   b   c

Fig. 4.11. Transverse impedance changes in a squid giant axon during the passage of an impulse. In a, the bridge was in balance during the resting state; in b and c it was out of balance during the resting state but was brought into balance by the impedance changes associated with the passage of the action potential. (After Cole and Curtis, 1939; redrawn.)

different times during the passage of the action potential and so to determine the resistance and capacitance of the membrane at these times. The results of these experiments showed that the membrane capacitance was 1.1 μF/cm^2 in the resting condition, and fell by about 2 per cent during activity. The resistance of the membrane, on the other hand, fell very markedly during activity, from its resting value of about 1000 Ω cm^2 to 25 Ω cm^2.*

THE PASSIVE ELECTRICAL PROPERTIES OF THE AXON MEMBRANE

From the results of the experiments so far described, it is evident that the axon membrane has a transverse resistance (r_m) and a transverse capacitance (c_m). If we wish to draw a circuit diagram to show the electrical properties of a length of axon, we must also include two other components, the longitudinal resistance of the external medium (r_o) and the longitudinal resistance of the axoplasm (r_i); the complete network is shown in the upper half of fig. 4.12. The suggestion that the axon can be represented in this way is known as the *core-conductor theory*, since it implies that the axon behaves as a poorly insulated cable. This representation is a simplification in some respects: (i) the membrane is a continuous structure, not a discontinuous series of elements; (ii) the resting potential is not represented – this could be incorporated by inserting a series of batteries in series with r_m, but since the object of the analysis is essentially to determine *changes* in membrane potential, this is an unnecessary refinement; and (iii) it omits the radial resistances of the external medium and axoplasm.

If we pass current through a part of the membrane so that a voltage (V_o) is set up across the membrane at that point, then the voltage across the membrane at some distant point $x(V_x)$ must be dependent on the distance x, and (because of the capacitances in the system) it must also vary with time, t. Hence it is necessary for us to know how V varies with x and t. We assume that the system is regular, i.e. that r_o, r_i, r_m and c_m do not change along the length of the axon, and that it is infinitely long. These quantities refer to unit lengths (e.g. 1 cm) of axon. Now the transverse current flowing through the membrane (i_m) will be the sum of the currents flowing through r_m and c_m, i.e.

$$i_m = \frac{V}{r_m} + c_m \frac{dV}{dt}. \tag{4.1}$$

* Notice that the membrane resistance must decrease with increase in area, hence it is measured in ohms *times* cm^2, whereas the membrane capacitance increases with increase in area and is therefore measured in microfarads *per* cm^2.

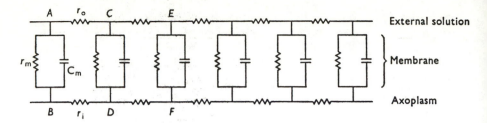

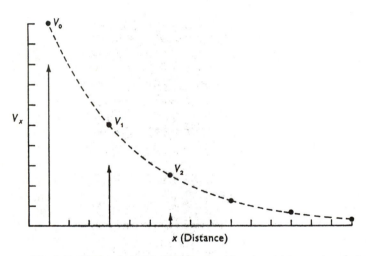

Fig. 4.12. Electrical model of the passive (electrotonic) properties of a length of axon. The graph shows the steady-state distribution of transmembrane potential along the model when points *A* and *B* on the model are connected to a constant current source. The vertical arrows on the graph are used to indicate that the potential at any point on the membrane rises gradually to its final value when the current is applied.

The current (i) flowing through the longitudinal resistances r_o and r_i must get progressively less as we move away from its source, since a constant fraction in each unit length is diverted through the membrane; thus it follows that

$$i = -\frac{dV}{dx}\left(\frac{1}{r_o + r_i}\right) \qquad (4.2)$$

and

$$i_m = -\frac{di}{dx}. \qquad (4.3)$$

Hence

$$i_m = \left(\frac{1}{r_o + r_i}\right)\frac{d^2 V}{dx^2}. \qquad (4.4)$$

Then, equating (4.4) with (4.1),

$$\frac{V}{r_m} + c_m \frac{\partial V}{\partial t} = \left(\frac{1}{r_o + r_i}\right)\frac{\partial^2 V}{\partial x^2},$$

or

$$V = \left(\frac{r_m}{r_o + r_i}\right)\frac{\partial^2 V}{\partial x^2} - r_m c_m \frac{\partial V}{\partial t}. \tag{4.5}$$

We now define two constants, a *space constant* λ, and a *time constant* τ, by

$$\lambda^2 = \frac{r_m}{r_o + r_i}$$

and

$$\tau = r_m c_m.$$

Substituting these definitions in (4.5), we get

$$V = \lambda^2 \frac{\partial^2 V}{\partial x^2} - \tau \frac{\partial V}{\partial t}. \tag{4.6}$$

This is the relation between V, x and t that we have been looking for. The solution of (4.6) requires the use of various transform methods (see Hodgkin and Rushton, 1946; Taylor, 1963; and Jack, Noble and Tsien, 1975); we shall merely consider some relatively simple results that follow from it.

First, consider the situation when a constant current has been applied for a long (effectively infinite) time, to set up the voltage V_0 at that point and the voltage V_x at a distance x from it. Then (4.6) simplifies to become

$$V = \lambda^2 \frac{\mathrm{d}^2 V}{\mathrm{d}x^2}.$$

The relevant solution of this equation is

$$V_x = V_0 e^{-x/\lambda}. \tag{4.7}$$

This means that the voltage across the membrane falls off exponentially with the distance from the point at which the current is applied. This feature is shown in the lower half of fig. 4.12. If we make x equal to λ in (4.7), V_x becomes $V_0 e^{-1}$; λ can thus be defined as the length over which the voltage across the membrane falls to $1/e$ of its original value.

Now consider the total charge (Q) on the membrane. This is given by

$$Q = c_m \int_0^\infty V \, \mathrm{d}x, \tag{4.8}$$

so that Q is obtained by integrating (4.6) with respect to x. If the applied current is constant, beginning at $t = 0$, then the solution of (4.8) is

$$Q_t = Q_\infty(1 - e^{-t/\tau}), \tag{4.9}$$

where Q_t is the charge at time t, and Q_∞ is the charge at infinite time (or, effectively, when $t \gg \tau$). If a current which has produced a charge Q_0 is suddenly switched off at $t = 0$, then the charge on the membrane decays according to the equation

$$Q_t = Q_0 e^{-t/\tau}. \tag{4.10}$$

If we put t equal to τ in (4.9) or (4.10) it becomes evident that τ can be defined as the time taken for the change in total charge on the membrane to reach $1/e$ of completion.

The equations describing the time course of the voltage change at any particular point are rather complex, but their general effect is that the potential rises and falls less rapidly the farther that point is from the point at which the current is applied. The responses to the onset and cessation of constant currents are, of course, limited by (4.7), (4.9) and (4.10). Fig. 4.13 shows the results of Hodgkin and Rushton's calculations of the voltage changes.

The core conductor equations we have examined so far have been based on quantities referring to a unit length of axon. For many purposes, however, it is useful to possess measurements of membrane properties in terms of unit area; the conversion from one to the other can easily be made if we know the radius of the axon. The relation between the membrane resistance of a unit length of axon, r_m, and the resistance of a unit area of membrane, R_m, is given by

$$r_m = R_m/2\pi a, \tag{4.11}$$

where a is the axon radius. Similarly, the membrane capacitance per unit length, c_m, is related to the membrane capacitance per unit area, C_m, by

$$c_m = 2\pi a C_m. \tag{4.12}$$

And the longitudinal resistance of the axoplasm in a unit length of axon, r_i, is related to the resistivity of the axoplasm, R_i, by

$$r_i = R_i/\pi a^2. \tag{4.13}$$

To take a numerical example, consider an axon of radius $25 \ \mu m$, membrane resistance (R_m) $2000 \ \Omega \ cm^2$, axoplasm resistivity (R_i) $60 \ \Omega \ cm$ and membrane capacitance (C_m) $1 \ \mu F/cm^2$. Then r_m (membrane resistance of a unit length) is $127\,000 \ \Omega \ cm$, r_i (internal longitudinal resistance) is $3\,060\,000 \ \Omega/cm$, and c_m (membrane capacitance per unit length) is $0.0157 \ \mu F/cm$.

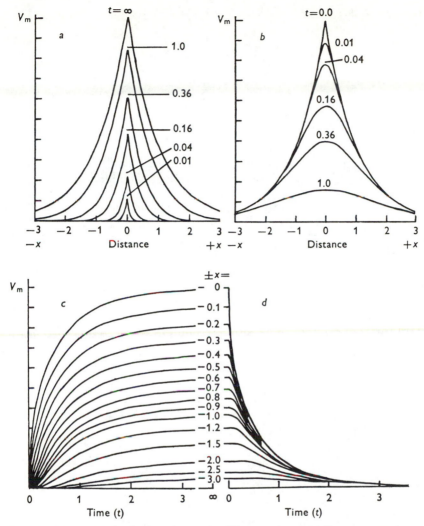

Fig. 4.13. Theoretical distribution of potential difference across a passive nerve membrane in response to onset (*a* and *c*) and cessation (*b* and *d*) of a constant current applied at the point $x = 0$. *a* and *b* show the spatial distribution of potential difference at different times, and *c* and *d* show the time courses of the potential change at different distances along the axon. Time (*t*) is expressed in units equal to the time constant, τ, and distance (*x*) is expressed in units equal to the space constant, λ. (From Hodgkin and Rushton, 1946, by permission of the Royal Society.)

These relations can usefully be inserted in some of the core-conductor equations. For simplicity we shall assume that the axon is in a large volume of external medium, so that r_o is very much less than r_i and can therefore be omitted from the equations. First, consider the space constant λ. If r_o is very small compared with r_i, then

$$\lambda = \sqrt{\frac{r_m}{r_i}}$$

$$= \sqrt{\frac{R_m/2\pi a}{R_i/\pi a^2}}$$

$$= \sqrt{\frac{a \cdot R_m}{2R_i}}.$$

Hence λ increases with increasing axon radius; if R_m and R_i remain constant, λ is proportional to the square root of the radius.

We have seen that the time constant is defined by

$$\tau = r_m c_m.$$

Substituting (4.11) and (4.12) in this we get

$$\tau = (R_m/2\pi a)(2\pi a C_m)$$

$$= R_m C_m.$$

Thus τ is essentially independent of the axon radius.

A further useful relation follows from (4.4); substituting (4.13) in this we get

$$i_m = \frac{\pi a^2}{R_i} \cdot \frac{d^2 V}{dx^2}.$$

Now the membrane current density per unit area, I_m, will be related to the membrane current per unit length, i_m, by

$$i_m = I_m \cdot 2\pi a.$$

Hence

$$I_m = \frac{a}{2R_i} \cdot \frac{d^2 V}{dx^2}. \tag{4.14}$$

(This equation, as we shall see later, is of considerable importance in the further analysis of nervous conduction.)

The core-conductor theory was experimentally tested by Hodgkin and Rushton (1946) on large axons from the walking legs of the lobster *Homarus*. Since they used extracellular electrodes, the voltage change measured was that on the outside of the membrane; this will be proportional to the change in potential across the membrane (the proportionality constant will be $r_o/(r_o + r_i)$) so that the core-conductor equations can still be applied. It was found that the observed potential changes were in accordance with (4.7), (4.9) and (4.10) and with equations derived from (4.6) describing the change of potential with time at different distances from the point of application of the current. This correspondence between fact and theory is obviously good evidence for the core-conductor theory. The next step was to evaluate the quantities r_o, r_i, r_m and c_m. Then, knowing the diameter of the axon, the resistance and capacitance of a square centimetre of membrane could be calculated. Average values from experiments on ten axons were $2300 \, \Omega \, \text{cm}^2$ for R_m, the membrane resistance (range 600 to $7000 \, \Omega \, \text{cm}^2$ – this quantity was rather variable), and $1.3 \, \mu\text{F/cm}^2$ for C_m, the membrane capacitance. These values are in quite good agreement with those obtained for squid axons by Cole and Curtis using the AC bridge method, a fact which further increases our confidence in the applicability of the core-conductor model.

THE LOCAL CIRCUIT THEORY

We are now in a position to consider how the action potential is propagated. Suppose a small length of axon is stimulated by a depolarizing pulse above threshold intensity, so that an action potential arises in the stimulated area. Regard this action potential as corresponding to the voltage V_0 in fig. 4.12 and assume that V_0 is about four times the threshold change in membrane potential. The potential across CD will now rise towards the point V_1, but at some point before reaching V_1 it will cross the threshold, so that an action potential appears across CD. This means that the potential across EF, originally moving towards V_2, now starts moving towards the potential level of V_1, and so it in turn crosses the threshold and an action potential now arises across EF. Thus an action potential has been propagated from AB through CD to EF, and will of course continue along the chain. Furthermore, unless there is any change along the length of the membrane in the threshold level or in the values of the component resistances and capacitances of the system, the conduction velocity will obviously be constant.

This hypothesis of the mechanism of conduction is known as the *local circuit theory*, since it postulates that conduction is dependent on the electrotonic currents across the membrane set up in front of the action potential. The local circuits are shown in fig. 4.14*a*. Notice that, in order to set up passive (electrotonic) currents at the beginning of the action potential, there must be some inward current flow at the peak of the action potential. This inward current flow is from negative to positive and is therefore analogous to the flow *inside* a battery; in other words, the electrical energy needed to cause the electrotonic currents (which flow from positive to negative) is derived from this inward movement of positive charge at the peak of the action potential. This concept is

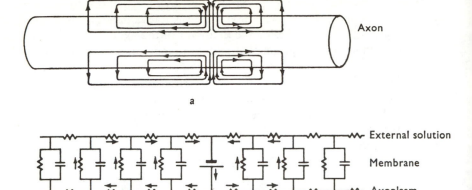

Fig. 4.14. *a*: Local circuit currents set up by a propagating action potential. *b*: Model system to show the local circuit currents (shown by arrows) set up by a battery inserted in the core-conductor model.

illustrated in fig. 4.14*b*. The 'battery'* is brought into action at any particular point when the membrane potential crosses the threshold at that point. The quantitative distribution of membrane current can be deduced from the form of the action potential by applying (4.4).

If the local circuit theory is correct, then it should be possible to observe the local currents associated with the action potential. This was done by Hodgkin (1937) by cooling a short length of nerve so that the action potential could not pass this region. On the other side of the block, small *extrinsic potentials* appeared (fig. 4.15), which diminished in size along the length of the nerve in the same way that electrotonic potentials do.

A less direct method of testing the theory is by consideration of the factors which affect conduction velocity. Reverting to fig. 4.12 again, it is evident that the conduction velocity must be dependent upon the time taken for the potential at a point a given distance in front of the action potential to cross the threshold. This time, in turn, will be governed by the values of the resistances in the system. Hence changes in r_o or r_i should produce changes in conduction velocity. Hodgkin (1939) showed that an increase in r_o, produced by immersing an axon in paraffin oil instead of Ringer's solution, decreased the conduction velocity. Conversely, a decrease in r_o, produced by laying the axon across a series of

---

* We shall see in the next chapter that this 'battery' is a sodium ion concentration cell and that the inward current it produces is an inward flow of sodium ions.

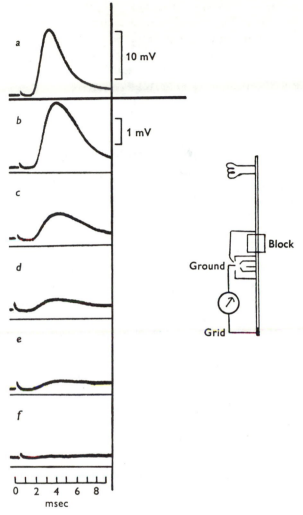

Fig. 4.15. Extrinsic potentials in frog sciatic nerve. A 3 mm length of nerve was blocked by cooling, and the electrical response of the nerve was recorded 2 mm proximal to the block (*a*) and at various distances distal to the block (*b–f*). Notice that the gain for records *b* to *f* is five times that for *a*. The distal records were made at the following distances from the block (mm): *b*, 1.4; *c*, 2.5; *d*, 4.1; *e*, 5.5; and *f*, 8.3. (From Hodgkin, 1937).

brass bars which could be connected by means of a mercury trough, resulted in an increase in conduction velocity. More recently, del Castillo and Moore (1959) have obtained large increases in conduction velocity by inserting a silver wire down the middle of an axon, so greatly reducing r_i.

Since r_i varies with axon radius, we would expect conduction velocity to vary with axon radius also. Theoretical arguments given by Hodgkin (1954) suggest that, other things being equal (a rather important proviso), conduction velocity is proportional to the square root of the axon radius. A simple argument why this should be so is as follows. For an action potential travelling at a constant velocity θ,

$$\frac{dV}{dx} = \frac{1}{\theta} \cdot \frac{dV}{dt}$$

and

$$\frac{d^2V}{dx^2} = \frac{1}{\theta^2} \cdot \frac{d^2V}{dt^2}.$$

Substituting this relation in (4.14), we get

$$I_m = \frac{a}{2R_i \cdot \theta^2} \frac{d^2V}{dt^2}. \qquad (4.15)$$

If the membrane properties remain the same in axons of different radii, then I_m and d^2V/dt^2 will be constant. In this case (4.15) becomes

$$\frac{a}{\theta^2} = \text{constant}$$

or

$$\theta = \text{constant} \times \sqrt{a},$$

i.e. the conduction velocity is proportional to the square root of the fibre radius.

An experimental investigation of this relation was carried out by Pumphrey and Young (1938), using the giant axons of *Sepia* and *Loligo*; the best fit for their results was given by conduction velocity being proportional to $(\text{radius})^{0.6}$, which is very near to the expected value. Other investigations have not always provided such good agreement with the square root relation, but it is generally observed that, in a group of axons of the same type, increase in radius is associated with increase in conduction velocity.

The conclusion to be drawn from the experiments and observations described in this section is that the action potential propagates by means of local circuits, as shown in fig. 4.14*a*. This analysis enables us to explain the shapes of action potentials as recorded by external electrodes. In fig. 4.16*a* two electrodes X and Y are placed a short distance apart on an axon. We shall assume that the input resistance of the recording system used to measure the potential difference between X and Y is very high. As the active region moves under electrode X, X becomes negative to Y, so producing an upward peak in the record. However, when the active region reaches electrode Y, X becomes

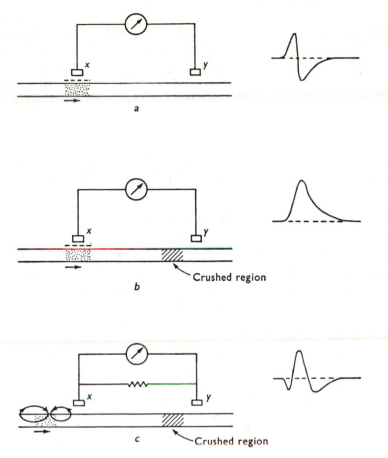

Fig. 4.16. The form of action potentials recorded from nerve axons with extracellular electrodes. The active region is stippled in each case. The records on the right show an upward deflection when electrode X is negative to electrode Y. Further explanation in the text.

positive to Y and so the potential record shows a downward peak. This type of recording is known as diphasic. In fig. 4.16b the axon is crushed between the electrodes so that the action potential is unable to reach Y, and therefore Y never goes negative; this gives a monophasic record which is similar in shape to the potential across the membrane at X. In fig. 4.16c there is a shunt resistance between the electrodes (such as might be produced by recording from a nerve in Ringer's solution or *in situ*) so that currents can flow between them. In this case the electrode X records the local currents associated with the action potential and the record is therefore triphasic.

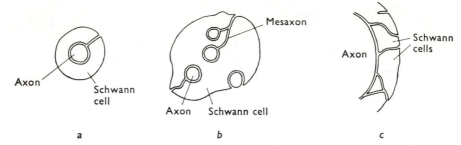

Fig. 4.17. The form of Schwann cells surrounding unmyelinated axons. *a* shows the basic arrangement, *b* the situation with many small axons, and *c* the situation with large axons.

SALTATORY CONDUCTION IN MYELINATED NERVES

All peripheral nerve axons are surrounded by accessory cells called *Schwann cells* (fig. 4.17). A connective tissue sheath, the *endoneurium*, which contains small endoneurial cells, surrounds the complex of Schwann cells and axon. Numbers of these Schwann cells and endoneurial cells are found distributed along the length of an axon. Transverse sections of large axons (such as squid giant axons, as shown in fig. 4.17*c*) shows that many Schwann cells are needed to surround them. When the axons are small, the reverse situation may be seen (fig. 4.17*b*), in which a single Schwann cell surrounds a number of axons. In all these cases the axon is separated from the Schwann cell by a space about 150 Å wide, which is in communication with the extracellular fluid via channels (mesaxons) between the membranes of the Schwann

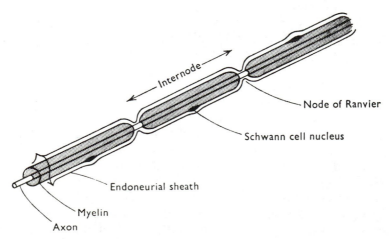

Fig. 4.18. Diagram to show the structure of a myelinated nerve fibre.

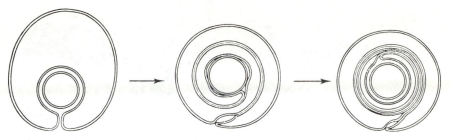

Fig. 4.19. The development of the myelin sheath by vertebrate Schwann cells. (From Robertson, 1960, by permission of Pergamon Press Ltd.)

cells. Examination of axons of this type by light microscopy does not reveal the presence of a fatty sheath surrounding the axon, and the nerve fibres (we shall regard a 'fibre' as comprising an axon plus its accessory cells) are therefore described as *unmyelinated*.

Many vertebrate and a few invertebrate axons (such as the large diameter axons of prawns; Holmes, 1942) are surrounded by a fatty sheath known as *myelin* and are described as being *myelinated*. The sheath is interrupted at intervals of the order of a millimetre or so, to form the *nodes of Ranvier* (fig. 4.18). Electron microscope studies by Geren (1954) and Robertson (1960) have shown that myelin is formed from many closely packed layers of Schwann cell membrane, produced by the mesaxon being wrapped round and round the axon (fig. 4.19). As a consequence of this arrangement, there are not extracellular channels between the axon membrane and the external medium in the internodal regions. Such contact is available at the nodes, as is shown in fig. 4.20.

It is obvious that the presence of a myelin sheath will considerably affect the electrical properties of the nerve fibre. The overlapping Schwann cell membranes act as chains of resistances and capacitances in series. This means that the myelin sheath will have a much higher transverse resistance and a much lower transverse capacitance than a normal cell membrane; the actual figures for frog fibres are about $160\,000\ \Omega\,cm^2$ and $0.0025\ \mu F/cm^2$. The values for the resting membrane at the node are quite different, being of the order of $20\ \Omega\,cm^2$ and $3\ \mu F/cm^2$. In view of these figures, we might expect that when current is passed across the membrane the current flow through the nodes would be greater than that at the internodes. If, in addition, the nodes are the regions at which excitation of the membrane occurs, then, applying the local circuit theory, we would expect conduction in a myelinated axon to be a discontinuous (or *saltatory*) process.

Verification of the saltatory conduction theory depends first of all on the ability of the experimenter to dissect out single fibres from a whole

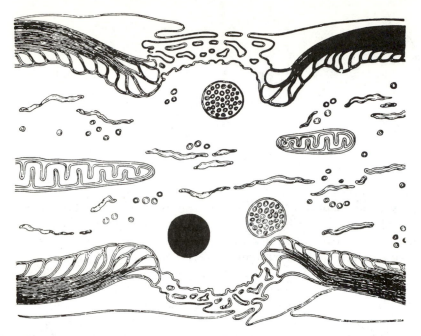

Fig. 4.20. Longitudinal section of a node of Ranvier, as seen by electron microscopy. (From Robertson, 1960, by permission of Pergamon Press Ltd.)

nerve trunk. This technique was first developed by Japanese workers in the nineteen-thirties. It was found that blocking agents such as cocaine or urethane were only effective when applied at the nodes, and that the threshold stimulus intensity was lowest at the nodes (fig. 4.21). These results accord well with the predictions of the saltatory theory, but it was possible to argue that the sensitivity of the nodal region was merely due to the exposure of the axon membrane at this site. More conclusive evidence that conduction is saltatory would be provided if it could be shown that inward flow of current is restricted to the nodes. Experiments to test this idea have been performed by Tasaki and Takeuchi (1942) and by Huxley and Stämpfli (1949); we shall now consider some of the details of these experiments.

Tasaki and Takeuchi's technique is shown in fig. 4.22. The nerve fibre is placed in three pools of Ringer's solution insulated from each other by air gaps. The two outer pools are earthed, and the middle pool is connected to earth via a resistance. With this arrangement, all the radial current flowing across the axon membrane and myelin sheath of that part of the fibre which is in the middle pool will flow through the resistance, and can therefore be measured by the potential of the middle pool with respect to earth. The results (fig. 4.22*b*) showed that inward

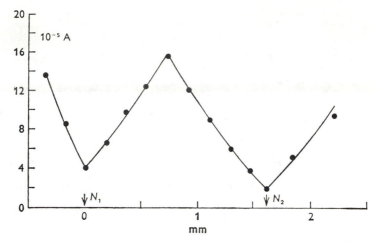

Fig. 4.21. The variation in threshold stimulus intensity along the length of a single myelinated fibre. N_1 and N_2 mark the positions of two nodes of Ranvier. (From Tasaki, 1953.)

currents occurred only when there was a node in the middle pool, and therefore inward current flow must be restricted to the nodal regions.

The technique used by Huxley and Stämpfli was rather different. The fibre was threaded through a fine hole in a glass slide, and the potential measured on each side of the hole during the passage of an impulse (fig. 4.23). If the resistance of the Ringer's solution surrounding the fibre in the hole is known, then the longitudinal current flow can readily be obtained from the measured potential by application of Ohm's law. By

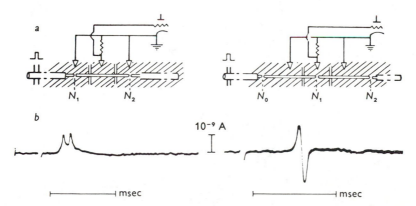

Fig. 4.22. The radial currents in a short length of a myelinated fibre during the passage of an action potential. *a*: The recording arrangement. *b*: The potential difference across the resistance *R* when the middle pool of Ringer does (right trace) or does not (left trace) contain a node. (From Tasaki and Takeuchi, 1942.)

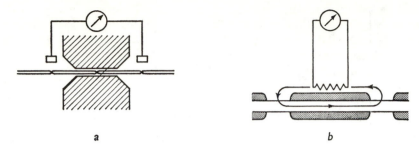

Fig. 4.23. The method used by Huxley and Stämpfli to measure longitudinal currents outside a short length of a myelinated fibre. *a* gives a schematic view of the recording arrangement, and *b* the equivalent circuit of the system. (After Huxley and Stämpfli, 1949.)

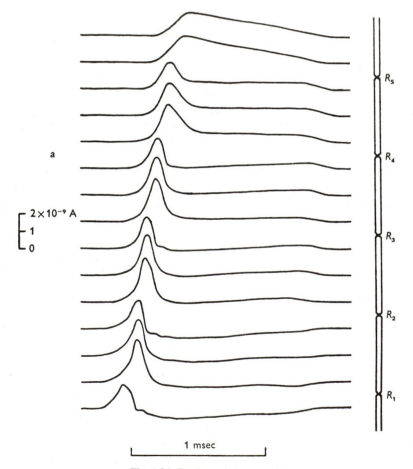

Fig. 4.24. For legend see opposite.

pulling the fibre through the hole, a series of such measurements can be made at intervals along its length, as shown in fig. 4.24a. The transverse current across the membrane is then given by (4.3), i.e. by the difference between the longitudinal currents at two adjacent recording sites. The calculated membrane currents are shown in fig. 4.24b, from which it is clear that here, again, inward current flow is restricted to the nodes. This implies that the transverse current in the internodal regions can be

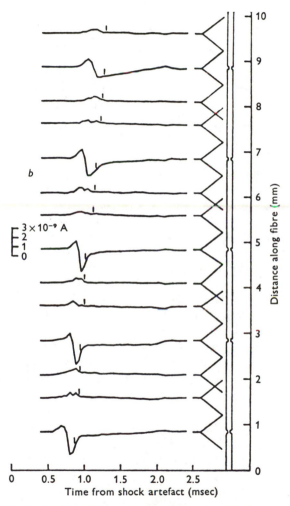

Fig. 4.24. Longitudinal and radial currents at different points along the length of a myelinated fibre during the passage of an action potential. a: Longitudinal currents, measured directly by the method shown in fig. 4.24. b: Radial (membrane) currents, obtained from the difference between the longitudinal currents at two points 0.75 mm apart. (From Huxley and Stämpfli, 1949.)

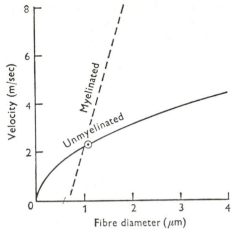

Fig. 4.25. Theoretical relations between fibre diameter and conduction velocity in myelinated and unmyelinated nerve fibres. (From Rushton, 1951.)

explained as a passive current through a resistance and capacitance in parallel, but there must be some 'active' component at the nodes of Ranvier.

The conduction velocity of myelinated fibres is normally greater than that of unmyelinated fibres of the same diameter. The reason for this is that the high resistance and low capacitance of the myelin sheath increases the longitudinal spread of the local currents involved in propagation. Theoretical arguments given by Rushton (1951) suggest that conduction velocity should be proportional to the fibre diameter, and not, as in unmyelinated axons, to the square root of fibre diameter. This appears to be so (Hursh, 1939; Tasaki, Ishi and Ito, 1943). This conclusion implies that the conduction velocity of very small myelinated fibres should be less than that of unmyelinated fibres of the same diameter (fig. 4.25). Rushton calculated that the critical diameter below which myelination confers no advantage should be about 1 μm, which is near that of the largest unmyelinated fibres in mammalian peripheral nerves. However Waxman and Bennett (1972) point out that myelinated fibres less than 1 μm in diameter are found in the central nervous system, and suggest that the critical diameter is as low as 0.2 μm.

ELECTRICAL STIMULATION PARAMETERS

Strength–duration relation

If an axon is stimulated with square constant current pulses, it is found that the threshold stimulus intensity rises as the pulse length is lessened.

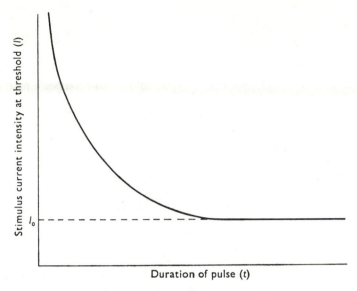

Fig. 4.26. The strength–duration curve.

This effect is known as the *strength–duration relation* (fig. 4.26). The curve in fig. 4.26 is quite well fitted by the empirical equation

$$\frac{I}{I_0} = \frac{1}{1 - e^{-t/k}} \tag{4.16}$$

where I is the intensity of the pulse, t is the length of the pulse, I_0 is the threshold stimulus intensity when t is large (the *rheobase*) and k is a constant (Lapique, 1907; see Hill, 1936). The pulse length when the threshold stimulus intensity is twice the rheobase is called the *chronaxie*.

Equation (4.16) is similar in form to that describing the change of voltage across a circuit consisting of a resistance and capacitance in parallel following the application of a constant current, i.e.

$$V = IR(1 - e^{-t/RC}) \tag{4.17}$$

where V is the voltage, I the current, t the time after application of the current, R the resistance and C the capacitance. If V is constant (for instance, if we regard V as the threshold membrane depolarization), this becomes

$$I \times \text{constant} = \frac{1}{1 - e^{-t/RC}}$$

which is equivalent to (4.16). But this equation involves a number of simplifications (see Noble, 1966). Firstly, the voltage across the passive

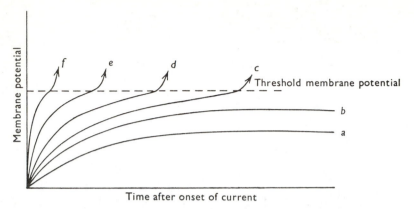

Fig. 4.27. Schematic diagram to show the effects of depolarizing currents on the membrane potential of a nerve axon. The current strengths increase in the order *a* to *f*. Responses *c* to *f* lead to excitation.

membrane cannot be described by (4.17); the much more complicated relations derived from (4.6) must be applied. Secondly, strong currents of short duration produce depolarizations which fall off rapidly with distance from the stimulating electrode, whereas the depolarizations produced by weaker currents of longer duration are more spread out, as will be evident from fig. 4.13. Since propagation is dependent on local circuits of sufficient intensity (involving movement of at least a minimal quantity of electric charge), the size of the active region is important, hence the threshold membrane depolarization is higher when smaller areas of membrane are depolarized, and therefore it is higher for brief constant current pulses than for longer ones. Finally, the existence of local responses (p. 44) means that (4.6) cannot be strictly applied for depolarizations greater than about half threshold. Thus, while it is easy to provide a qualitative explanation for the strength–duration relation (depolarization is more rapid with stronger currents; see fig. 4.27), a quantitative formulation of the various factors involved (Noble and Stein, 1966) must necessarily be very complex.

Latency

The time between the onset of a stimulus and the peak of the ensuing action potential is called the *latency* of the response. It is clear from fig. 4.27 that the latency decreases with increasing current strengths.

Latent addition

For a short time after a brief subthreshold cathodal stimulus, the membrane potential is nearer to the threshold membrane potential than usual. During this time, less current is required to depolarize the

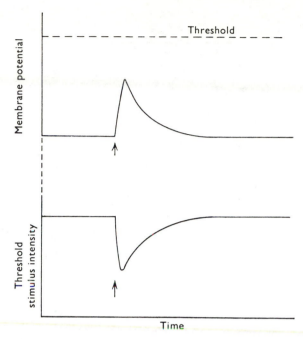

Fig. 4.28. To illustrate latent addition. A subthreshold depolarizing current pulse is applied to an axon at the time shown by the arrow. The upper trace shows the consequent change in membrane potential. The lower trace, which is proportional to the difference between the upper trace and the threshold membrane potential, shows the changes in threshold stimulus intensity which this produces.

membrane by enough to cause excitation, so that stimuli which are normally of subthreshold intensity may be sufficient to elicit an action potential (fig. 4.28). Conversely, if an anodal shock is applied, a cathodal stimulus soon afterwards has to be of a higher intensity than usual in order to cause excitation. These phenomena constitute *latent addition.* They can be used to investigate the time course of subthreshold responses. Katz (1937), for instance, demonstrated indirectly the presence of local responses in this way; his results were very similar to those obtained by Hodgkin (1938) from direct observations on crab nerves.

Accommodation

If a constant subthreshold depolarizing current is passed through an axon membrane, it is found that the threshold slowly rises during the passage of the current, and then falls again after its cessation. Conversely, when hyperpolarizing current is used, the threshold falls. This delayed dependence of the threshold on membrane potential is known

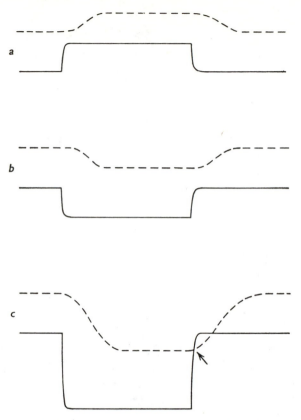

Fig. 4.29. Accommodation in response to constant currents. In each case the continuous line shows the membrane potential and the dotted line the threshold membrane potential. *a*: Depolarizing current; *b* and *c*; hyperpolarizing current. In *c* the current is strong enough to cause 'anode break excitation' after it is switched off, at the point marked by the arrow.

as accommodation (fig. 4.29). With hyperpolarizing currents of sufficient strength, the threshold may fall beyond the resting potential, so that when the current is switched off, the membrane potential is temporarily above threshold. This leads to the initiation of an action potential (fig. 4.29c), and the phenomenon is therefore known as *anode break excitation*.

If an axon is stimulated by a linearly rising cathodal current, the membrane potential change follows an approximately linear course, and, because of accommodation, the threshold membrane potential also changes slowly. Consequently, the threshold current intensity is lower for rapidly rising currents than it is for slowly rising ones. When the rate of rise of current is sufficiently low, the change in membrane potential is

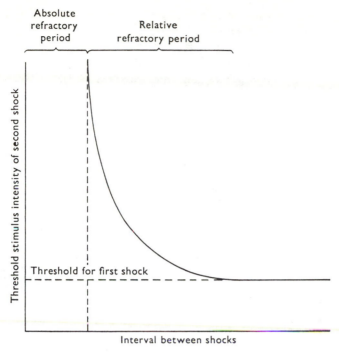

Fig. 4.30. Refractory periods.

too slow to be able to gain on the change in threshold level, and therefore no excitation occurs.

The refractory period

For a short time after the passage of an action potential, it is not possible to elicit a second action potential, however high the stimulus intensity is. This period is known as the *absolute refractory period* (fig. 4.30). Following the end of the absolute refractory period, there is a period during which it is possible to elicit a second action potential, but the threshold stimulus intensity is higher than usual. During this *relative refractory period* the second action potential, if it is recorded near the stimulating electrodes, may be reduced in size. The existence of refractoriness places an upper limit on the frequencies at which axons can conduct nerve impulses.

COMPOUND ACTION POTENTIALS

The isolated sciatic nerve of the frog has been much used in the study of nervous phenomena. If such a nerve is arranged so that it can be

TABLE 4.1. *Fibre groups in bullfrog sciatic nerve, with typical values for their conduction velocities at room temperature.* (Based on Erlanger and Gasser, 1937.)

Fibre group		Conduction velocity (m/sec)
	α	41
A	β	22
	γ	15
B		4
C		0.7

stimulated at one end and recorded from at the other, action potentials can be recorded in the usual manner. These action potentials differ from those of isolated single fibres in that their size is, over a restricted range, proportional to the size of the stimulus. The reason for this is that the recorded action potential is the result of simultaneous activity in a large number of axons (each of which itself obeys the all-or-nothing law) which have different thresholds.

When the distance between the stimulating and recording electrodes is large, it is found that the monophasic compound action potential (recorded as in Fig. 4.16*b*) consists of a number of potential waves; this is due to the nerve trunk containing fibres with different conduction velocities. Erlanger and Gasser (1937) classified the groups of nerve fibres in the trunk according to their conduction velocities, as is shown in table 4.1.

5 The ionic theory of nervous conduction

In chapter 3, we saw that the resting potential is determined by the ionic concentration gradients across the cell membrane and the relative permeabilities of the membrane to the different ions in the system. The aim of this chapter is to show that the electrical excitability of the axon membrane, as described in the previous chapter, is dependent upon changes in its ionic permeability; in particular, upon changes in the permeability to sodium and potassium ions.

According to the ionic theory of nervous conduction, we can interpret the properties of the axon membrane in terms of a conceptual model, the electrical circuit shown in fig. 5.1. In this model, we assume that the concentration gradient of any particular ion in the system acts as a battery, whose electromotive force is given by the Nernst equation for

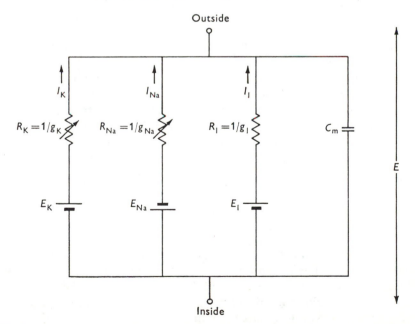

Fig. 5.1. Conceptual model of a patch of electrically excitable membrane. (After Hodgkin and Huxley, 1952d.)

71

that ion. For example, there is a 'potassium battery', E_K, whose electromotive force is given by

$$E_K = \frac{RT}{F} \log_e \frac{[K]_o}{[K]_i}.$$ (5.1)

In squid axons, E_K is usually about -75 mV (inside negative). Similarly, the electromotive force of the 'sodium battery', E_{Na}, is given by

$$E_{Na} = \frac{RT}{F} \log_e \frac{[Na]_o}{[Na]_i}.$$ (5.2)

Since $[Na]_o$ is usually greater than $[Na]_i$, E_{Na} is positive, typical values being in the region of $+55$ mV. The other ions in the system, of which chloride is usually the most important, contribute to a third battery, E_l, which may produce a 'leakage current'; the value of E_l is near the resting potential, at about -55 mV. In series with each battery is a resistance (R_K, R_{Na} and R_l in fig. 5.1); but it is more convenient to talk about the conductances, g_K, g_{Na} and g_l (conductance is the reciprocal of resistance). These conductances represent the ease with which ions can pass through the membrane; they are related to the permeability coefficients mentioned in chapter 3, but of course the units of measurement are different. We assume that the sodium and potassium conductances are variable. These three ionic channels are arranged in parallel, the current flowing through each channel being represented by I_K, I_{Na} and I_l. Finally, the membrane capacitance is represented by a fourth unit C_m.

A moment's thought will show that the net potential across the membrane, E, will, in the absence of externally applied current, be determined by the relative values of the ionic conductances. If g_K is much higher than g_{Na}, or g_l, for instance, E will be near to E_K; if g_{Na} is then increased, E will move towards E_{Na}, and so on. We must now consider the evidence that this conceptual model provides an adequate representation of the properties of the axon membrane.

THE 'SODIUM THEORY' OF THE ACTION
POTENTIAL

The suggestion that the resting potential is due to the selective permeability of the membrane to potassium ions was first made by Bernstein in 1902. He also suggested that the action potential is brought about by a breakdown in this selectivity, so that the membrane potential falls to zero. However the discovery that the membrane potential 'overshoots' the zero level during the action potential, so that the inside becomes positive, implied that some other process must be involved. Hodgkin and Katz (1949) suggested that this process was a rapid and specific increase in the permeability of the membrane to sodium ions. In terms of the model shown in fig. 5.1, g_{Na} becomes temporarily very much larger than g_K, so that E moved towards E_{Na}.

In order to test the validity of the 'sodium theory' as this suggestion is known, Hodgkin and Katz measured the height of action potentials from squid giant axons placed in solutions containing different concentrations of sodium ions. No action potentials could be produced in the absence of sodium ions in the external medium. If the axon was placed in a solution with reduced sodium ion concentration (prepared by mixing sea water with an isotonic glucose solution), the action potential was reduced in height (fig. 5.2). Furthermore the slope of the curve relating

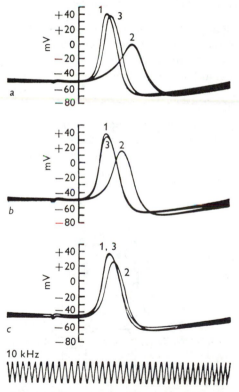

Fig. 5.2. The effect of reducing the external sodium ion concentration on the size of the action potential in a squid giant axon. In each set of records, record 1 shows the response with the axon in sea water, record 2 in the experimental solution, and record 3 in sea water again. Experimental solutions were made by mixing sea water and isotonic glucose solutions, the proportions of sea water being *a*, 33 per cent; *b*, 50 per cent and *c*, 70 per cent. (From Hodgkin and Katz, 1949.)

the height of the action potential to $\log_{10} [Na]_o$ was close to $58 \, mV$ per unit, except at very low sodium ion concentrations where failure of conduction was immanent; this is just what one would expect from (5.2). It was also found that the height of the action potential was increased after addition of sodium chloride to the external solution.

Similar results have been obtained from a variety of other electrically excitable cells, including frog myelinated axons (Huxley and Stämpfli, 1951), insect axons (Narahashi, 1963), frog 'fast' skeletal muscle fibres (Nastuk and Hodgkin, 1950) and mammalian heart muscle fibres (Draper and Weidmann, 1951). The 'sodium theory' thus seems to be of fairly general application, although certain arthropod muscle fibres provide an exception (see chapter 13).

IONIC MOVEMENTS DURING ACTIVITY

Direct measurements of ionic movements, using radioactive isotopes as tracers and various accurate methods of chemical analysis, are obviously relevant to a study of the nature of the action potential. A large number of studies of this kind have been made in recent years, but we will here consider just one set of experiments, by Keynes (1951), on the movement of radioactive sodium and potassium ions across the membrane of the giant axons of the cuttlefish *Sepia*. The principle of the technique is shown in fig. 5.3. The axon, held in two pairs of forceps, could be loaded with the radioactive isotope by immersion in a pot of sea water containing the isotope. In order to observe its radioactivity, it was transferred to a chamber containing flowing non-radioactive sea water and placed over a Geiger counter. Stimulating pulses could be applied via one pair of

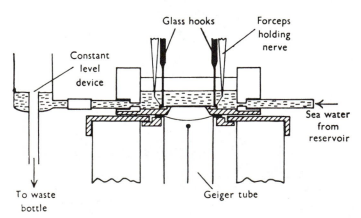

Fig. 5.3. The method used by Keynes to measure the radioactivity of *Sepia* giant axons. (From Keynes, 1951.)

forceps, and action potentials could be recorded from the other pair so as to provide a check on the excitability of the axon.

Fig. 5.4 shows the results of one of Keynes's experiments on the movement of radioactive potassium ions. The axon was initially immersed in ^{42}K-sea water. After 15 minutes it was removed to the counting chamber and its radioactivity measured at intervals during the next 40 minutes; when plotted on a logarithmic scale, these values fell on a straight line, and so the radioactivity at 15 minutes could be estimated by producing this line back. From this value, knowing the potassium ion concentration of the sea water, and its specific activity, the rate of entry of potassium could be calculated. Then the axon was transferred to the radioactive sea water again and stimulated for ten minutes; on returning to the measuring chamber its increase in radioactivity could be calculated as before, from which the extra influx on stimulation could be determined. The rate of loss of radioactivity in the resting condition could be obtained from the slope of the lines through the experimental points in fig. 5.4, and the rate of loss on stimulation could be found by stimulation in the non-radioactive sea water in the

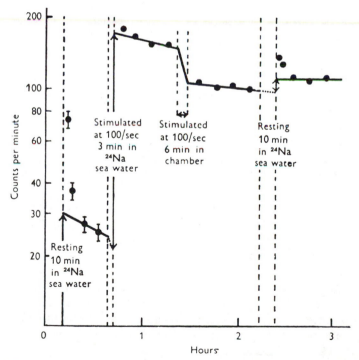

Fig. 5.4. Results of an experiment to determine potassium ion movements in a *Sepia* giant axon. (From Keynes, 1951.)

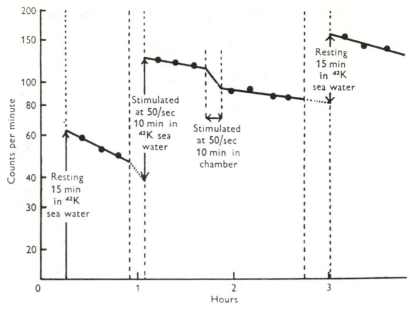

Fig. 5.5. Results of an experiment to determine sodium ion movements in a *Sepia* giant axon. The values recorded immediately after a period of immersion in radioactive sea water probably included radioactivity carried over in the external medium; the straight lines are therefore drawn through the later points only. (From Keynes, 1951.)

counting chamber. By making plausible assumptions as to the internal potassium concentration, these figures could be converted into absolute values of potassium efflux.

Similar experiments were performed with sodium ions (fig. 5.5). The main results of the investigation are shown in graphical form in fig. 5.6. The net effect of each impulse was to produce a net entry of 3.7 pmoles/cm^2 of sodium ions and a net exit of 4.3 pmoles/cm^2 of potassium ions.

Entry of sodium ions will cause a build-up of positive charge on the inside of the axon membrane (making the membrane potential more positive), and exit of potassium ions will remove it. Hence it is reasonable to suggest, from Keynes's results, that the rising phase of the action potential is brought about by inward movement of sodium ions, and the falling phase by outward movement of potassium ions. If this suggestion is correct, we should be able to show that the quantity of sodium ions entering the axon is sufficient to cause the observed changes in membrane potential. The charge (Q, measured in coulombs) on a capacitance (C, measured in farads) is given by

$$Q = CV$$

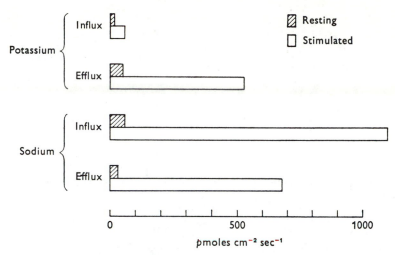

Fig. 5.6. Fluxes of sodium and potassium ions in *Sepia* giant axons. Shaded columns: resting. White columns: stimulated at 100 shocks/sec. (Based on data of Keynes, 1951.)

where V is the voltage across the capacitance. If this charge is produced by a univalent ion, the number of moles of the ion moved from one side of the capacitance to the other is given by

$$n = CV/F$$

where F is Faraday's constant. In the case of an axon, C is $1\ \mu\text{F/cm}^2$ and V (the height of the action potential) is about $110\ \text{mV}$. Hence

$$n = \frac{10^{-6} \times 0.11}{10^5}\text{moles/cm}^2$$

$$= 1.1\ \text{pmoles/cm}^2.$$

It follows that the net flow of sodium ions is more than sufficient to account for the change in membrane potential during the action potential.

Isotopic experiments provide a very good measure of the gross ionic exchanges occurring on stimulation, but their time resolution is naturally extremely poor in relation to the duration of the action potential. A more detailed picture of ionic movements during activity can only be obtained by using electrical methods, which we must now consider.

VOLTAGE CLAMP EXPERIMENTS

If the 'sodium theory' is correct, then the initial depolarization which produces an action potential must result in an increase in sodium

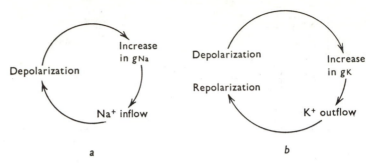

Fig. 5.7. Effects of depolarization on the sodium and potassium ion conductances of the axon membrane. *a*: Sodium: this is a positive feedback loop. *b*: Potassium: this is a negative feedback loop.

conductance, and this increase in sodium conductance will itself produce a further depolarization. Thus we are dealing with a positive feedback system, as shown in fig. 5.7*a*, and such systems are very difficult to analyse. These difficulties have been overcome by means of a special technique known as the *voltage clamp method*, in which the effect of sodium conductance on membrane potential is eliminated by passing current through the nerve membrane so as to hold the membrane potential constant at any desired value. These experiments were first performed by Hodgkin and Huxley at Cambridge (Hodgkin, Huxley and Katz, 1952; Hodgkin and Huxley, 1952*a–d*), and led to a remarkably complete analysis of the events occurring during the action potential.

A schematic diagram of the experimental arrangement is shown in fig. 5.8. Two thin silver wire electrodes *a* and *b* are passed down the middle of a squid giant axon. The axon is placed in a trough passing through pools of sea water separated by insulated partitions; on the outside of these pools is an earthed electrode *e*. Current is passed through the nerve membrane by means of a generator connected to the internal

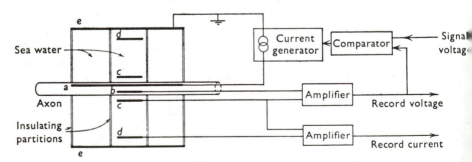

Fig. 5.8. Schematic diagram of the method used to determine membrane currents in a squid axon under voltage clamp. (Based on Hodgkin, Huxley and Katz, 1952.)

electrode *a* and the earth electrode *e*. In the middle pool, this current must pass through the resistance provided by the sea water between electrodes *c* and *d*, and thus (applying Ohm's law) the voltage between *c* and *d* is proportional to the current flowing through the axon membrane in contact with the middle pool. The voltage across this part of the membrane is recorded by means of electrode *c* and the internal electrode *b*. This voltage, after amplification, is fed into a comparator which is also supplied by a signal voltage. The output of the comparator is passed to the current generator* so as to increase or decrease the current flowing through the membrane, which makes the voltage across the membrane (after amplification) equal to the signal voltage. Hence this arrangement constitutes a negative feedback control system in which the voltage across the membrane is determined by the externally applied signal voltage. The presence of the outer pools, with part of the current between *a* and *e* passing through them, ensures that the membrane potential is constant over the length of axon in the middle pool.

The current flowing through the axon membrane is assumed to consist of two components, a capacity current (i.e. changes in charge density at the inner and outer surfaces of the membrane) and an ionic current (i.e. passage of ions through the membrane). Hence the total current, *I*, is given by

$$I = C_m \cdot \frac{\mathrm{d}E}{\mathrm{d}t} + I_i \qquad (5.3)$$

where C_m is the membrane capacitance, E is the membrane potential and I_i is the ionic current. Thus when the voltage is held constant ('clamped') $\mathrm{d}E/\mathrm{d}t = 0$, and so the record of current flow gives a direct measure of the total ionic flow.

Fig. 5.9*a* shows the type of record obtained by Hodgkin and Huxley. Notice that the voltage measured, *V*, corresponds to the difference between the resting potential (E_r) and the clamped membrane potential (E); the reason for this is that it is difficult to obtain accurate absolute values for the membrane potential with silver wire electrodes. The current record shows three components. First, there is a brief 'blip' of outward current; this is caused by discharge of the membrane capacitance. After this the current is inward for about 1 msec, and finally the current becomes outward and climbs to a steady level which is maintained while the clamp lasts. Now, applying the 'sodium theory' of the action potential, we might expect that the initial inward current during

---

* In practice, a single high-gain differential DC amplifier acts both as the comparator and the current generator.

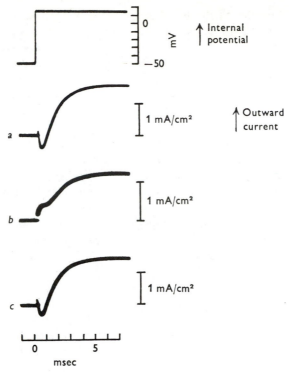

Fig. 5.9. Typical records of the membrane current during a voltage clamp experiment. *a* and *c*: In sea water; *b*: in a sodium-free choline chloride solution. (From Hodgkin, 1958, after Hodgkin and Huxley, 1952*a*.)

the clamp is due to sodium ions flowing inwards. If this is so, then it should disappear if the axon is placed in a solution containing no sodium ions. In fact, it was found (by using choline chloride as a substitute for sodium chloride) that the inward current is replaced by an outward current under these conditions (fig. 5.9*b*). This is understandable if we assume that depolarization causes a brief increase in sodium conductance, since the internal sodium concentration is now very much greater than that outside, and we would therefore expect sodium ions to flow outwards.

The direction of sodium ion flow must be dependent upon the membrane potential as well as on the concentration gradient. When the membrane potential is equal to the sodium equilibrium potential, E_{Na} (given by (5.2)), there will be no net flow of sodium ions, so that if the membrane potential is clamped at E_{Na} (produced by a depolarization of V_{Na}, where $V_{Na} = E_{Na} - E_r$) there should be no sodium current. If the depolarization does not reach V_{Na}, sodium current will be inwards, and

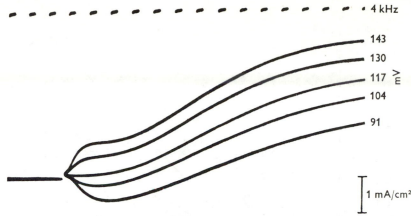

Fig. 5.10. Membrane currents at large depolarizations. Values of V are shown at the right of each record. (From Hodgkin, 1958, after Hodgkin, Huxley and Katz, 1952.)

if it is greater than V_{Na} it will be outwards. This effect is shown in fig. 5.10, in which it is evident that on this interpretation V_{Na} is approximately 117 mV. An alternative way of eliminating the sodium current is to alter the external sodium ion concentration, so changing E_{Na}; trace *b* in fig. 5.12 was obtained in this way.

These results imply very strongly that the initial current flow under voltage clamp conditions is caused by movement of sodium ions. In order to clinch the matter, it is necessary to show that the potential at which the initial current reverses really is the sodium equilibrium potential, E_{Na}. This could not be done directly in the original experiments since the internal sodium ion concentration was not known exactly, and also because of the uncertainties involved in translating the measured potentials (V) into membrane potentials (E). However, it was possible to test the hypothesis by measuring *changes* in E_{Na} (measured as changes in V_{Na} plus a small correction for any resting potential change) which occurred when the external sodium ion concentration was changed. The observed changes never differed from those expected by more than 3 per cent.

The slowly rising, maintained outward current is only very slightly affected by changes in the external sodium ion concentration, and is therefore caused by movement of some other ion, probably (it was assumed) potassium. Direct evidence that this was so was provided by an experiment in which the efflux of ^{42}K from the region of an axon membrane under a cathode was measured (Hodgkin and Huxley, 1953). The increase in potassium ion efflux was linearly proportional to the current density with a slope equal to Faraday's constant (fig. 5.11), and hence the current was carried by potassium ions.

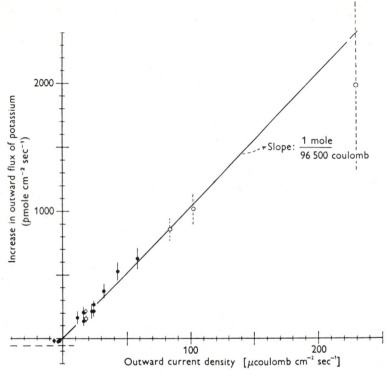

Fig. 5.11. Efflux of potassium under a cathode applied to a *Sepia* axon. (From Hodgkin and Huxley, 1953.)

It was now possible to analyse the ionic current following depolarization into two components, due to the flow of sodium and potassium ions. In fig. 5.12, trace *b* represents the current produced by an axon in 10 per cent sea water (90 per cent isotonic choline chloride solution) when V was equal to V_{Na}. Trace *a* is the current produced by the same depolarization (56 mV in this case) with the same axon in sea water. Now since there is no sodium current (I_{Na}) when V is equal to V_{Na}, trace *b* must represent the potassium current I_K) plus a small constant component, the leakage current (I_l). Hence the difference between trace *a* and trace *b* must be equal to the sodium current; this is drawn as trace *c* in fig. 5.12. The problem of separating the sodium and potassium currents is a little more complicated when the clamped depolarization in the low sodium solution was not equal to V_{Na}, but, by assuming that the sodium current in the low sodium case was always a constant fraction of that in the high sodium case, it is possible to calculate the sodium and potassium currents in each case.

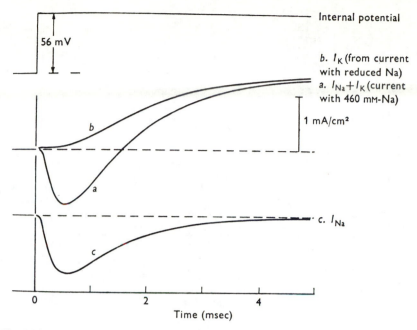

Fig. 5.12. Analysis of the ionic current in a *Loligo* axon during a voltage clamp. Trace *a* shows the response to a depolarization of 56 mV with the axon in sea water. Trace *b* is the response with the axon in a solution comprising 10 per cent sea water and 90 per cent isotonic choline chloride solution. Trace *c* is the difference between traces *a* and *b*. Further explanation in the text. (From Hodgkin, 1958, after Hodgkin and Huxley, 1952*a*.)

The next step in the analysis was to determine the conductance of the membrane to sodium and potassium ions during a clamped depolarization. Applying Ohm's law again, the ionic current flow through the membrane is equal to the product of the conductance and the electromotive force. In the case of sodium ions, for instance, the electromotive force is given by the difference between the membrane potential and the sodium equilibrium potential, and so the sodium current is given by

$$I_{Na} = g_{Na}(E - E_{Na})$$
$$= g_{Na}(V - V_{Na}). \tag{5.4}$$

Thus, from trace *c* in fig. 5.12, and knowing V_{Na} it is a simple matter to calculate g_{Na} throughout the course of the clamp. By a similar process, the potassium conductance can be calculated from trace *b* and the relation

$$I_K = g_K(V - V_K). \tag{5.5}$$

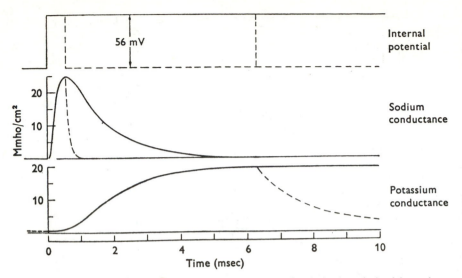

Fig. 5.13. Ionic conductance changes during a clamped depolarization, derived from the current curves shown in fig. 5.12. The broken curves show the effects of repolarization. (From Hodgkin, 1958, by permission of the Royal Society.)

The results of these calculations are shown in fig. 5.13. Fig. 5.14 shows the conductance changes resulting from depolarizations of different magnitudes.

The results of this investigation at this stage can be summarized as follows. Depolarization produces three effects: (1) a rapid increase in sodium conductance, followed by (2) a slow decrease in sodium conductance (known as the sodium inactivation process) and (3) a slow increase in potassium conductance. The extent of these conductance changes increases with increasing depolarization. If the membrane potential is suddenly returned to its resting level, these changes are reversed, as is shown by the broken curves in fig. 5.13.

Calculation of the form of the action potential

If the conceptual model of the membrane shown in fig. 5.1 is correct, then it should be possible to calculate the form of the action potential if we know the values of the fixed components in the system (E_K, E_{Na}, E_l, R_l and C_m) and if equations describing the behaviour of the variables g_K and g_{Na} can be obtained. From their voltage clamp experiments, Hodgkin and Huxley (1952d) were able to provide a series of equations describing the changes of g_K and g_{Na} with depolarization and time (from which the continuous curves in fig. 5.14 were calculated), and so were able to undertake this rather laborious calculation. The results were

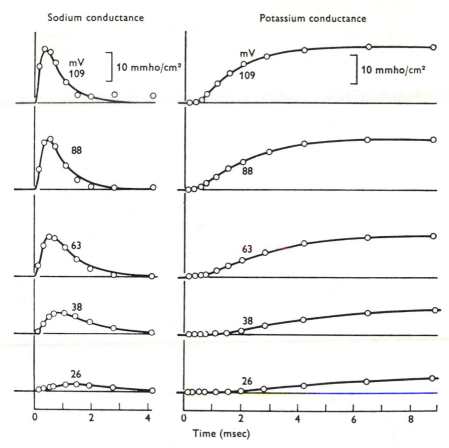

Fig. 5.14. Conductance changes brought about by clamped depolarizations of different extents. The circles represent values derived from the experimental measurements of ionic current, and the curves are drawn according to the equations used to describe the conductance changes. (From Hodgkin, 1958, after Hodgkin and Huxley, 1952*d*.)

dramatically accurate, with only slight differences between the predicted action potentials and those actually observed (fig. 5.15).

Fig. 5.16 shows the theoretical solution for a propagated action potential, and the associated changes in g_{Na} and g_K. It is instructive to follow the potential and conductance changes during its course. Initially, g_K is small but g_{Na} is much smaller, so that the resting potential is near E_K. Then the presence of an action potential approaching along the axon draws charge out of the membrane capacitance by local circuit action, so causing a depolarization. When the membrane has been depolarized by about 10 mV, the sodium conductance begins to increase, so that a small number of sodium ions cross the membrane,

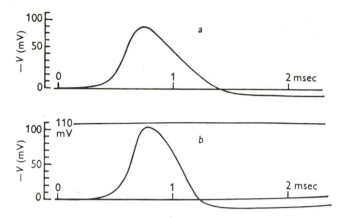

Fig. 5.15. Comparison of computed (*a*) and observed (*b*) propagated action potentials in squid axon at 18.5 °C. The calculated velocity of conduction was 18.8 m/sec; the observed velocity was 21.2 m/sec. (From Hodgkin and Huxley, 1952*d*.)

flowing down their electrochemical gradient into the axon. This transfer of charge results in a further depolarization, which causes a further increase in g_{Na}, so more sodium ions cross the membrane, and so on. The result of this regenerative action between depolarization and sodium conductance is that the sodium battery is relatively much more important than the potassium battery in determining the membrane potential, and so the membrane potential goes racing up towards the sodium equilibrium potential, E_{Na}. But now the two slower consequences of depolarization, sodium inactivation and increase in potassium conductance, begin to take effect. This means that the potassium battery becomes more important and the sodium battery less important in determining the membrane potential (because sodium ion inflow declines and potassium ion inflow increases), and so the membrane potential begins to fall. This repolarization further reduces g_{Na} (it also reduces g_K, but more slowly) so that the membrane potential is brought rapidly back to its resting level. At this point, although g_{Na} is extremely low, g_K is still considerably higher than usual, so that the membrane potential passes the resting level and moves even nearer to the potassium equilibrium potential. Finally, as g_K declines to its normal low value, the membrane potential returns to its resting level.

The equations

Now let us have a brief look at the equations used by Hodgkin and Huxley (1952*d*) in the calculation of the form of the action potential. The first step was to find empirical equations to describe the conductance changes seen, for example, in fig. 5.14. The potassium conduc-

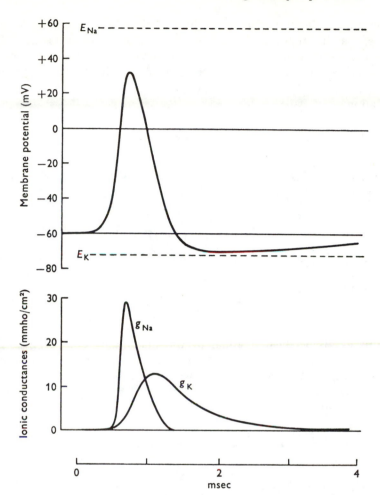

Fig. 5.16. Calculated changes in membrane potential (upper curve) and sodium and potassium conductances (lower curves) during a propagated action potential in a squid giant axon. The scale of the vertical axis is correct, but its position may be slightly inaccurate; it has been drawn here assuming a resting potential of $-60\,\mathrm{mV}$. The positions of E_{Na} and E_K are correct with respect to the resting potential. In the original calculations, voltages were measured from the resting potential, as in fig. 5.15. (After Hodgkin and Huxley, 1952d; redrawn.)

tance is given by

$$g = \bar{g}_K n^4$$

where $\bar{g}_K$ is a constant equal to the maximum value of g_K. The idea behind this equation was that potassium ions might be let through the membrane when four charged particles moved to a certain region of the membrane under the influence of the electric field, the quantity n being the probability that one of these particles is in the right position. This idea may or may not be correct; the question is unimportant from the point of view of providing a mathematical description of the conductance changes. The variation of the quantity n with time is given by the equation

$$\frac{dn}{dt} = \alpha_n(1-n) - \beta_n n,$$

in which α_n and β_n are rate constants which, at $6\,°C$, vary with voltage according to the equations

$$\alpha_n = \frac{0.01(V+10)}{\exp\left[(V+10)/10\right]-1}$$

and
$$\beta_n = 0.125 \exp\left(V/80\right).$$

The sodium conductance is given by

$$g_{Na} = \bar{g}_{Na} m^3 h, \tag{5.6}$$

where $\bar{g}_{Na}$ is a constant equal to the maximum value of g_{Na}. This equation is based on the idea that the sodium channel can be opened by movement of three particles, each with a probability m of being in the right place, and inactivated by an event of probability $(1-h)$. m is given by the equation

$$\frac{dm}{dt} = \alpha_m(1-m) - \beta_m m, \tag{5.7}$$

where, at $6\,°C$,
$$\alpha_m = \frac{0.1(V+2.5)}{\exp\left[(V+25)/10\right]-1}$$

and
$$\beta_m = 4 \exp\left(V/18\right).$$

h is given by the equation

$$\frac{dh}{dt} = \alpha_h(1-h) - \beta_h h, \tag{5.8}$$

where, at $6\,°C$,
$$\alpha_h = 0.07 \exp\left(V/20\right)$$

and
$$\beta_h = \frac{1}{\exp\left[(V+20)/10\right]+1}.$$

The total membrane current, I, is given by (5.3), which is expanded to give

$$I = C_m \cdot \frac{\mathrm{d}V}{\mathrm{d}t} + I_K + I_{Na} + I_l$$

$$= C_m \cdot \frac{\mathrm{d}V}{\mathrm{d}t} + g_K(V - V_K) + g_{Na}(V - V_{Na}) + g_l(V - V_l)$$

$$= C_m \cdot \frac{\mathrm{d}V}{\mathrm{d}t} + \bar{g}_K n^4 (V - V_K) + \bar{g}_{Na} m^3 h (V - V_{Na}) + (V - V_l).$$

If I is known, it is possible to work out V from this equation by numerical integration. The simplest case occurs when an appreciable stretch of the membrane is excited simultaneously by means of an internal silver wire electrode; there are here no local circuit currents and the net current flow through the membrane is zero, the ionic current being equal and opposite to the capacitance current. This is known as a 'membrane' action potential. In the case of the propagated action potential, the situation is a little more complicated; we have seen in (4.15) that

$$I = \frac{a}{2R\theta^2} \cdot \frac{\mathrm{d}^2 V}{\mathrm{d}t^2},$$

where a is the radius of the axon, R is the resistivity of the axoplasm, and θ is the conduction velocity. The right value of θ has to be found by trial and error; wrong values lead to infinite voltage changes.

Other predictions

Besides predicting the form of the action potential, the equations derived from analysis of the voltage clamp experiments can account for a number of other features of the physiology of the axon.

By adding g_K and g_{Na}, the impedance change during the course of the action potential can be obtained; the result is very similar to that observed by Cole and Curtis (fig. 4.11). Knowing the time course of the ionic currents during the action potential it is possible, by integrating with respect to time, to calculate the net flow of sodium and potassium ions across the membrane during an impulse. For a propagated action potential the calculated net sodium entry was 4.33 pmoles/cm^2 and the calculated net potassium loss was 4.26 pmoles/cm^2. These values are very close to those obtained by Keynes from radioisotope measurements on *Sepia* axons.

By calculating the responses to instantaneous depolarizations of varying extent, curves very similar to those of fig. 4.7 were obtained. Thus the local responses produced by membrane potential displacements just less than threshold are caused by small increases in g_{Na} which are soon swamped by increase in g_K. The threshold for production of an action potential is the point at which the increase in g_{Na} is just large and rapid enough to avoid this effect.

At the end of an action potential, g_K is higher than usual, and the sodium inactivation process is well developed, so that g_{Na} cannot be much increased by depolarization. The membrane is thus inexcitable for a time; this corresponds to the duration of the absolute refractory period. A little later, g_K and the extent of the sodium inactivation process have fallen somewhat, so that a submaximal regenerative response can be initiated by sufficient depolarization. This period corresponds to the relative refractory period, where the threshold is higher than normal and the ensuing action potential may be reduced in size. Finally g_K and the sodium inactivation process return to their normal resting levels, and the membrane shows its normal excitability.

Accommodation is explained by the increase in the sodium inactivation process and potassium conductance, which is produced by a slowly rising membrane potential, so raising the threshold. Conversely, anode break excitation is due to the lowering of the threshold by means of a decrease in sodium inactivation and potassium conductance.

It has long been known that the conduction velocity of an axon increases with increase in temperature. This is because the time course of the electrical changes constituting the action potential is much faster at higher temperatures, and these electrical changes are of course themselves dependent upon the time courses of the conductance changes produced by depolarization. In terms of the Hodgkin–Huxley equations, the rate constants for the conductance changes (the alphas and betas in the equations) increase threefold for every 10 °C rise in temperature.

Direct measurement of the conductances during an action potential

Bezanilla, Rojas and Taylor (1970) have provided a nice verification of the Hodgkin–Huxley analysis. Advances in electronic design enabled them to subject an axon membrane to a voltage clamp in the middle of an action potential and measure the immediate value of the resulting ionic current. Fig. 5.17 shows the results of one such experiment, in which the membrane was clamped at the potassium equilibrium potential at different times during a 'membrane' action potential. The current traces therefore measure the sodium current, from which the sodium conductance can be calculated as in (5.4). In each case it can be seen that the sodium conductance falls rapidly, as one would expect at this membrane potential, but its initial value will be the same as it was immediately before the application of the voltage clamp. (This conclusion arises from the kinetics of sodium conductance changes as

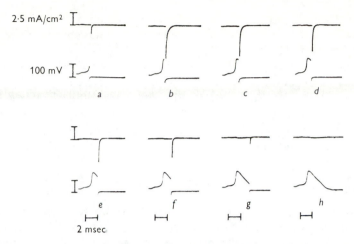

Fig. 5.17. An experiment to determine the sodium conductance at different times during the 'membrane' action potential in a squid (*Dosidicus*) giant axon. Lower traces show the membrane potential, the voltage being clamped at progressively later times (*a* to *h*) during the action potential. Upper traces show membrane ionic current flow; the capacitance current is over within 30 μsec of the onset of the clamp. (From Bezanilla, Rojas and Taylor, 1970.)

described by the Hodgkin–Huxley equations: the alphas and betas in these equations do change instantaneously with change in voltage, but the quantities m and h do not, as is evident in (5.7) and (5.8), where dm/dt and dh/dt always have finite values.) So by applying the clamp at different times it is possible to determine the time course of the sodium conductance during the action potential, with the result shown by the open circles in fig. 5.18.

Similarly, by clamping the membrane at the sodium equilibrium potential it is possible to measure the initial potassium currents at

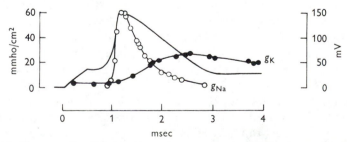

Fig. 5.18. Time course of the ionic conductance changes during a membrane action potential. The sodium conductance (open circles) was determined from an experiment such as that of fig. 5.17; the potassium conductance (black circles) was determined by clamping the membrane at the sodium equilibrium potential at different times. (From Bezanilla, Rojas and Taylor, 1970.)

different times. The time course of the potassium conductance during the action potential is then readily calculated, and is shown by the black circles in fig. 5.18.

Comparison of fig. 5.18 with fig. 5.16 shows that the agreement between the experimental curves and those predicted by Hodgkin and Huxley is very good. A sharp eye will detect some differences: the fall of sodium conductance is a little more rapid in the experimental determination, and the time course of the potassium conductance change is rather slower. But in general the similarities are much more impressive than the differences.

Myelinated nerves

The membrane potentials of myelinated nerve fibres, which are usually less than 15 μm in diameter, cannot be measured with the ease with which those of squid giant axons are. It is not normally possible to insert electrodes into the axon, and so indirect methods have to be used. These depend essentially upon using the axoplasm itself as the intracellular electrode by insulating the cut end of an internode from the external fluid surrounding an adjacent node. The results (reviewed by Hille, 1971) show that the properties of the nodal membrane are remarkably similar to those of squid axon membrane. The currents produced during a voltage-clamped depolarization consist of an initial sodium current and a delayed potassium current, and can be described by relatively minor modifications of the Hodgkin–Huxley equations.

FIXED CHARGES AND THE INVOLVEMENT OF CALCIUM IONS

Several investigations in recent years have suggested that there are fixed negative charges on the outer and inner sides of the axon membrane (see for example Hille, Woodhull and Shapiro, 1975, and Gilbert, 1971). We would expect such charges to occur at the polar ends of the membrane lipids and possibly also on the protein molecules. Their existence would modify the electrical potential field in the membrane and in its immediate vicinity, perhaps in the manner shown in fig. 5.19a. We would also expect that the excitability mechanisms in the membrane (which must presumably be dependent upon the electric field in the membrane) would be affected by any alterations in the density of these fixed charges. Such alterations would be produced by changes in the divalent ion concentration, pH or ionic strength of the solutions on either side of the membrane. As an example of these effects, let us look at the action of calcium ions.

Reduction of the calcium ion concentration in the external medium produces an increase in the excitability of axons, seen as a reduction in

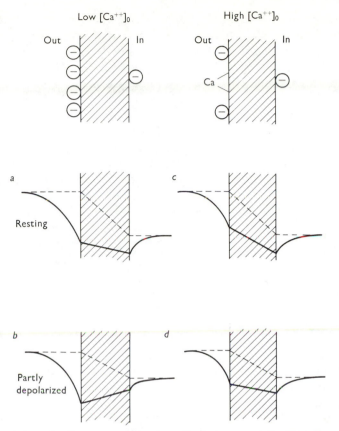

Fig. 5.19. Schematic diagrams to show how fixed charges might affect the potential distribution (heavy lines) in the membrane and in its immediate vicinity. The broken lines show the potential distribution expected in the absence of such fixed charges. The diagrams show low (*a* and *b*) and high (*c* and *d*) external calcium ion concentrations, with membrane potentials at the resting level (*a* and *c*) and partly depolarized (*b* and *d*). Remember that the fixed charges only influence the potential field in the immediate vicinity of the membrane: they cannot be measured with microelectrodes, which are much too large for the job.

threshold and a tendency towards spontaneous and repetitive activity. Frankenheuser and Hodgkin (1957), applying the voltage clamp technique, concluded that these effects were due to an increase in the sodium and potassium conductances of the membrane, such that a fivefold decrease in external calcium ion concentration produces effects similar to a depolarization of 10 to 15 mV.

Fig. 5.19 shows one way in which these results can be interpreted in terms of fixed charges. With low external calcium ion concentrations, there are many more fixed negative charges on the outer surface of the membrane than on its inner surface. These charges will not affect the total potential across the membrane as measured at short distances away from its surfaces, but they will have marked effects on the detailed form of the potential gradient within the membrane and in its immediate vicinity. In particular, the greater number of negative charges on the outer surface ensures that the gradient of the potential field within the membrane will be less than it would otherwise be (fig. 5.19*a*). Depolarization by a few tens of millivolts will therefore reverse the direction of the field within the membrane (fig. 5.19*b*) even though the total field is still inside-negative. It is an attractive speculation that the voltage-dependent ionic conductances are activated by such a reversal. If this is so, we can see how an increased external calcium ion concentration could raise the threshold: by reducing the negative charge density on the outer surface of the membrane it would increase the gradient of the potential field within the membrane and hence raise the membrane potential at which reversal of this field occurred (fig. 5.19*c, d*).

In addition to their action on the sodium and potassium conductances, there is a small flow of calcium ions into the axon during the action potential (Hodgkin and Keynes, 1957). The size of this calcium ion flow is only about 1/700 of the sodium ion flow, and so it plays a negligible part in the form of the action potential. Its time-course has been investigated by Baker, Hodgkin and Ridgeway (1971) using voltage clamp techniques and the bioluminescent protein aequorin, which emits light in the presence of calcium ions. One of their results is shown in fig. 5.20, from which it is apparent that calcium inflow occurs in two phases: a rapid inflow during the first 200 μsec which is abolished by tetrodotoxin, and a slower inflow which is not. The timing and tetrodotoxin-sensitivity of the rapid phase suggest that the calcium ions

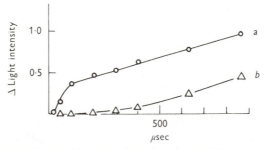

Fig. 5.20. Calcium entry in squid axon under voltage clamp. The axon was injected with the protein aequorin, which emits light in the presence of calcium ions. The curves show the increase in light intensity produced by 80 mV depolarizing pulses of different lengths. *a*, under normal conditions; *b*, in the presence of tetrodotoxin. (From Baker, Hodgkin and Ridgeway, 1971.)

are here entering through the sodium channel. One might then guess that the late entry occurs through the potassium channel, but in testing this idea it was found that tetraethylammonium ions (which block the potassium channel) had no effect on the late calcium entry. It looks, therefore, as though there must be a third type of potential-dependent ionic channel which is specifically permeable to calcium ions.

This calcium entry into nerve axons on depolarization is of more general interest for two reasons: it appears to be important in the release of chemical transmitter substances from nerve endings (p. 146), and there are some other systems, including vertebrate heart muscle, in which calcium ions may carry an appreciable part of the current flow during an action potential (see Reuter, 1973).

EXPERIMENTS ON PERFUSED GIANT AXONS

It has been possible to test and extend some of the conclusions of the ionic theory of nervous conduction by means of the technique of intracellular perfusion (Baker, Hodgkin and Shaw, 1962a; Tasaki and Shimamura, 1962). In the method used by Baker *et al.*, the axoplasm is squeezed out of a squid axon, leaving a flattened tube consisting mainly of the axon membrane and the Schwann cells and endoneurial sheath. This tube is then reinflated by filling it with a perfusion fluid isotonic with sea water.

When the perfusion fluid is an isotonic potassium chloride solution, the resting potential is about -55 mV. Isotonic solutions of other potassium salts, such as sulphate or isethionate, produce resting potentials a few millivolts greater, indicating that the membrane is slightly permeable to chloride ions (Baker *et al.* 1962b). If the internal solution is now replaced by an isotonic sodium chloride solution, the membrane potential falls to near zero. By gradually increasing the internal potassium ion concentration, it is possible to see how far the membrane behaves as a potassium electrode; it was found that the membrane potential tended to reach a saturating value of -50 to -60 mV at an internal potassium ion concentration of about 300 mV. This relation is compatible with (3.8) if we assume that P_K decreases with increasing (more negative) membrane potentials. If the solutions inside and outside the membrane were identical in composition, the membrane potential was within 1 mV of zero. Finally, when the internal solution was isotonic sodium chloride and the external solution isotonic potassium chloride, the inside became positive to the outside by 40 to 60 mV. All these results provide further confirmation of the theory of the resting potential discussed in chapter 3.

In the experiment of Hodgkin and Katz shown in fig. 5.2, the action potential was reduced in height by reducing the external sodium ion concentration, so reducing the sodium equilibrium potential, as given by (5.2). The intracellular perfusion method makes it possible to reduce the sodium equilibrium potential by the alternative method of increasing the internal sodium ion concentration. When this was done it was found that the overshoot of the action potential was reduced as expected.

The voltage clamp technique coupled with internal perfusion has been used by Atwater, Bezanilla and Rojas (1969) to measure sodium fluxes during clamped depolarizations. They placed an axon in sea water containing ^{22}Na and perfused it with a potassium fluoride solution. By measuring the radioactivity of this perfusion solution after it had passed through the axon it was possible to calculate the sodium influx. First the sodium influx at rest was measured. Then the axon was subjected to a few thousand clamped depolarizations at a rate of 10/sec, each pulse lasting about 3 msec. The radioactivity of the emerging perfusion solution rose to reach a new steady level within a few minutes, from which the influx associated with each clamped depolarization could be calculated. The corresponding inward current was calculated from the difference between clamped currents measured in the presence and absence of tetrodotoxin, a substance which blocks the early inward current (p. 99).

Fig. 5.21 summarizes the results obtained in these experiments. At a number of different membrane potentials the sodium influx almost

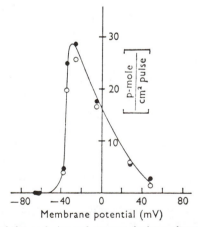

Fig. 5.21. The nature of the early inward current during voltage clamp in a squid giant axon. Open circles show the sodium ion influx as measured with a radioactive tracer method. Black circles show the charge movement (coulombs divided by Faraday's constant) attributable to the early inward current, measured from the difference between ionic current flow in the presence and absence of tetrodotoxin. (From Atwater, Bezanilla and Rojas, 1969.)

entirely accounts for the inward current flow, in most convincing agreement with Hodgkin and Huxley's analysis.

Chandler and Meves (1965) carried out voltage clamp experiments on giant axons perfused internally with different solutions. When an axon was perfused with a 300 mM potassium chloride solution, they found that the initial inward current (which is attributed entirely to sodium ions in the Hodgkin–Huxley analysis) became an outward current at a reversal potential in the region of +70 mV. But under these conditions, with no internal sodium ions, the sodium equilibrium potential is infinite and so a purely sodium ion current should always be inward. Chandler and Meves concluded that potassium ions can flow through the sodium channel. By applying the constant field equation (3.8) it is possible to estimate the relative permeabilities of the channel to sodium and potassium ions. Thus with no internal sodium and no external potassium, and assuming that the channel is impermeable to anions, (3.8) becomes

$$E = \frac{RT}{F} \log_e \left(\frac{P_{Na}[Na]_o}{P_K[K]_i} \right). \tag{5.9}$$

At 0 °C, with 320 mM potassium ions inside and 472.5 mM sodium ions outside, the mean reversal potential for the early current was +67.8 mV. Substituting these values in (5.9) we get

$$67.8 = 54.2 \log_{10} \left(\frac{472.5 \, P_{Na}}{320 \, P_K} \right)$$

whence $\qquad\qquad P_{Na}/P_K = 12.0$.

By repeating this type of experiment with other monovalent cations it was possible to estimate their permeabilities through the sodium channel. They fell in the order Li > Na > K > Rb > Cs, with relative values of 1.1, 1, 1/12, 1/40 and 1/61.

The potassium channel seems to be rather more selective, and indeed its permeability to potassium ions is reduced in the presence of other monovalent cations: sodium, rubidium and especially caesium cause reductions in the potassium current during a voltage clamp. Ammonium ions can pass through both the sodium and potassium channels, their permeability being about a third of that of the normal ion in each case (Binstock and Lecar, 1969).

The effect of internal anions on membrane function was investigated by Tasaki, Singer and Takenaka (1965). They found that some ions were much better than others in maintaining excitability, in the order F > HPO_4 > glutamate > aspartate > SO_4 > acetate > Cl > NO_3 > Br > I > SCN. Hence in recent years it has become quite common to use fluoride as the internal anion in perfusion experiments.

Dilution of the internal medium with a solution of sucrose or some other nonelectrolyte results in a reduced resting potential, as we would expect. But a similar reduction in excitability does not occur: large and often prolonged action potentials may be seen under these conditions. Using the voltage clamp technique it was possible to show that the relation between sodium conductance and membrane potential was moved to a less negative position; the sodium inactivation relation was moved similarly. In other words, the sodium channel behaved as though the membrane potential was more negative than it actually was. Chandler, Hodgkin and Meves (1965) explained this in terms of fixed charges on the inside of the membrane; when the ionic strength of the internal solution is low such charges will have a much larger effect on the

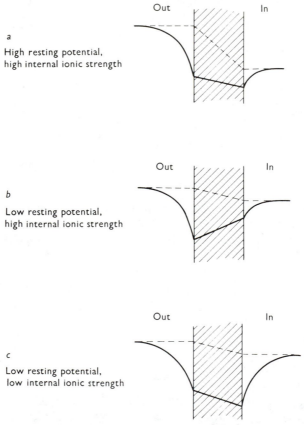

a
High resting potential,
high internal ionic strength

b
Low resting potential,
high internal ionic strength

c
Low resting potential,
low internal ionic strength

Fig. 5.22. Schematic diagrams to show how the gradient of potential across the membrane at low membrane potentials may be altered by internal perfusion with a solution of low ionic strength. Such a solution allows the fixed charges on the inner face of the membrane to exert a much greater effect. Compare with fig. 5.19.

potential field within the membrane. Fig. 5.22 gives an indication of how this system might work.

THE ACTIONS OF DRUGS

The actions of drugs on nervous conduction are interesting both in themselves and in relation to their possible use in medicine. They have also become much used in recent years as tools in the study of the mechanisms of membrane activity (see Narahashi, 1974, Hille, 1970, and Armstrong, 1974).

Tetrodotoxin is a virulent nerve poison found in the tissues of the Japanese puffer fish *Spheroides rubripes*. Narahashi, Moore and Scott (1964), using voltage-clamped lobster axons, found that tetrodotoxin in the external solution blocks the increase in sodium conductance that occurs on depolarization but has no effect on the increase in potassium conductance (see fig. 5.23*a*). This provides further evidence in favour of the idea that sodium and potassium ions flow through separate channels. *Saxitoxin*, a poison produced by the dinoflagellate protozoan *Gony-aulax*, also blocks the sodium channel. Both these substances have no effect on nervous conduction if they are injected into the axoplasm.

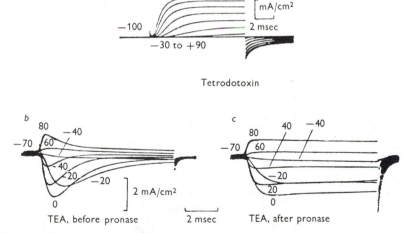

Fig. 5.23. Separation of the ionic currents seen in the voltage clamped squid axon by treatment with various agents. Each diagram shows a number of superimposed traces of the current flow produced by depolarization to different membrane potentials. *a*, Treatment with tetrodotoxin; notice the absence of any inward (sodium) current; *b*, internal perfusion with a solution containing tetraethylammonium (TEA) ions; this blocks the late outward (potassium) current; *c*, the same after adding the proteolytic enzyme pronase to the internal perfusion fluid; notice that the sodium current is now maintained rather than transient as in *b*. (From Armstrong, 1974.)

Hence their site of action must be at or near the external surface of the axon membrane.

Tetraethylammonium (TEA) ions produce prolongation of the action potential when injected into squid axons, but external application is ineffective. Voltage clamp experiments have shown that this prolongation is due to a blockage of the potassium channel, as in fig. 5.23*b*.

A particularly interesting experiment was performed by Armstrong, Bezanilla and Rojas (1973). They perfused a squid axon internally with TEA (so eliminating the potassium current) and pronase, a proteolytic enzyme, and discovered that the sodium current was not inactivated (fig. 5.23*c*). This means that the sodium inactivation process really is a separate process, as is implied in Hodgkin and Huxley's analysis. In the absence of TEA it is evident that the sodium and potassium currents are both high in the later stages of a clamped depolarization, and hence pronase perfusion has no effect on the potassium channels. This important result shows that the rise in potassium conductance is not linked to the fall in sodium conductance and therefore that we cannot suppose that sodium channels change into potassium channels during maintained depolarizations.

Local anaesthetics such as procaine are effective in their function because they block nervous conduction. Voltage clamp experiments (e.g. Taylor, 1959) have shown that they reduce both sodium and potassium currents. They are effective when applied to the outside of the axon, but since they are readily soluble in lipids they would easily be able to cross the membrane for action at internal sites.

Some insecticides act on the conductance changes involved in nervous conduction. Thus DDT, which prolongs the action potential and may cause repetitive firing, blocks the sodium inactivation process and reduces the potassium conductance. Allethrin, derived from the natural insecticide pyrethrum, reduces not only these two processes but the increase in sodium conductance as well.

Veratridine, one of the veratrum alkaloids found in a group of plants of the lily family, increases the sodium permeability of the resting membrane. This produces an initial increase in excitability, but prolonged depolarization may make the membrane inexcitable, probably as a result of the sodium inactivation process.

Less direct effects are produced by drugs which interfere with the maintenance processes of the cell, so that the ionic gradients across the membrane are altered. Cardiac glycosides, such as ouabain, prevent sodium extrusion directly; metabolic inhibitors do so by removing the energy supply for the process. As one would expect from the ionic theory of nervous conduction, poisoning of the sodium extrusion system does not immediately affect the electrical properties of the axon. In an experiment by Hodgkin and Keynes (1955), over 200 000 impulses were elicited from a *Sepia* axon which had been poisoned with dinitrophenol.

GATING CURRENTS

We have seen that the Hodgkin–Huxley equations were based on the idea that the ionic channels are opened by the movement of charged particles within the membrane. Indeed it seems almost inevitable that a voltage-dependent change should be initiated by some sort of movement of charge. There should therefore be some current flow in the membrane just prior to the increase in ionic permeability. The search for such 'gating currents' (so called because they would open the 'gates' for ionic flow) has recently been successful (Armstrong and Bezanilla, 1973, 1974; Keynes and Rojas, 1974; see also Armstrong, 1974, and Rojas, 1976).

The main difficulty in detecting these gating currents is that they are very small in comparison with the ionic and capacity currents. The ionic currents can be blocked pharmacologically (by using external tetrodotoxin to block the sodium current and internal caesium fluoride to block the potassium current, for example). The capacity current is set aside by subtracting the response to a hyperpolarizing voltage clamp pulse from the response to a depolarizing pulse of the same magnitude; the difference is then the gating current. It is pertinent to consider why the

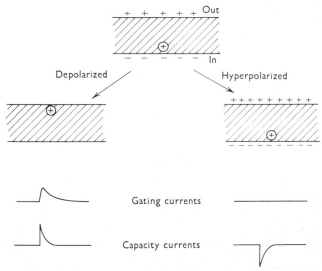

Fig. 5.24. Schematic diagram showing one way in which gating currents could arise. The capacity currents are much larger than the gating currents.

gating current should be asymmetrical in this way. Suppose the charges that are moved during the gating current are positively charged particles that are held near to the inside of the membrane by the resting potential (fig. 5.24). During a hyperpolarization they will not be able to move any further towards the inside of the membrane simply because they are at the limit of their possible movement. There will therefore be no gating current on hyperpolarization. On depolarization, however, the particles will move towards the outer side of the membrane and this movement will constitute the gating current. (It is possible, of course that the gating particles will reach the inner limits of their movement at membrane potentials more negative than the resting potential, but it would seem unlikely that the difference would be very great. It is for this reason that many measurements of gating currents have been made from 'holding potentials' of -100 mV or so, rather more negative than the resting potential.) Notice that the arguments are similar in principle if the gating particles are negatively charged and move inwards or if they are rotating dipoles.

The upper trace in fig. 5.25 shows the gating current of a squid axon for a clamped depolarization of 70 mV, and the lower trace shows the ionic current obtained from the same axon in sea water. Notice that the

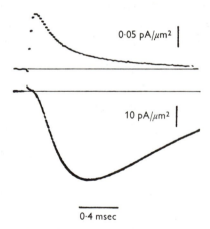

0·05 pA/μm^2

10 pA/μm^2

0·4 msec

Fig. 5.25. The upper trace shows the gating current in a squid giant axon when subjected to a depolarization of 70 mV. It was produced by summing the non-ionic currents produced by an equal number of hyperpolarizations and depolarizations of 70 mV. Ionic currents were eliminated by placing the axon in 'sea water' in which Tris replaced sodium (to eliminate the sodium current) and perfusing it internally with a caesium fluoride solution (to eliminate the potassium current). The lower trace shows the current (mainly sodium current) recorded from the same axon in sea water with potassium fluoride internally, in response to a clamped depolarization of 70 mV. The membrane potential before the onset of clamped pulses was held at -70 mV in both experiments. (From Armstrong, 1974.)

gating current is outward and has a much faster time course than the ionic current. The total quantity of charge moved (obtained by integrating the current with respect to time) is about 30 nC/cm^2 in a squid axon, which corresponds to about 2000 electronic charges per square μm of membrane (Keynes and Rojas, 1974). We shall return to this figure shortly.

On repolarization at the end of a voltage clamp pulse there is an inward gating current in which the total charge movement is equal and opposite to that at the beginning of the pulse. Internal perfusion with zinc chloride produces proportional reductions in both the gating current and the ionic sodium current; this observation provides support for the view that the gating current is indeed concerned with opening the sodium channels. But it is interesting that tetrodotoxin can block the sodium channels without affecting the gating current. This supports the idea that the sodium channel is a transmembrane pore with a movable gate inside and a tetrodotoxin receptor outside.

COUNTING THE CHANNELS

Clearly it is of considerable interest to estimate how many ionic channels there are in an area of axon membrane. One general way of doing this for sodium channels is to measure the binding of tetrodotoxin or saxitoxin to the nerve. There are technical difficulties in performing such experiments, and a number of different estimates have been produced in recent years. Careful experiments by Ritchie, Rogart and Strichartz (1976), using radioactively labelled saxitoxin, suggest that some of the

TABLE 5.1. *Estimates of sodium channel density in nonmyelinated axons.*

Preparation	Method	Channels/μm^2	Reference
Garfish olfactory nerve	Radioactive STX binding	35	Ritchie, Rogart and Strichartz, 1976
Lobster walking leg nerve	Radioactive STX binding	90	Ritchie, Rogart and Strichartz, 1976
Rabbit vagus	Radioactive STX binding	110	Ritchie, Rogart and Strichartz, 1976
Squid giant axon	Gating current divided by 3.9	483	Keynes and Rojas, 1974
	Radioactive TTX	553	Levinson and Meves, 1975
	Kinetics of TTX binding	300–600	Keynes, Bezanilla, Rojas and Taylor, 1975

STX, Saxitoxin; TTX, tetrodotoxin

earlier measurements gave results that were too low. The later measurements vary from 35 sodium channels per μm^2 for the thin fibres of the garfish olfactory nerve to about $500/\mu m^2$ for squid giant axons (see table 5.1).

Sodium channel densities in squid axons can also be measured from the magnitude of the gating currents. Thus Keynes and Rojas (1974) found that the average maximum gating charge displacement was 1882 charges/μm^2. To utilize this figure we need to know how many gating charges move per sodium channel. From the Hodgkin–Huxley equations we assume that the gating current is carried by three particles; it is then necessary to estimate the charge (or effective valency, z') on each of these particles. Keynes and Rojas were able to calculate z', defined as the valency of the particles multiplied by the fraction of the membrane field over which they move, from an equation similar to (3.2):

$$E = \frac{25}{z'} \log_e \left(\frac{\text{gating charge density inside}}{\text{gating charge density outside}} \right).$$

The estimated value of z' was 1.3, and so the gating charge per channel is 3.9. With a maximum gating charge displacement of 1882 charges/μm^2, therefore, the sodium channel density is $483/\mu m^2$.

We can now calculate the conductance of a single sodium channel. The maximum possible sodium conductance, occurring if all the sodium channels were open at once (i.e. $\bar{g}_{Na}$ in (5.6)), was estimated by Hodgkin and Huxley (1952d) to be 120 mmho/cm^2, or 1200 pmho/μm^2. If there are 500 channels/μm^2, then each channel will have a conductance of 2.4 pmho.

Is there an optimum density for the sodium channels? Clearly, conduction velocity will increase with increasing sodium conductance. But it will decrease with increasing membrane capacitance, and each channel, with its little store of movable gating charges, must act as a small capacitance. At very high sodium channel densities, therefore, the effective membrane capacitance will be large enough to reduce the conduction velocity. Hence there is an optimum sodium channel density at which conduction velocity is maximal. This argument is developed by Hodgkin (1975), who calculates that the optimum density is in the range 500 to 1000 channels/μm^2, in pleasing agreement with the figures derived from experiments on squid axons.

Why should the sodium channel density be lower in axons of smaller diameter? Perhaps it is simply to reduce the quantity of sodium entering per impulse, all of which must later be pumped out using metabolic energy. This is much more of a problem for small diameter fibres because of their higher surface-to-volume ratio.

An important corollary of the differences in channel density in different axons is that calculations which assume that membrane characteristics are the same for fibres of different diameters (as on p. 56) are, to say the least, oversimplified.

Gating currents have recently been measured in frog myelinated nerves (Nonner, Rojas and Stämpfli, 1975). The calculated sodium channel density in the nodal membrane is about $5000/\mu m^2$, ten times as high as in squid axons.

It is interesting to notice that the density of sodium channels in unmyelinated axons is about one-tenth of that of sodium pumping sites (see p. 26). This ratio is presumably related to the much greater rate of flow of sodium ions through the channels than through the pumping sites.

6 The neuromuscular junction

Having dealt with the propagation of impulses along the axons of nerve cells, we now have to consider how the information contained in nerve impulse sequences is passed on to other nerve cells and muscle cells, which brings us to the problem of synaptic transmission. This process is most fully understood in the case of the neuromuscular junction in the sartorius muscle of the frog; consequently this chapter is mainly devoted to experiments on this preparation and on other vertebrate neuromuscular junctions with similar physiological characteristics.

THE PROBLEM OF SYNAPTIC TRANSMISSION

When the neuron theory had become generally accepted, at about the beginning of this century, the problem arose as to how excitation was transmitted across the gap between two neurons. Du Bois Reymond, in 1877, suggested that the presynaptic cell could influence the post-synaptic cell either by electrical currents or by chemical mediators. We now know that transmission at the majority of synapses is brought about by release of a chemical *transmitter substance* from the presynaptic cell, although there are some cases of electrical transmission. The history of the establishment of this generalization provides a fascinating example of the development of a scientific theory, and we shall therefore briefly examine a few of the main steps in the story.

The first tentative evidence for the chemical transmission theory came from the work of Elliott 1904. He showed that those mammalian muscles which are innervated by the sympathetic nervous system are responsive to adrenaline, whereas muscles which are not so innervated are not. In order to explain these observations he suggested that sympathetic nerves act by release of adrenaline from the nerve endings when a nerve impulse arrives at the periphery.

In 1914, Dale investigated the effects of the substance acetylcholine, originally isolated from ergot (a fungal infection of cereals) on various body functions. He found that it lowered the blood pressure in the cat, inhibited the heart beat in the frog, and caused contractions of frog intestinal muscles. He suggested that acetylcholine occurred naturally in

106

the body, possibly acting as an antagonist to the effects of adrenaline, and, as an inspired guess, suggested that it was normally rapidly broken down by a hydrolytic enzyme; this would account for the fact that it has so far been impossible to isolate acetylcholine from the body.

The next stage in this story came from a classic experiment by Loewi (1921) on the frog heart. The heart normally beats spontaneously, but it can be inhibited by stimulation of the vagus nerve. Loewi found that the perfusion fluid from a heart which was inhibited by stimulation of the vagus would itself reduce the amplitude of the normal beat in the absence of vagal stimulation. Perfusion fluid from a heart beating normally did not have this effect. This means that stimulation of the vagus results in the release of a chemical substance (initially called the *Vagusstoff*), presumably from the nerve endings. The inhibitory action of vagal stimulation and of the vagusstoff was prevented in the presence of the alkaloid atropine. In this and other respects the pharmacological properties of the vagusstoff seemed to be similar to those of acetylcholine. It was later shown that the actions of the vagusstoff and acetylcholine could be potentiated by the alkaloid eserine, and that this substance acted by preventing the hydrolysis of acetylcholine by the enzyme acetylcholinesterase.

These experiments were leading to the conclusion that acetylcholine and the vagusstoff were one and the same substance; all that was needed for the general acceptance of this conclusion was a demonstration that acetylcholine does in fact occur in the body. This was provided by Dale and Dudley (1929), who isolated it from the spleens of cows and horses. Further advances came with the demonstrations that acetylcholine is released on stimulation from the sympathetic ganglia (Feldburg and Gaddum, 1934) and from skeletal muscles (Dale, Feldburg and Vogt, 1936).

However, as the evidence for the chemical transmission theory accumulated, some difficulties arose in its application to all synapses. A number of neurophysiologists remained convinced that many cases of synaptic transmission must involve electrical current flow across the synapse. The principal objections to the chemical transmission theory arose from the rapidity of synaptic action (it was difficult to see how the transmitter substance could be released, produce its action and be destroyed all within the space of a few milliseconds) and from the conflicting evidence as to the action of acetylcholine in the central nervous system. These doubts were not finally resolved until the early nineteen-fifties, when it became possible to insert intracellular electrodes into the postsynaptic region (Fatt and Katz, 1951; Brock, Coombs and Eccles, 1952), as we shall see in the following pages.

Finally, a somewhat ironic footnote on the controversy over the electrical and chemical theories of synaptic transmission is provided by the discovery of some synapses in which the transmission process *is* by means of electric currents.

We must now turn to the main subject of this chapter, the structure and physiology of the neuromuscular junction in the 'twitch' fibres of vertebrate skeletal muscle. It is important to realize that the situation here differs in some details from that found at other neuromuscular junctions and at synapses between neurons. Nevertheless, the detailed study of this particular synapse provides an excellent introduction to the physiology of chemically transmitting synapses in general

THE STRUCTURE OF THE NEUROMUSCULAR JUNCTION

A vertebrate 'twitch' skeletal muscle fibre, such as is found in the sartorius muscle of the frog, is a long cylindrical multinucleate cell, which is innervated at one or sometimes two points along its length by branches of a motor nerve axon. The structure of the neuromuscular junction as determined by light microscopy, used in conjunction with special staining techniques (see Couteaux, 1960), is shown in figs. 6.1 and 6.2. In the region of contact, the muscle fibre is modified to form the motor end-plate, which contains numbers of mitochondria and nuclei. The motor nerve axon loses its myelin sheath near the ending, and the terminal branches of the axon are partly sunken into 'synaptic gutters' as they spread over the surface of the end-plate. Schwann cell (teloglia)

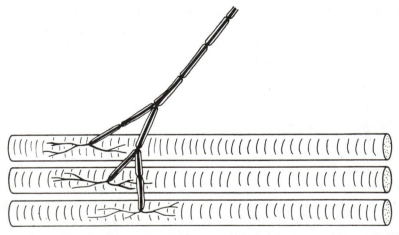

Fig. 6.1. Diagrammatic picture of a vertebrate 'twitch' muscle motor nerve terminal. In most cases a single motor axon innervates many more muscle fibres than the three shown here.

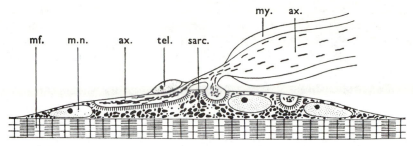

Fig. 6.2. The fine structure of a vertebrate motor end-plate and nerve ending, seen in longitudinal section, as determined by light microscopy. *mf*, myofibril; *m.n.*, muscle nucleus; *ax*, axon; *tel*, Schwann cell (teloglia) nucleus; *sarc*, sarcoplasm; *my*, myelin sheath. (From Couteaux, 1960.)

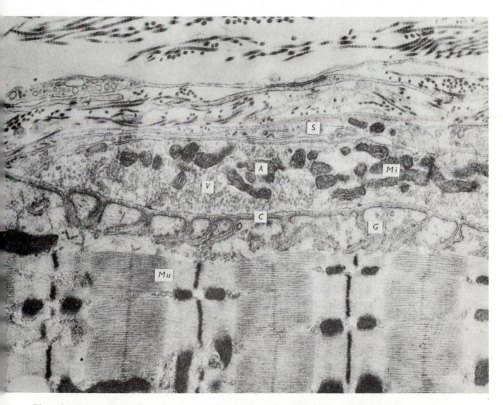

Plate 6.1. Electron micrograph of a neuromuscular junction in the diaphragm of a rat. The axon terminal (*A*) contains numerous mitochondria (*Mi*) and synaptic vesicles (*V*), and is covered with a Schwann cell (*S*). The muscle fibre (*Mu*) is separated from the axon terminal by the synaptic cleft (*C*), which contains some darkly staining material. The postsynaptic membrane of the muscle fibre is indented to form synaptic gutters (*G*). Magnification 12 500 times. (Photograph kindly supplied by Professor R. Miledi.)

nuclei can be seen in association with the unmyelinated terminals, but the precise distribution of their cytoplasm cannot be determined with the light microscope. A series of 'subneural lamellae' project into the end-plate cytoplasm under the terminal branches of the axon.

These observations have been extended by the use of electron microscopy (Robertson, 1956; Birks, Huxley and Katz, 1960), which shows conclusively that there is a space about 500 Å wide, the *synaptic cleft*, between the nerve terminal and the muscle cell (plate 6.1). The 'subneural lamellae' are seen to be infoldings of the postsynaptic membrane, and have since become known as 'synaptic gutters'. There is some extracellular material in the cleft, especially in the synaptic gutters. The Schwann cell covers the nerve terminal in places where it is not in contact with the muscle cell. The terminal branches of the axon contain large numbers of small vesicles about 500 Å in diameter, which may be concentrated in regions opposite the folds in the postsynaptic membrane.

THE RELEASE OF ACETYLCHOLINE BY THE MOTOR NERVE ENDINGS

In a very elegant series of experiments, Dale, Feldburg and Vogt (1936) showed that acetylcholine is released when the motor nerves of various vertebrate skeletal muscles are stimulated. Their experiments on the cat's tongue serve to illustrate some of their methods and conclusions. In most cases the superior cervical ganglia had been removed some weeks previously, so that the hypoglossal nerve would now contain no fibres from the sympathetic system, the only efferent fibres being the motor axons of the tongue muscles. The blood vessels to the tongue were now prepared so that a perfusion fluid (Locke's solution saturated with oxygen at 37 °C) could be passed into the lingual artery via the carotid arteries and collected from the jugular vein after its passage through the capillaries of the tongue. A small quantity of eserine was added to the perfusion fluid so as to prevent the breakdown of acetylcholine by acetylcholinesterase.

The quantities of acetylcholine released in an experiment of this kind are too small to be identified and measured by ordinary chemical methods, and so some type of bioassay has to be used. Acetylcholine in very dilute concentrations causes contraction of the dorsal longitudinal muscles of leeches (after sensitization by eserine) and produces a fall in the blood pressure of anaesthetized cats. Thus if the perfusate also produced these effects, and particularly if the same concentration of acetylcholine was needed to mimic the effects of the perfusate in

Fig. 6.3. Contractions of eserinized leech muscle in response to acetylcholine and perfusates from cat tongue muscle under various conditions. Explanation in text. (From Dale, Feldburg and Vogt, 1936.)

different tests, then there is a strong indication that the perfusate contained acetylcholine.

Now consider fig. 6.3, which shows the results of one of the experiments on the cat's tongue. Each record in the diagram is a smoked drum trace of the contraction of a strip of leech muscle in response to perfusate from the tongue (diluted by 50 per cent) or a solution containing a known concentration of acetylcholine. Trace *a* shows the absence of response to perfusate obtained before stimulation of the hypoglossal nerve. Trace *b* shows the contraction produced by acetylcholine at a concentration of 10^{-8} w/v. Trace *c* shows the response to perfusate obtained during a period of stimulation at 5 shocks/sec, traces *d* and *e* the responses to perfusate obtained 2 and 20 minutes later. Trace *f* shows the response to perfusate obtained during a second period of stimulation; trace *g*, 20 minutes later. Finally trace *h* shows a smaller response to perfusate obtained from a third period of stimulation, which can be compared with trace *j*, the contraction produced by acetylcholine at a concentration of 2.5×10^{-9} w/v.

This experiment shows that a substance which causes leech muscle to contract is released on stimulation of the nerves supplying the cat's tongue. It was also found that the perfusate from a period of stimulation reduced the blood pressure in the cat, and that this effect was increased in the presence of eserine and abolished by atropine; in all these respects the released substance was similar to acetylcholine. Furthermore, the concentrations of acetylcholine required to mimic the effects of the perfusate on cat blood pressure and on the leech muscle were the same. It follows that acetylcholine must have been present in the perfusion fluid obtained during a period of stimulation.

Where does this acetylcholine come from? At this stage of the investigation there were three possibilities: the acetylcholine could be released from the motor nerves, from the sensory nerves, or from the muscle fibres. In order to resolve this question, Dale and his colleagues used a similar perfusion technique on leg muscles of the dog and the frog. Stimulation of the sensory nerves via the dorsal roots (which were cut proximally to the electrodes so as to avoid reflex excitation of the motor nerves) did not induce release of acetylcholine, whereas stimulation of the motor nerves via the ventral roots did. Direct stimulation of the muscles also produced acetylcholine release, but of course it is impossible to stimulate a normal muscle without also stimulating the motor nerve endings in the muscle. However, stimulation of a muscle which had been denervated ten days previously, or stimulation of a muscle after transmission had failed following prolonged stimulation of the motor nerves, did not result in release of acetylcholine. Hence acetylcholine is released from the motor nerve endings on stimulation.

This conclusion was very interesting in itself, but, in terms of the chemical transmission theory, it had to be shown that the release of acetylcholine was relevant to the excitation of the muscle, and was not merely some incidental by-product of nervous activity. Earlier work had shown that some contraction could be obtained from muscles after injection of acetylcholine, but only when the amount injected was very high. Brown, Dale and Feldburg (1936) reinvestigated the problem, paying particular attention to the details of the perfusion technique, so as to ensure that the administered doses of acetylcholine came into contact with all the fibres in the muscle as rapidly as possible. In this way they were able to show that injections of as little as 2 μg of acetylcholine could produce contractions reaching as high as that in a twitch produced by stimulation of the motor nerves. This response was abolished in the presence of curare, a South American arrow poison which blocks neuromuscular transmission. Injections of eserine caused the contractions produced by single stimuli to the motor nerves to be enhanced and prolonged, as in repetitive stimulation. Hence it was concluded that the action of acetylcholine on the muscle fibre is an essential step in the transmission process.

THE END-PLATE POTENTIAL

We have now to seek an answer to the question: what is this action of acetylcholine on the muscle fibre? Electrical activity in the region of the end-plate, produced by nervous stimulation, was first observed by Göpfert and Schaefer in 1938 and was subsequently investigated by a number of other workers. The general conclusion to be reached from these initial experiments was that the first postsynaptic electrical event is a depolarization (recorded as a negative potential with external electrodes) of the muscle fibre membrane, known as the *end-plate potential* (EPP). This initial depolarization is restricted to the end-plate region of the muscle fibre, but is normally of sufficient size to elicit an all-or-none action potential which propagates along the length of the fibre.

The end-plate potential was first investigated with intracellular electrodes by Fatt and Katz (1951), using the frog sartorius muscle. It is much easier to study the nature of the EPP if it is reduced in size by partial block with curare, so that the complicating effects of the propagated action potential are absent. Under these conditions, the EPP is a brief depolarization with a rapid rising phase and a much slower falling phase, as is shown in the uppermost record of fig. 6.4.

When the electrode is inserted at increasing distances from the end-plate, the electrical change becomes reduced in size and its time course is lengthened (fig. 6.4). This is what one would expect if the EPP is a result of a brief 'active' phase of ionic current flow across the membrane

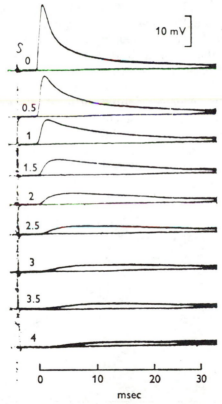

Fig. 6.4. The end-plate potential in curarized frog sartorius muscle, recorded by means of an intracellular microelectrode inserted at different distances from the neuromuscular junction. The figures to the left of each record indicate the distance (mm) between the electrode and the motor nerve ending; the time scale at the bottom is in msec. (From Fatt and Katz, 1951.)

at the junctional region, followed by a passive electronic spread and decay of the charge (or rather, deficiency of charge, since the membrane is depolarized) across the membrane. That is to say, the form of the EPP should be in accordance with the equations used by Hodgkin and Rushton (p. 49) in the application of the core-conductor theory to nerve axons. A relatively simple way of testing this idea is to see if the total charge on the membrane (which is proportional to the integral of voltage with respect to distance) decays exponentially, as is predicted by (4.10). Fig. 6.5 shows that it does. A more elaborate test was provided by assuming that the idea was correct, calculating the resistance and capacitance of the membrane from the form of the EPP, and then comparing the values obtained with those derived from measurements using applied current pulses; similar estimates were obtained in each case.

This analysis also allows the amount of charge displaced during an EPP to be calculated. The peak value in fig. 6.5 is 3.2×10^{-3} V cm; this will give the peak charge displacement in coulombs if multiplied by the membrane capacitance of a unit length of muscle fibre. With a membrane capacitance of $6 \ \mu F/cm^2$ and fibre diameter of $135 \ \mu m$, this works out as $2.54 \times 10^{-7} \ \mu F/cm$, giving a charge transfer of about 8×10^{-10} coulombs, which is equivalent to a net movement of about 8×10^{-15} moles of univalent ions. These calculations refer to the EPP of curarized muscle, so that in the normal condition the action of the neuromuscular transmitter must result in an ionic movement of at least three times this amount at the end-plate.

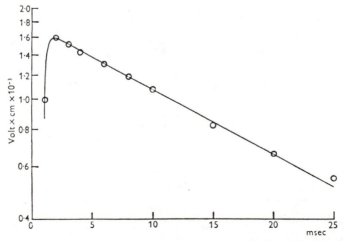

Fig. 6.5. Diagram to show the exponential decay of charge associated with the end-plate potential. The ordinate shows the integral of voltage with respect to distance (derived from the records of fig. 6.4), plotted on a logarithmic scale. (From Fatt and Katz, 1951.)

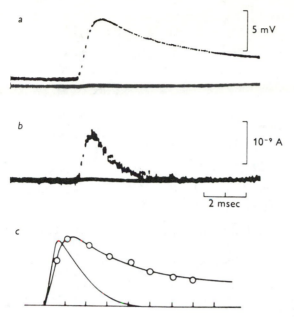

Fig. 6.6. Voltage clamp analysis of the end-plate potential in a curarized frog muscle. Trace *a* shows the EPP, recorded in the usual manner. Trace *b* shows the EPC recorded from the same end-plate with the membrane potential clamped at its resting level. The continuous lines in *c* show the actual EPP and EPC superimposed; the circles show the expected EPP as calculated from the EPC, assuming that the muscle fibre has a time constant of 25 msec and an effective resistance of 320 kΩ. (From Takeuchi and Takeuchi, 1959.)

The ionic current at the end-plate has since been measured directly by Takeuchi and Takeuchi (1959), using a voltage clamp technique. The results (shown in fig. 6.6) are fully in agreement with the scheme suggested by Fatt and Katz.

The initiation and form of the action potential

If a frog sartorius fibre is stimulated electrically so as to depolarize the membrane sufficiently, an action potential arises in the depolarized region and is propagated along the length of the fibre.* This action potential is similar to that of nonmyelinated axons (although the time course is longer) in that it shows the familiar phenomena of threshold, all-or-nothing response and refractoriness, and is produced by a specific regenerative increase in the permeability of the cell membrane to sodium ions (Nastuk and Hodgkin, 1950).

* It is important to realize that vertebrate 'twitch' muscle fibres are rather specialized in this respect; excitation does not result in an all-or-nothing propagated action potential in most other types of muscle fibre (see chapter 13).

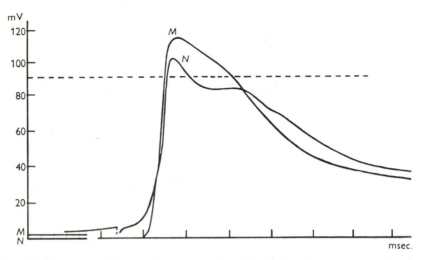

Fig. 6.7. Muscle action potentials recorded intracellularly from the end-plate region in response to direct electrical stimulation of the muscle fibre (trace *M*) and in response to stimulation of the motor nerve (record *N*). The voltage scale shows the change with respect to the resting potential; the broken line shows where the potential across the membrane is zero. (From Fatt and Katz, 1951.)

In the normal muscle, the depolarization produced by the EPP is sufficient to cross the threshold for production of an action potential. However, Fatt and Katz showed that the form of an action potential produced in response to nervous stimulation and recorded at the end-plate (the *N* response) is different from that recorded at distances away from the end-plate (fig. 6.8). It is also different from responses to direct stimulation of the muscle fibre (*M* responses) recorded at the end-plate (fig. 6.7). Fig. 6.8 shows the transitional stages between the two types that occurs at points near the end-plate. What is the reason for these differences? It is obvious that the inflection in the rising phase of the *N* response at about -50 mV is caused by the action potential 'taking off' from the rising phase of an EPP as the membrane potential crosses the threshold. The later differences between *N* and *M* responses were at first more difficult to interpret; the peak of the *N* response is lower than that of the *M* response and there is a hump on the falling phase of the *N* response. Fatt and Katz suggested that these phenomena were due to the effect of the conductance change at the end-plate on the form of the normal action potential.

The ionic basis of the end-plate potential

Del Castillo and Katz (1954*a*) found that the EPP increased in size if the muscle cell membrane was hyperpolarized at the end-plate region,

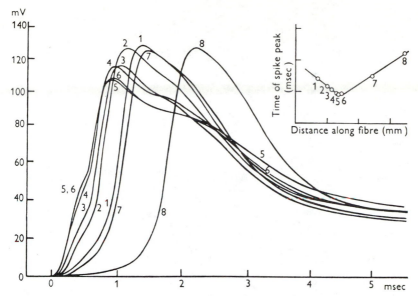

Fig. 6.8. The initiation of the muscle action potential at the motor end-plate. Traces obtained by inserting a microelectrode at different distances (shown in the inset) along the length of a frog muscle fibre, and recording the response to stimulation of the motor nerve. Membrane potentials recorded as changes from the resting level. (From Fatt and Katz, 1951.)

and decreased if it was depolarized. This observation provides further evidence for the view that the EPP is produced by an increase in membrane conductance. But it was not clear whether or not this conductance increase occurred for all ions. The problem was solved by Takeuchi and Takeuchi (1960), using a voltage clamp method on frog sartorius muscle fibres. It is not possible to clamp the membrane potential of a whole muscle fibre since one cannot insert an intracellular wire electrode along its length. However, using glass intracellular micro-electrodes, the membrane potential can be clamped at the end-plate region, as long as no large currents flow through the adjacent unclamped membrane. As we have seen (fig. 6.6), this technique allows the end-plate current (EPC) to be measured. The principle of the Takeuchis' experiments was to determine the relations between the size of the EPC and the membrane potential in solutions containing different ionic concentrations.

First, consider fig. 6.9a. In a solution containing 3×10^{-6} g/ml curare, the EPC varies linearly with membrane potential over the measured range of -50 to -120 mV (it is not possible in these experiments to clamp the membrane at depolarizations greater than about -50 mV). When the line drawn through these points is extrapolated, it cuts the

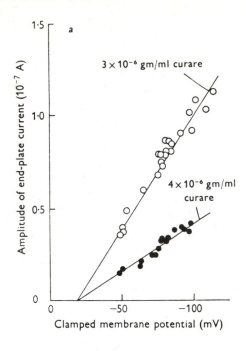

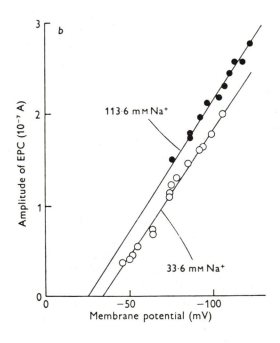

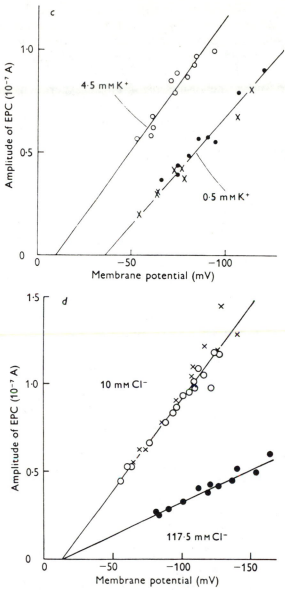

Fig. 6.9. The effect of membrane potential on the end-plate current of a frog muscle fibre, determined by means of a voltage clamp technique, under various conditions. *a*: Effect of different curare concentrations. *b*: Effect of different external sodium ion concentrations. *c*: Effect of different external potassium ion concentrations. *d*: Effect of different external chloride concentrations. Notice particularly the changes in the reversal potential brought about by changes in sodium and potassium ion concentrations, and the absence of such changes when the curare or chloride concentrations are altered. (From Takeuchi and Takeuchi, 1960.)

membrane potential axis at -15 mV. This means that the EPC would be zero at this value, and negative (i.e. in the opposite direction) beyond it; hence this point (-15 mV) is called the *reversal potential*. If the curare concentration is increased, the values of EPC are reduced, but the reversal potential is not altered.

Fig. 6.9*b* shows the results of a similar experiment, in which the relation between the EPC and the membrane potential in a curarized muscle fibre was compared in two solutions containing different sodium ion concentrations. A number of such experiments gave the relation between sodium ion concentration and reversal potential in fig. 6.10*a*. It

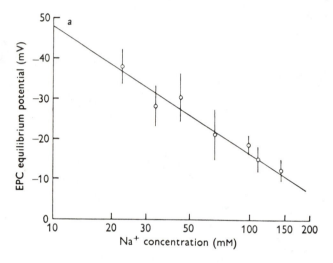

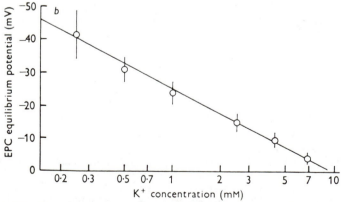

Fig. 6.10. The effects of different external sodium (*a*) and potassium (*b*) ion concentrations on the reversal potential of the end-plate current in frog sartorius muscle fibres. (From Takeuchi and Takeuchi, 1960.)

is evident that reduction of the sodium ion concentration makes the reversal potential more negative, and thus it follows that the sodium conductance of the end-plate membrane must increase during the EPC. Similar experiments were carried out on the effect of variations in potassium ion concentration (figs. 6.9c and 6.10b). Again, it was found that the reversal potential is dependent upon the potassium ion concentration, and therefore the potassium conductance of the end-plate membrane must also increase during the EPC. However, with drastic changes in chloride concentration, using glutamate as a substitute, there was no change in reversal potential, and therefore there is no change in the chloride conductance of the end-plate membrane during the EPC (fig. 6.9d).*

Thus it would seem that the transmitter substance, acetylcholine, acts by increasing the permeability of the subsynaptic membrane to cations, so that sodium ions flow inward and potassium ions flow outward. Thus the EPC is the sum of the currents due to sodium and potassium ion flow, or

$$I_{EPC} = I_{Na} + I_K.$$

Now
$$I_{Na} = \Delta g_{Na}(E - E_{Na}),$$

where Δg_{Na} is the change in the sodium conductance, E is the membrane potential, and E_{Na} is the sodium equilibrium potential, given by the Nernst equation for sodium ions, (5.2). Similarly

$$I_K = \Delta g_K(E - E_K).$$

Therefore
$$I_{EPC} = \Delta g_{Na}(E - E_{Na}) + \Delta g_K(E - E_K).$$

At the reversal potential, the end-plate current is zero, therefore

$$\Delta g_{Na}/\Delta g_K = (E - E_K)/(E_{Na} - E).$$

Thus, in normal Ringer solution, assuming that at the reversal potential $E = -15$ mV, $E_K = -99$ mV and $E_{Na} = 50$ mV,

$$\Delta g_{Na}/\Delta g_K = 84/65 = 1.29,$$

so that the end-plate membrane is slightly more permeable to sodium than it is to potassium during the action of the transmitter substance. Notice that, in contrast with the situation in a propagated action potential, the *ratio* of the sodium and potassium conductances does not vary with time, and moreover the conductances are not initiated by changes

---

* Although the reversal potential is unchanged in fig. 6.9d, it is obvious that the EPCs recorded were much higher in the absence of chloride. The reason for this is that the glutamate solution contained a higher calcium concentration, which would increase the amount of transmitter released (p. 135), and also a lower curare concentration.

in membrane potential, so that the potential change constituting the EPP is not regenerative.

THE ACTION OF ACETYLCHOLINE ON THE END-PLATE MEMBRANE

If the transmitter substance at the neuromuscular junction really is acetylcholine, then it should be possible to show that acetylcholine produces electrical changes at the end-plate similar to those following nervous stimulation. But how is the acetylcholine to be applied to the end-plate? As we have seen, the normal action of the transmitter substance is a brief, impulsive event lasting no more than a few milli-seconds, yet it is not possible to limit the action of externally applied acetylcholine to less than a few seconds when using normal perfusion techniques. The answer to this problem was provided by an ingenious technique developed by Nastuk (1953) and further used by del Castillo and Katz (1955) and others. Acetylcholine ionizes in solution so that the acetylcholine ion is positively charged. Consequently it will move elec-trophoretically if placed in an electric field. Thus after filling a glass micropipette (of the same order of size as a glass microelectrode) with acetylcholine, it is possible to eject it from the tip of the pipette for very brief times by passing a pulse of electrical current outward through the pipette. This technique is known as *ionophoresis*. It is usually necessary to pass an inward 'braking' current through the pipette in order to prevent outward diffusion of acetylcholine in the absence of an ejecting pulse.

The experimental arrangement used by del Castillo and Katz is shown in fig. 6.11. When the acetylcholine pipette was brought near the end-plate, an ejecting pulse caused a depolarization of the muscle fibre membrane, which could initiate propagated action potentials if it was sufficiently large (fig. 6.12). On the other hand, if the pipette was

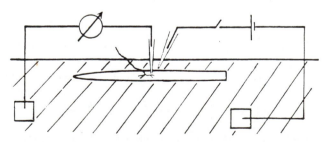

Fig. 6.11. Arrangement for recording membrane potential changes in response to iono-phoretic application of acetylcholine to the end-plate of a muscle fibre. (From del Castillo and Katz, 1955.)

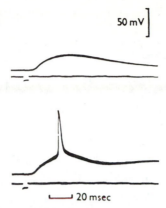

Fig. 6.12. Responses of a muscle fibre to ionophoretic application of acetylcholine to the end-plate. The upper traces show the membrane potential of the fibre, the lower traces monitor the current in the ionophoresis circuit. (From del Castillo and Katz, 1955.)

inserted into the muscle fibre, the acetylcholine ejecting pulse merely caused an electrotonic potential change, which is what would occur even if there were no acetylcholine in the pipette. This is well shown in fig. 6.13, which gives the successive responses to ejecting pulses as the pipette was withdrawn from a fibre.

These experiments show that acetylcholine, when applied to the outside of the end-plate membrane, causes a depolarization similar to that produced during neuromuscular transmission. The longer time course of the rising phase of the acetylcholine potential was ascribed to the much longer distance from the sensitive sites: a few microns instead of 500 Å. The acetylcholine is thought to act via combination with particular molecular sites (known as receptors) which occur in the end-plate region on the outside of the postsynaptic membrane. The combination between an acetylcholine molecule and a receptor thus

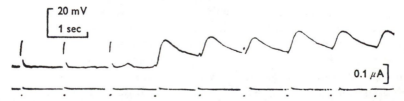

Fig. 6.13. The effect of withdrawing an acetylcholine-filled ionophoresis pipette from the motor end-plate. The lower trace shows the current pulses applied to the pipette so as to eject acetylcholine, the upper trace is a record of the membrane potential. At the beginning of the record, the pipette was inside the muscle fibre. It was then progressively withdrawn, leaving the fibre between the third and fourth current pulses. (From del Castillo and Katz, 1955.)

momentarily opens a channel through which sodium and potassium ions can flow.

Localization of the acetylcholine receptors

Responses to acetylcholine in ionophoresis experiments can only be obtained when it is applied to the end-plate region of the muscle fibre. This suggests that the acetylcholine receptors are localized in or near the subsynaptic region. Just how localized they are has been shown in some technically superb experiments by Kuffler and Yoshikami (1975a). They used the proteolytic enzyme collagenase to free the nerve terminals from the muscle fibres, so enabling acetylcholine to be applied directly to the subsynaptic membrane. They found, as expected, that the subsynaptic membrane had a high sensitivity to acetylcholine; but the sensitivity fell to as little as two per cent of this within a distance of 2 μm once the ionophoresis pipette was moved over the extrasynaptic membrane.

The localization of acetylcholine receptors has also been determined histochemically. A most useful substance for this is α-bungarotoxin, a polypeptide constituent of the venom of the Formosan snake *Bungarus multicinctus*, which causes neuromuscular block by binding tightly to the acetylcholine receptors (see Lee, 1970). Using radioactive toxin (made by acetylating the toxin with ^{3}H-acetic anhydride) it is a simple matter to show by autoradiography that the toxin rapidly becomes attached to the postsynaptic membrane at the end-plate regions (Barnard, Wieskowski and Chiu, 1971, Porter and Barnard, 1975). By counting the grains of silver produced in the autoradiograph, it is then possible to calculate the number of toxin molecules that have been bound and from this to estimate the number of receptors present. The results suggest that there are about 3×10^7 binding sites per end-plate in mammals, corresponding to an average density in the region of 10^4 sites per μm^2. Closer examination (by Fertuck and Salpeter, 1974) reveals that the binding sites occur only on that part of the postsynaptic membrane which is opposed to the presynaptic membrane, and not in the synaptic gutters. In these regions, therefore, their density must be very high indeed – perhaps over 3×10^4 sites per μm^2. The number of ionic channels will be lower than this of course, if there are a small number of binding sites associated with each channel. Even so, the density is appreciably larger than that of sodium channels in unmyelinated axon membranes.

Cooperativity

Werman (1969) pointed out that many dose–response curves for drug–receptor reactions are not in accordance with the simple idea that one drug molecule combines with one receptor molecule, all such combinations being independent of one another. He suggested instead that there may be cooperative interactions between individual drug binding sites, there being more than one such site in each receptor. This idea has received some support from recent experiments on the dose–response curves for the ionophoretic application of acetylcholine to end-plates: Dreyer and Peper (1975) suggest that three molecules combine with each receptor, whereas Dionne and Stevens (1975) conclude that the number is two, or perhaps one or two. There is also some evidence for such cooperative action from biochemical investigations on isolated receptors (see Changeux *et al.*, 1976).

A likely explanation for these effects is that each receptor site consists of an ionic channel in the middle of a small number of receptor molecules which are tightly bound to each other. The channel is then only opened when acetylcholine molecules are combined with some or all of these units. Another possibility is that combination of acetylcholine with one receptor site slightly alters the characteristics of neighbouring sites so that they combine more readily with acetylcholine. This could only happen if the receptor sites were closely packed together in the membrane, as indeed they are.

The molecular nature of acetylcholine receptors

A number of laboratories have made progress in chemically isolating and characterizing the acetylcholine receptor (see, for example, Raftery, Vandlen, Reed and Lee, 1976, Karlin, Weill, McNamee and Valderrama, 1976, and Changeux *et al.*, 1976). The experiments have been done with the electric organs of fishes such as the electric eel and electric ray. We shall examine the functioning of these organs in chapter 13; for our purposes here it is sufficient to notice that they are much richer in receptors than are muscle cells. Isolation of receptors usually involves combination with a highly specific binding agent from snake venom (such as α-bungarotoxin), which may be radioactively labelled, and treatment with detergents so as to free them from the membrane in which they are bound.

Such isolated receptors are proteins with a high proportion of hydrophobic (lipophilic) amino acid residues. This is just what we would expect if they are firmly embedded in the cell membrane. The molecules consist of three or four different types of protein chains, one of which binds to acetylcholine. We assume that the act of binding the acetylcholine molecule results in an alteration of the configuration of the protein chains of the receptor so as to open a channel permitting the flow of cations across the membrane.

The time course of the conductance change during the EPP

For a long time it was supposed that the conductance changes occurring during the EPP were totally independent of the membrane potential. More recently, however, it has been found that the end-plate current decays more rapidly at less negative membrane potentials. For example, Kordas (1969) found that the half-time of decline of the end-plate current was 1.6 msec when the membrane potential was clamped at -120 mV, but only 0.6 msec at $+40$ mV.

Similar results have been obtained by Magleby and Stevens (1972*a*, *b*), who have produced a quantitative theory to describe the time course of the end-plate currents. We need not examine the detail of this theory here, but its basic ideas provide a useful picture of what may be happening at chemically transmitting synapses. It is assumed that each molecule of the transmitter substance T combines with a molecule of the receptor substance R to form a complex TR and that this then undergoes a conformational change (to become TR^*) which is associated with the opening of an ionic channel. These reactions are reversible and are governed by different rate constants (k_1, k_2, α and β) to give an overall reaction scheme

$$T + R \underset{k_2}{\overset{k_1}{\rightleftharpoons}} TR \underset{\alpha}{\overset{\beta}{\rightleftharpoons}} TR^*. \tag{6.1}$$

It is also assumed that k_1 and k_2 are much faster than α and β, so that the conformational change is the rate-limiting step. α and β are affected by the membrane potential, presumably because the conformational change involves some small redistribution of charge within the receptor molecule.

What will happen to the acetylcholine released into the synaptic cleft by a presynaptic nerve impulse? Most of it will combine with receptors to produce TR and then TR^*, so opening the ionic channels in the postsynaptic membrane and hence resulting in current flow across it. Some transmitter will diffuse out of the cleft, and some will be hydrolyzed by acetylcholinesterase (p. 147). The activated complex TR^* will revert to TR and, since the concentration of T is now low, this in turn will dissociate to free more T which will itself diffuse away or be hydrolyzed. The decay of the end-plate current is therefore largely determined by α, the rate constant for conversion of TR^* to TR; indeed Magleby and Stevens assume that the rate constant for the exponential decay of the end-plate current is identical with α. Further support for this view comes from measurements on acetylcholine-induced 'noise', as we shall see later.

Desensitization

If a muscle is soaked in a solution containing acetylcholine, the initial period of excitation is followed by neuromuscular block which can only be removed by washing away the acetylcholine. Thesleff (1955) showed that such application causes an initial depolarization which then dies away, and introduced the term *desensitization* to describe this effect. The situation was further investigated by Katz and Thesleff (1957*a*) using the ionophoretic application technique. Both barrels of a double-barrelled pipette were filled with an acetylcholine solution; steady ('conditioning') currents were passed through one and brief pulses of current through the other. When acetylcholine was continuously applied by switching on the 'conditioning' current, the resulting depolarization gradually faded, and the response to 'pulses' of acetylcholine also declined, as is shown in fig. 6.14. Cessation of the conditioning current was followed by a recovery in the response to acetylcholine pulses.

What is the cause of this desensitization? Katz and Thesleff suggested that the receptor sites become inactive after combination with acetylcholine for some time. Some recent studies on purified receptor protein by Changeux and his colleagues (1976) are in accordance with this view. They find that acetylcholine combines initially with receptor protein

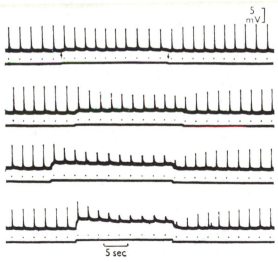

Fig. 6.14. A series of records showing desensitization to acetylcholine in a frog muscle fibre. In each case the lower trace monitors the ionophoretic current and the upper trace shows the membrane potential of the muscle fibre. Acetylcholine was applied iono-phoretically from a double-barrelled micropipette. Brief 'test' pulses were applied through one barrel (producing the dots on the lower traces and the vertical deflections on the upper traces), and constant 'conditioning' currents were applied through the other barrel at the time indicated. The strength of the 'conditioning' current is lowest in the top record and highest in the bottom one. (From Katz and Thesleff, 1957*a*.)

with a low affinity, but that this slowly changes to a high affinity complex which, it is thought, does not possess an open ionic channel in the membrane.

Acetylcholine noise

The depolarization produced by acetylcholine is the sum of the effects of a very large number of random collisions between acetylcholine molecules and receptor molecules. We would expect there to be random fluctuations in the rate of these molecular collisions, which would produce corresponding fluctuations – or 'noise' – in the resulting depolarization. Katz and Miledi (1970, 1972) found that such fluctuations are indeed detectable, as can be seen in fig. 6.15. The phenomenon has also been investigated by Anderson and Stevens (1973) with the voltage clamp technique; this allowed the noise (measured as fluctuations in current) to be measured at different membrane potentials, independently of the acetylcholine concentration. Let us examine some of their results.

It would be inappropriate here to delve into the mathematics of fluctuation analysis. Suffice it to say that it is possible to resolve a fluctuating current record into a series of sine and cosine waves of different frequencies. A *spectral density curve* (or 'power density spectrum') can then be constructed from these; the density at any particular frequency is the sum of the squared amplitudes of the sine and cosine components of the noise at that frequency. (It will be clear that the calculations in this Fourier analysis are best done by computer.) The spectral density curve thus shows the contributions of different frequencies to the noise, as is shown in fig. 6.16. It can also – and this is its really useful property – give us some information about the time characteristics of the individual events which cause the fluctuations.

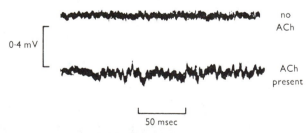

Fig. 6.15. End-plate 'noise' produced by acetylcholine. The upper trace shows a record at high gain of the membrane potential from a frog sartorius muscle fibre, in the end-plate region with no acetylcholine (ACh) present. The lower trace is a similar record obtained when acetylcholine was applied by diffusion from a nearby micropipette. (From Katz and Miledi, 1972.)

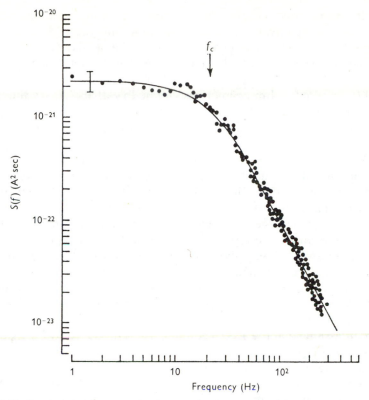

Fig. 6.16. Spectral density curve (or 'power density spectrum') of current fluctuations produced at the end-plate by acetylcholine. The membrane potential was clamped at −60 mV. The curve is drawn according to (6.2), with $\alpha = 0.132 \, \text{msec}^{-1}$. (From Anderson and Stevens, 1973.)

(Should there be any readers who regard the last sentence as nonsensical magic, I invite them to consider the Analogy of the Fleas. Suppose the platform of a small rapid-reading balance is placed in a box in which there are a large number of fleas jumping about at random. Each flea remains still for a short period of time (for an average of 100 msec, let us say) between jumps. Clearly the number (and hence weight) of fleas on the platform will fluctuate, and so the continuous record of weight will be 'noisy'. If we measure the weight of fleas on the platform every second we will find that successive measurements are essentially independent of one another because almost all the fleas on the platform at one particular time will have jumped off when we look again one second later. Hence the variance of the difference between successive weights recorded at one-second intervals will be maximal. The same will apply for measurements taken at 10-second intervals. Another way of describing this situation is to say that the low frequency components of the noise are large. However if we measure at intervals of 1 msec, each successive measurement will be very close to the previous one, because only about one hundredth of the fleas will have jumped off in the meantime. Hence the variance of the differences between successive weights recorded at 1 msec intervals will be low, that is to say the high frequency components of the noise are small. At intermediate frequencies we will see corresponding intermediate values for the components of the noise, and so a graph of these components against frequency might well look like that in fig. 6.16. Notice that the contribution of any particular high frequency to the noise depends upon the time for which the fleas are still between jumps. If we warm them up, for example, so that this time is reduced to 10 msec, then measurements at 1 msec intervals will show larger differences between successive measurements because now about one tenth of the fleas will have jumped off in the meantime instead of only about one hundredth. Clearly there must be information in the frequency characteristics of the noise which can tell us about the length of time between jumps. For fleas read acetylcholine molecules, for platform read postsynaptic membrane and for weight read current. Now read on.)

Let us assume that the acetylcholine noise is produced by fluctuations in the number of ionic channels opened by combinations of acetylcholine molecules with receptors, and that each channel is either open or closed. It can then be shown that the form of the spectral density curve should be given by

$$S(f) = \frac{S_0}{1 + (2\pi f/\alpha)^2} \tag{6.2}$$

where f is the frequency, S_0 is a constant (the value of $S(f)$ when $f = 0$) and α is the rate constant for the closing of the ionic channels (as in (6.1)). Now let f_c be the frequency at which the spectral density falls to half its maximum value. Then

$$S(f_c) = \frac{S_0}{2} = \frac{S_0}{1 + (2\pi f_c/\alpha)^2}.$$

Hence $$(2\pi f_c/\alpha)^2 = 1$$

and so $$f_c = \alpha/2\pi.$$

In fig. 6.16, for example, the value of f_c is 21 Hz, giving a rate constant of 0.13 msec^{-1}. The rate constant is the reciprocal of the time constant, which is also the mean of an exponentially distributed population of times for which the channels are open. In the experiment of fig. 6.16, therefore, the average time for which a channel is open is 7.6 msec.

Anderson and Stevens found that f_c became larger at more positive membrane potentials and at higher temperatures. They were also able to show that the values of α derived from noise measurements were identical with those measured from the decay of current during neurally-evoked end-plate currents (as in Magleby and Stevens' experiments, p. 126), as is shown in fig. 6.17.

It is not clear what the reason for the change in α with change in membrane potential is, or whether it serves any functional purpose. It may simply be an accidental consequence of the enormous electric field across the membrane. It is certainly quite possible that the interaction with acetylcholine involves some slight redistribution of electric charge in the receptor molecule, and an intense electric field might well affect the rate of such a redistribution.

Direct observation of the current flow through single acetylcholine receptor channels has been achieved by Neher and Sakmann (1976). It has been known for some time (e.g. Axelsson and Thesleff, 1959) that

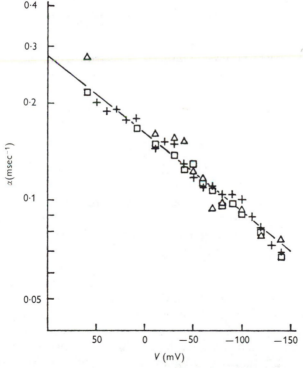

Fig. 6.17. Variation of α, the rate constant of end-plate current decay, with membrane potential. α is determined in three independent ways: from the rate of decay of the end-plate current (+) and of miniature end-plate currents ($\triangle$) and from fluctuation analysis of acetylcholine noise ($\square$). (From Anderson and Stevens, 1973.)

denervation results in a remarkable change in vertebrate muscle fibres: the whole fibre surface becomes sensitive to acetylcholine, as if the occurrence of receptors outside the end-plate region were normally inhibited in some way by the presence of the motor axon. Neher and Sakmann pushed the tip of a microelectrode containing a low concentration of acetylcholine against the membrane of a denervated muscle fibre and were thus able to measure the activity of the channels in a small patch of membrane. Since the density of the receptors is relatively low in the non-end-plate region, only a few hundred channels would be covered by the electrode. They found that each channel produces a square pulse of current lasting a few milliseconds. Of course, the characteristics of these extra-junctional channels may be somewhat different from the normal ones (for example Brockes, Berg and Hall, 1976, find that the extra-junctional receptors bind curare less tightly than do junctional ones); nevertheless the results are in pleasing agreement with the conclusions from noise analysis.

THE QUANTAL NATURE OF TRANSMITTER RELEASE

Miniature end-plate potentials

With a microelectrode inserted into a frog sartorius muscle fibre in the end-plate region, and using a high-gain low-noise amplifier, Fatt and Katz (1952) observed the occurrence, in the resting muscle, of small fluctuations in membrane potential of much the same shape as normal EPPs, but only about 0.5 mV in height (fig. 6.18). These *miniature end-plate potentials* could only be found at the end-plate region, were reduced in size by curare, and increased in the presence of prostigmine (an anticholinesterase similar to eserine in its action). They were not found in muscle that had been denervated two weeks previously. Statistical analysis of the time intervals between successive miniature EPPs showed that their occurrence was randomly distributed in time. These observations suggested that the miniature EPPs were caused by the action of acetylcholine released spontaneously from the motor nerve ending.

How much acetylcholine is involved in the production of a single miniature EPP? The most obvious suggestion is perhaps just one molecule. Fatt and Katz argued that this could not be so, however, for the following reasons. First, if one molecule could produce a depolarization of about 0.5 mV, then not more than 1000 at the most would be needed to produce an EPP of 50 mV, and therefore the amount of acetylcholine released per impulse would be no more than 10^{-21} moles. In fact it is much larger than this (Krnjević and Mitchell, 1961, estimate that each nerve terminal in rat diaphragm muscle releases about 10^{-17} moles of

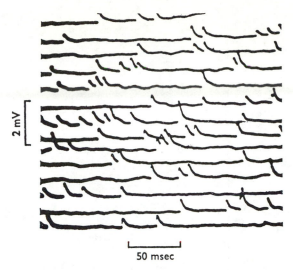

2 mV

50 msec

Fig. 6.18. A series of membrane potential records from a frog neuromuscular junction showing miniature end-plate potentials. (From Fatt and Katz, 1952.)

acetylcholine per impulse), so that each miniature EPP must be produced by several thousand molecules of acetylcholine. Secondly, the response to externally applied acetylcholine is a smooth depolarization, not a series of unitary events about 0.5 mV in height.

Kuffler and Yoshikami (1975b) have recently provided a more direct estimate of the number of molecules of acetylcholine per miniature EPP. They used preparations in which the subsynaptic membrane was exposed, by removing the nerve terminals from frog or snake muscles which had been treated with the enzyme collagenase so as to loosen their connective tissue sheaths. They then measured the size of the responses to ionophoretic applications of acetylcholine. The amount of acetylcholine ejected from the pipette during each ionophoretic pulse was determined by means of ingenious bioassay: several thousand such pulses were passed into a droplet of Ringer solution which was then applied to an end-plate under a layer of fluorocarbon oil, and its depolarizing effect was compared with that of a droplet of the same volume of acetylcholine solution of known concentration. They found that 6000 molecules of acetylcholine were required to produce a depolarization of 1 mV. The average size of a miniature EPP in their experiments was about 1.5 mV, and hence somewhat less than 10 000 molecules were involved in its production.

A most interesting property of these miniature EPPs is that their frequency can be altered by procedures which we would expect to change the membrane potential of the presynaptic nerve terminals.

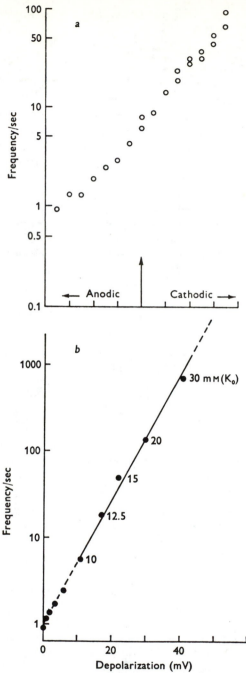

Fig. 6.19. The effect of the presynaptic membrane potential on the frequency of miniature end-plate potentials in a rat diaphragm muscle fibre. a: The effect of applying various electrical currents (measured in arbitrary units) to the nerve ending. b: The effect of increasing the external potassium ion concentration; the abscissa shows the estimated depolarizations of the presynaptic nerve membrane produced by this treatment. (a from Liley, 1956; b from Katz, 1962, derived from results given by Liley, 1956.)

Liley (1956) has investigated these effects at the neuromuscular junctions in the rate diaphragm. If depolarizing currents were passed through an external electrode placed near the nerve terminals, the frequency of discharge increased; conversely, it decreased if the current was hyperpolarizing (fig. 6.19*a*). The frequency also increased if the external potassium ion concentration was raised above 10 mM; in this case the increase can be related to an estimate of the presynaptic membrane potential, which cannot be measured directly since the terminals are too fine (fig. 6.19*b*).

We have seen that the size of the miniature EPPs can be altered by curare and anticholinesterases; it also changes with the postsynaptic membrane potential in the same way as does the EPP. In the frog sartorius, the size of miniature EPPs varies inversely with the diameter of the muscle fibre; thus, since large fibres have a smaller total membrane resistance than small fibres, it would seem that the currents associated with the miniature EPPs (and therefore the amounts of acetylcholine released) are more or less constant (Katz and Thesleff, 1957*b*). The discharge frequency can be altered by changing the osmotic pressure of the bathing solution or (in mammals) the magnesium or calcium ion concentrations, as well as by polarization of the nerve terminals. These observations lead to an important generalization: as Katz (1962) puts it, 'the *frequency* of the miniature potentials is controlled entirely by the conditions of the *pre*synaptic membrane, while their *amplitude* is controlled by the properties of the *post*synaptic membrane'.

The quantal nature of the end-plate potential

It has long been known that an excess of magnesium ions blocks neuromuscular transmission. Del Castillo and Engbaek (1954) showed that this effect was due to a reduction of the quantity of acetylcholine released per impulse, so that the EPP was very much smaller. Calcium ions antagonized this action, i.e. the degree of reduction of acetylcholine release was dependent upon the ratio of the magnesium and calcium ion concentrations. When the block produced by this treatment was intense, it was found that the size of successive EPPs may fluctuate in a step-wise manner (fig. 6.19). This led to the suggestion that acetylcholine is released from the motor nerve terminals in discrete 'packets' or *quanta*, that the normal EPP is the response to some hundreds of these quanta, and that miniature EPPs are the result of spontaneous release of single quanta.

An extension of the quantal release hypothesis is that there is a large population of quanta in each nerve terminal, each one of which has a small probability of being released by a nerve impulse. This idea was tested by del Castillo and Katz (1954*b*) by means of a statistical analysis

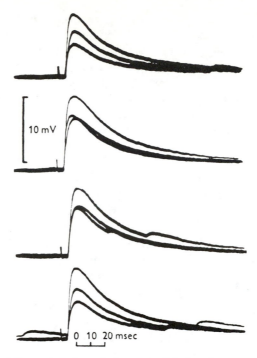

Fig. 6.20. Quantal fluctuations in end-plate potential size after magnesium poisoning. Spontaneous miniature potentials can be seen in some of the records. (From del Castillo and Katz, 1954*b*.)

applied to a large number of nervous stimuli. The experiments were performed on a frog toe muscle in a solution containing a high magnesium ion concentration so as to reduce the quantal content of each response. Let x be the number of quanta comprising any one response, and let m be the mean of all values of x. Let N be the total number of events measured, and let n_0, n_1, n_2, etc., be the number of events in the classes $x = 0$, 1. 2, etc.* Then the relative frequencies (n/N) of the different values of x should be given by the terms of a Poisson distribution, i.e.

$$\frac{n_x}{N} = e^{-m} \cdot \frac{m^x}{x!}. \tag{6.3}$$

The easiest way of finding m in this equation is to take the case when

* An analogy may help the non-mathematical reader with the terminology here. Suppose a large class of students attends a long series of lectures, and each student has a constant low probability of sneezing during a lecture. Then x is the number of sneezes in any particular lecture, m is the mean of x, N is the total number of lectures, n_0 the number of lectures with no sneezes, n_1 the number with one sneeze, and so on.

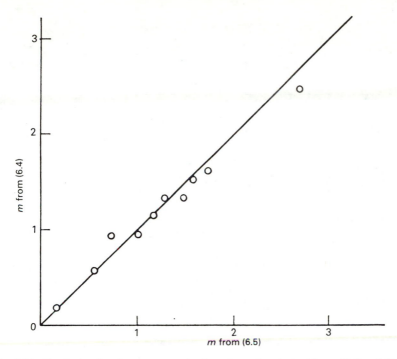

Fig. 6.21. Graph showing the agreement in the values of m derived from (6.4) and (6.5). Explanation in text. (From del Castillo and Katz, 1954*b*.)

$x = 0$, when we get

$$\frac{n_0}{N} = e^{-m}$$

or $$m = \log_e (N/n_0), \tag{6.4}$$

i.e. m is given by the natural logarithm of the reciprocal of the proportion of events in which transmission fails. There is also another way of finding m. Since we suspect that each miniature EPP corresponds to one quantum, the mean number of quanta per response should be given by

$$m = \frac{\text{mean amplitude of EPP}}{\text{mean amplitude of miniature EPPs}}. \tag{6.5}$$

Thus we have two independent ways of determining m, from (6.4) and (6.5), and if the hypothesis is correct, they should agree. Fig. 6.21 shows that they do.

Now consider the situation when x is 1 or more. The relative frequencies of EPPs of different sizes should be described by a Poisson distribution, as in (6.3). A slight complication arises here, however,

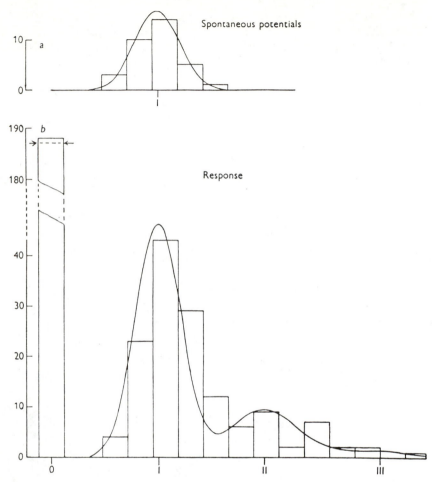

Fig. 6.22. Statistical analysis of the quantal components of transmission at a frog neuromuscular junction. *a*: Histogram showing the variation in size of spontaneous miniature end-plate potentials; the continuous curve is a normal distribution curve. *b*: Frequency distribution histogram of the sizes of responses to nervous stimulation in the same muscle fibre as *a*, after magnesium block. The ordinate shows the number of occurrences of potentials in a particular size range; the abscissa shows the size of the potentials, the unit of measurement being the mean size of the spontaneous potentials, from *a*. The arrows (to show the expected number of failures of transmission) and the continuous curve were calculated from a Poisson distribution, modified to take account of the variability in size of the quantal units as seen in *a*. (From del Castillo and Katz, 1954*b*.)

since the size of the spontaneous miniature EPPs varies somewhat (fig. 6.22*a*), and this has to be taken into account when calculating the frequency distribution of EPP amplitudes. Fig. 6.22*b* shows that the curve calculated in this way is in good agreement with the actual distribution of EPP amplitudes.

Does this analysis apply to normal EPPs, where m is large? Martin (1955) showed that it does, as long as (6.5) is modified to take account of the non-linear addition of quantal changes. For example, it follows from Ohm's law that a change in conductance which produces a depolarization of 0.5 mV when the membrane potential is -95 mV will only produce a depolarization of three-quarters of that amount (i.e. 0.375 mV) at -75 mV (assuming that the reversal potential is -15 mV). From these considerations, Martin calculates that a normal EPP of about 50 mV is produced by the action of about 250 quanta.

We can now see the importance of the experiments showing that the frequency of spontaneous discharge of miniature EPPs rises when the presynaptic terminal is depolarized. If the line in fig. 6.19b is extrapolated to, say, 90 mV (so that, to put it another way, we are assuming that the probability of discharge of a quantum of actylcholine is logarithmically proportional to the depolarization of the presynaptic terminal membrane), the discharge frequency would be about 10^6 per second. Since the presynaptic action potential only lasts for a fraction of a millisecond, this is about the right order of magnitude to account for the release of the number of quanta needed to produce an EPP.

But why should acetylcholine be discharged from the nerve ending in quantal units of 10^3 to 10^4 molecules? Plate 6.1 shows that the nerve terminal contains large numbers of vesicles about 500 Å in diameter. There vesicles were first observed by de Robertis and Bennett (1954) and Palade and Palay (1954), and have since been found in the pre-synaptic terminals of all chemically transmitting synapses where they have been looked for. Del Castillo and Katz (1956) suggested that they contain the transmitter substance, and that the discharge of the contents of one vesicle into the synaptic cleft corresponds to the release of one quantum of the transmitter. This suggestion is known as the 'vesicle hypothesis'.

Contents of the vesicles

If the vesicle hypothesis is correct it should be possible to isolate vesicles and show by chemical means that they actually do contain acetylcholine. Much work on this problem has been done by Whittaker and his colleagues. Homogenization of a suitable tissue (such as guinea-pig brains) followed by centrifugation enables a subcellular fraction containing fragments of nerve endings to be isolated. The fragments are known as synaptosomes and contain synaptic vesicles. The vesicles can be released from the synaptosomes by rupturing them in a hyposmotic solution. Such extracts from homogenized brains contain quantities of a variety of transmitter substances, including acetylcholine (Whittaker, Michaelson and Kirkland, 1964).

More recently it has been possible to isolate nearly pure fractions of acetylcholine-containing vesicles by preparing them from the electric organ of the electric ray, *Torpedo*, which is very richly innervated (Whittaker and Zimmermann, 1974). The acetylcholine content of each vesicle is in the region of 66 000 molecules. Electric organ vesicles are larger than those of the neuromuscular junction (the outside diameter is about 84 nm instead of 50 nm and so the volume they contain – allowing for a membrane thickness of 8 nm – is nearly seven times larger). So we might expect neuromuscular junction vesicles to contain about 10 000 molecules of acetylcholine, perhaps fewer if we take the osmotic concentration of the tissues into consideration. This figure is in good agreement with the estimates of quantal content from physiological experiments.

The vesicles also contain adenosine triphosphate (ATP), about one molecule for every five molecules of acetylcholine (Dowdall, Boyne and Whittaker, 1974). There is as yet no clear evidence as to what the physiological function of the ATP is, although we do know that it is released into the synaptic cleft along with the acetylcholine, as is shown by collecting perfusates from rat muscle during stimulation (Silinsky, 1975).

Preparations of vesicles also contain two types of protein. Some of them are of fairly high molecular weight and are membrane-bound and so are probably the intrinsic protein of the vesicle membrane. However one protein is of low molecular weight (about 10 000) and is not membrane-bound; it has been named 'vesiculin'. Amino acid analysis shows it to be rich in acidic residues. Whittaker, Dowdall and Boyne (1972) suggest that these negatively charged residues serve to neutralize the positively charged acetylcholine ions and that acetylcholine is replaced by cations such as sodium or potassium when the vesicle discharges. There is some evidence that vesiculin is not discharged along with the acetylcholine and ATP: vesicles from electric organs which have been depleted of acetylcholine by prolonged stimulation have lowered acetylcholine and ATP contents, whereas their vesiculin content is normal.

FACILITATION, DEPRESSION AND POST-TETANIC POTENTIATION

If a curarized rat diaphragm muscle is repetitively stimulated through its motor nerve, it is found that successive EPPs decline in size, as is shown in fig. 6.23*a*. This phenomenon is known as *neuromuscular depression*. However, if the calcium ion concentration is lowered, the reverse effect, known as *neuromuscular facilitation*, occurs (fig. 6.23*b*). The

Fig. 6.23. Facilitation and depression in curarized rat diaphragm muscle, stimulated at 120 shocks/sec. *a*: External calcium ion concentration 2.5 mM, showing depression. *b*: External calcium ion concentration 0.28 mM, showing facilitation. (From Lundberg and Quilisch, 1953.)

mechanisms of these effects were investigated by del Castillo and Katz (1954*c*), using a frog toe muscle. In solutions with high magnesium concentrations, it was found that the proportion of failures of transmission decreased during a train of stimuli. If the stimuli were paired, the number of failures following the second stimulus was less than that following the first stimulus. These experiments indicate that facilitation is a presynaptic phenomenon; it is produced by an increase in the probability of discharge of acetylcholine quanta. When transmission was blocked by curare (which, of course, does not itself reduce the number of quanta released per impulse), depression was observed, and it was found that the later EPPs in a train showed fluctuations in amplitude. Here, again, it is evident that depression is mainly a presynaptic phenomenon, although it is possible that desensitization may be important in some cases after prolonged stimulation.

It has been known for some time that monosynaptic spinal reflex responses are increased for a time after a period of repetitive stimulation (Lloyd, 1949). This phenomenon is known as *post-tetanic potentiation*. Hubbard (1963) has investigated this phenomenon at the neuromuscular junction in rat diaphragm muscle. Let us examine the results of one of his experiments, shown in fig. 6.24. Neuromuscular transmission was partially blocked by magnesium, the motor nerve was stimulated 50 times at a frequency of 200 impulses/sec, and the size of the EPP measured at different times after the end of the period of stimulation. Immediately after the cessation of stimulation, the EPP was increased in size; this effect occurred after single impulses as well as after long trains of stimuli, and we can therefore describe it as facilitation. Following this, about one second after the end of the stimulation period, was a brief period of depression. This was then followed by a prolonged period in which EPPs were increased in size. This second period of increased responses did not occur after single impulses and could only be brought about by repetitive stimulation, hence we can describe it as post-tetanic potentiation.

We are now faced with the problem of providing an explanation for the occurrence of these time-dependent changes in the amount of

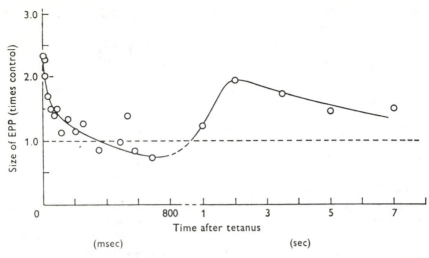

Fig. 6.24. Post-tetanic potentiation in magnesium-paralyzed rat diaphragm muscle. The graph shows the height of the end-plate potentials at various times after the end of a period of 50 stimuli at 200/sec. (After Hubbard, 1963.)

transmitter released following prior stimulation. Some progress in this respect can be made if we assume that only part of the transmitter present in the presynaptic terminal (or, to be more specific, only a proportion of the population of presynaptic vesicles) is available for release on the arrival of a nerve impulse. We can then regard depression as a consequence of the depletion of this readily available fraction by the discharge following previous impulses; this idea would fit with the observation that depression does not occur during magnesium block. Application of the concept to facilitation necessitates the suggestion that the size of the readily available fraction is temporarily increased by a presynaptic impulse. With post-tetanic potentiation we are on rather firmer ground. Hubbard and Schmidt (1963) discovered that repetitive stimulation increases the size of the presynaptic action potential (as recorded with an extracellular electrode), possibly by means of an accumulated after-hyperpolarization; this would of course increase the size of the EPP.

PRESYNAPTIC EVENTS PRODUCING TRANSMITTER RELEASE

We have seen that the arrival of a nerve impulse at the motor nerve terminal results in the release of a number of quantal packets of acetylcholine from the terminal. The next question to be raised is the nature of the connecting links between these two phenomena.

Synaptic delay

When an impulse arrives at the motor nerve ending, there is a short delay before the EPP arises in the muscle fibre. This delay is a general property of chemically transmitting synapses, and has recently been investigated at the frog sartorius neuromuscular junction by Katz and Miledi (1965*a*, *b*). The essence of their technique was to use an extracellular microelectrode applied closely to a point on the terminal so as to be able to make 'focal recordings' of the presynaptic action potential and the postsynaptic EPP from localized points at the junction.

One possible explanation of synaptic delay was that the action potential in the motor nerve axon does not invade the terminal endings, so that the depolarization there would be a slower process, caused by electrotonic spread from the myelinated region of the axon. However, Katz and Miledi found that the presynaptic spike recorded from the myelinated terminals occurred at later times after the stimulus as the recording electrode was moved distally, and also that antidromic* impulses could be set up in the axon by focal stimulation of the terminals. Hence it was concluded that the action potential in the axon does invade the terminals.

The focal recording technique was then applied to measuring accurately the extent of the delay at any particular point; the method of Katz and Miledi used for this is extremely ingenious. The preparation was kept in a calcium-free Ringer solution containing 1 mM magnesium ions; this completely blocked neuromuscular transmission but did not prevent presynaptic impulse conduction. The recording microelectrode was filled with a 0.5 M solution of calcium chloride, so that diffusion of calcium ions from the electrode (which could be controlled electrophoretically) allowed transmission to occur at, and only at, the point of recording. Thus the focal electrode recorded the presynaptic action potential at a point on the nerve terminal and the EPP produced immediately below it; there could be no interference from transmitter action at distances away from this point, where presynaptic depolarization would occur at different times. Fig. 6.25*a* shows the form of the record obtained by this means, and fig. 6.25*b* shows a number of successive actual records. The size of the EPPs produced fluctuates in a step-wise manner and may fail completely, as is to be expected from the quantal release hypothesis. More interesting is the fact that the synaptic delay is shown to be variable. By measuring the delays of a large number of responses it was possible to measure the probability of transmitter

---

* An *antidromic* impulse is one travelling in the opposite direction to that which normally occurs in the animal. An *orthodromic* impulse is one which travels in the same direction as naturally occurring impulses.

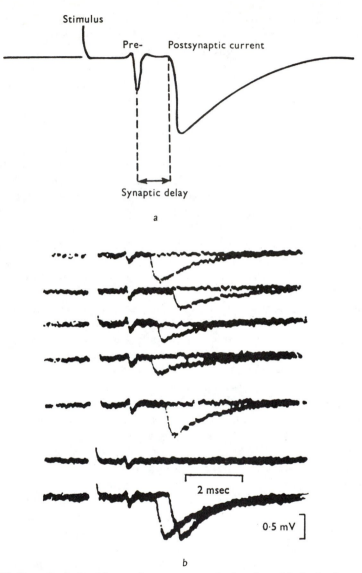

Fig. 6.25. Records obtained from a frog neuromuscular junction with the focal recording technique, details of which are given in the text. A downward deflection indicates negativity at the focal electrode. *a*: Form of a typical record, showing the components of the response and the method of measuring the synaptic delay. *b*: Part of a series of such responses; two traces are superimposed on each record. (From Katz and Miledi, 1965*b*, by permission of the Royal Society.)

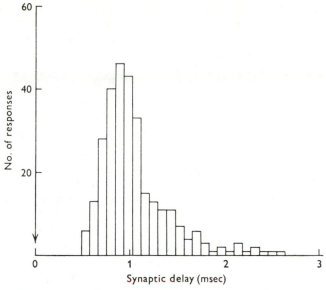

Fig. 6.26. Frequency distribution of synaptic delays in a large number of responses of the type shown in fig. 6.25. (From Katz and Miledi, 1965*b*, by permission of the Royal Society.)

quanta being released at various times after the presynaptic impulse. The results of this experiment are given in fig. 6.26, showing that the minimum synaptic delay is about 0.5 msec at 17 °C.

We now have the problem of what the nature of the synaptic delay is. Katz and Miledi considered three possibilities: delay between depolarization of the axon terminals and release of acetylcholine, the time taken for the acetylcholine to diffuse across the synaptic cleft, and the time between the impact of acetylcholine on the postsynaptic membrane and the subsequent depolarization. They showed that the time needed for acetylcholine molecules to diffuse the distance of 500 Å across the synaptic cleft must be very small, and certainly less than 0.05 msec. When acetylcholine was applied ionophoretically to the end-plate, the consequent postsynaptic depolarization began within 0.15 msec of the start of the ejecting current pulse. Hence it was concluded that at least 0.3 msec, the major part of the synaptic delay, is needed for the release of acetylcholine from the presynaptic nerve terminals.

The depolarization of the presynaptic terminal

As mentioned in chapter 5, the puffer fish poison tetrodotoxin blocks nervous conduction. A frog sartorius nerve-muscle preparation treated with tetrodotoxin will not therefore show the normal transmission process when the nerve is stimulated in the normal manner, because the axon and terminal membranes have been made inexcitable. However, Katz and Miledi (1967*a*) showed that if the terminals of such a preparation are briefly depolarized by a current pulse applied via a microelectrode, an end-plate potential is elicited in the muscle fibre. Thus the depolarization of the terminal membrane, however it is produced, appears to be the effective agent in producing transmitter release.

Further experiments with the tetrodotoxin-treated preparation (Katz and Miledi, 1967*b*) showed that the number of quanta released increased as the strength and duration of the applied current were increased. A most interesting finding here was that the latency of the postsynaptic response was greater with longer current pulses. This effect was explained by suggesting that transmitter release is dependent upon the entry of a positive charged substance (such as calcium ions) into the axon terminal; this substance would then be held back during the application of the depolarizing current, but would be free to enter after its cessation. In other words, it is suggested, depolarization of the terminal results in an increase in the conductance of the terminal membrane to this positively charged substance so that it can move down its electrochemical gradient into the cytoplasm of the terminal.

The role of calcium ions

If this suggestion is correct, and if the substance involved is calcium ions (or possibly a positively charged calcium ion complex), then one would expect transmission to be markedly affected by the external calcium ion concentration at the time when the terminal membrane is depolarized. This can be tested by applying calcium ions electrophoretically to the terminals of a tetrodotoxin-treated preparation in calcium-free Ringer solution at various times before and after the application of depolarizing pulses (Katz and Miledi, 1967*c*). It was found that application of calcium ions before the depolarizing pulse (the interval between the two pulses could be as little as 50 μsec) resulted in transmission, whereas application of calcium ions after the depolarizing pulse did not.

This result suggests that entry of calcium ions during depolarization of the terminal membrane is indeed an essential prerequisite for the liberation of transmitter. The existence of synaptic delay suggests that, once inside the terminal, the calcium must become involved in some further chemical reaction before transmitter release can occur. Possibly it forms an intermediate, CaX (or perhaps Ca_4X – see Dodge and Rahamimoff, 1967), which is needed for the release process. If this is so, the antagonistic action of magnesium ions can be explained by suggesting that they compete for X molecules to form a complex MgX which is ineffective in the release process.

Berl and his colleagues (1973) have detected a most interesting pair of proteins in the synaptosomal fraction of brain tissue. Named neurin and stenin, they are similar in many ways to the actin and myosin of muscle, described in chapter 11. Thus they will readily combine to form a complex (called neurostenin), and neurin stimulates the magnesium-activated ATPase activity of both stenin and muscle myosin. When the synaptosomal fraction is subjected to osmotic shock and then separated to give a vesicle fraction and a synaptic membrane fraction, stenin is found with the vesicles and neurin with the membranes. This leads to the suggestion that the neurin–stenin combination is an essential stage in the release of the contents of a vesicle from the presynaptic terminal. Perhaps the role of calcium ions is to activate the neurin–stenin combination.

ACETYLCHOLINESTERASE

The enzyme acetylcholinesterase hydrolyses acetylcholine to form choline and acetic acid. Non-specific ('pseudo') cholinesterases occur in the bloodstream; these are not as active in hydrolyzing acetylcholine as is 'true' cholinesterase, and can be distinguished by the fact that they are able to hydrolyze butyrylcholine. Acetylcholinesterase occurs in very high concentration at the neuromuscular junction. The first evidence for this came from the work of Marnay and Nachmansohn (1938), using a manometric estimation technique on frog sartorius muscle. A portion of the tissue under investigation was ground up in Ringer solution, and acetylcholine was added. The acetic acid formed by the enzymic hydrolysis of the acetylcholine released carbon dioxide from the bicarbonate in the Ringer, and the volume of this carbon dioxide was then measured manometrically. They found that portions of the muscle from regions which contained no nerve endings would hydrolyze 0.14 mg of acetylcholine per 100 mg of tissue per hour, whereas in the region with the greatest number of nerve endings the rate was about 0.8 mg. Since the neuromuscular junctions would only constitute a very small proportion of the latter sample, it is evident that the acetylcholinesterase concentration there must be very high. It was further shown that the concentration in the sciatic nerve was not sufficient to account for this high concentration at the junctional region, which must therefore be a specific property of the junction. Marnay and Nachmansohn calculated that there was sufficient acetylcholinesterase at each neuromuscular junction to hydrolyze about 10^{-14} moles of acetylcholine in 5 msec; this is more than enough to deal very rapidly with the amount of acetylcholine released per impulse.

The localization of acetylcholinesterase can be determined much more precisely by histochemical means. The essentials of this technique, as developed by Koelle and Friedenwald (1949), are as follows. Fresh-frozen material (or, in Couteaux's modification, material lightly fixed with formaldehyde) is incubated in a medium containing acetyl-thiocholine in the presence of magnesium ions and copper glycinate. The enzyme hydrolyses the acetylthiocholine to thiocholine and acetic acid, and the thiocholine then complexes with the copper glycinate. On treatment of this complex with ammonium sulphide, copper sulphide is deposited at the enzymic site, and can be easily seen under the micro-scope. The results show that acetylcholinesterase is located in the terminal gutters of the postsynaptic membrane.

7 Synapses between neurons

Synapses between neurons are of two physiologically distinct types: those in which transmission is chemical in nature as at the neuromuscular junction, and those in which the presynaptic cell directly excites the postsynaptic cell by means of electric current. We shall first consider the properties of some representative chemically transmitting synapses.

The structure of chemically transmitting synapses (fig. 7.1) shows some variation in detail, but two particular features are common to all of them: synaptic vesicles are found in the presynaptic nerve ending, and the pre- and postsynaptic cells are separated by an extracellular space (the synaptic cleft) about 200 Å across. In addition, the pre- and postsynaptic membranes are usually thickened in places, and it seems probable that these thickened regions are particularly associated with the transmission process. A variety of other subcellular components may occur in different types of synapse (see Gray and Guillery, 1966; Gray, 1974), but their functional significance is in most cases not yet clear.

SYNAPTIC EXCITATION IN MAMMALIAN SPINAL MOTONEURONS

Motoneurons are neurons which directly innervate skeletal muscle fibres. In mammals, the cell bodies of the motoneurons innervating the limb muscles lie in the ventral horn of the spinal cord, and their axons pass out to the peripheral nerves via the ventral roots. The cell body, or soma, is about 70 μm across, and extends into a number of fine branching processes, the dendrites. The surface of the soma and dendrites is covered with small presynaptic nerve endings (terminal boutons), showing the typical features of a chemically transmitting synapse. Many of these terminal boutons are the endings of group Ia fibres from stretch receptors (muscle spindles) in the muscle which the motoneuron innervates. Stretching the muscle excites these group Ia fibres, which may then excite the motoneurons supplying the muscle so that it contracts. This system is known as a *monosynaptic reflex* (fig. 7.2).

149

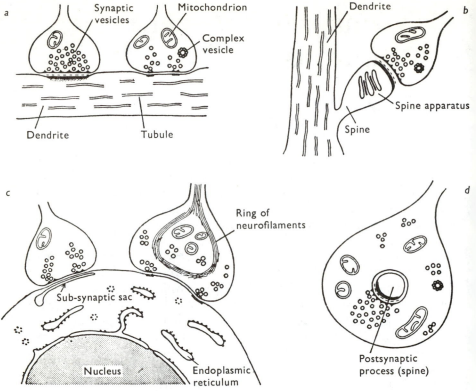

Fig. 7.1. Diagrams to show the variety of structure found in chemically transmitting synapses in the mammalian central nervous system. *a*: Two types of synapse between axon terminals and a dendrite. *b*: Synapse between an axon terminal and a dendritic spine. *c*: Two types of synapse between axon terminals and a neuron soma. *d*: A synapse whose postsynaptic spine is invaginated into the presynaptic terminal. Type 1 synapses (p. 165) are shown in *b* and on the left in *a*; type 2 synapses are shown in *c* and on the right in *a*. (From Whittaker and Gray, 1962.)

Most of our knowledge of the synaptic responses of motoneurons comes from investigations in the lumbar region of the spinal cord of the cat, using intracellular electrodes. This technique was first developed by Brock, Coombs and Eccles (1952), and has since been much used in Eccles's laboratory and elsewhere. The cat is anaesthetized and its spinal cord transected in the thorax. The spinal cord is exposed, a patch of the tough sheath round the cord removed, and the microelectrode inserted through this region until it reaches a motoneuron. Maps of the positions of various motoneuron groups are available, prepared by observing the chromatolytic changes following section of the nerve supplying a particular muscle (e.g. Romanes, 1951), so that the microelectrode can be placed in approximately the desired position. The

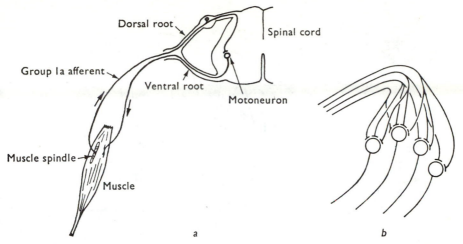

Fig. 7.2. Anatomical organization of the monosynaptic stretch reflex system. *a* is grossly simplified; there are very many stretch receptors and afferent and efferent neurons associated with each muscle. *b* is slightly less simplified, showing how the afferent fibres branch to synapse with different members of the motoneuronal pool.

identity of a motoneuron is accurately established by stimulating the motor nerve fibres of a particular muscle and observing the monosynaptic response in the motoneuron. In many experiments the ventral roots are cut so that stimulation of a peripheral nerve does not result in antidromic stimulation of the motoneurons. Fig. 7.3 shows some of the muscles whose motoneurons are mentioned in the following pages.

Motoneurons have a resting potential of about −70 mV. Depolarization of the membrane by about 10 mV results in the production of an action potential which propagates along the axon to the nerve terminals. The results of a number of experiments on the effects of injection of various ions into motoneurons via microelectrodes indicate that the ionic basis of the resting and action potentials is much the same as in squid axons. That is to say, the resting potential is slightly less than the potassium equilibrium potential, the action potential is caused primarily by a regenerative increase in sodium permeability, and the ionic concentration gradients necessary for these potentials are produced by a Donnan equilibrium system in conjunction with an active extrusion of sodium ions.

Excitatory postsynaptic potentials

The responses shown in fig. 7.4 were recorded by means of a microelectrode inserted into a motoneuron supplying fibres in the biceps or semitendinosus muscles. They were produced by stimulating the nerve supplying these muscles with single shocks. The size of the mass

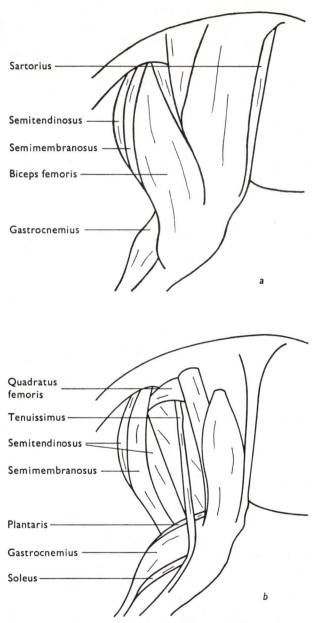

Fig. 7.3. Some of the thigh muscles of the cat. *a*: Superficial muscles, seen from the outer side. *b*: Deeper muscles.

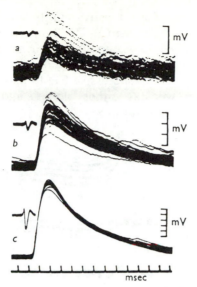

Fig. 7.4. Excitatory postsynaptic potentials recorded intracellularly from a cat biceps–semitendinosus motoneuron in response to stimuli of increasing intensity (from *a* to *c*) applied to the group Ia afferent fibres from the muscle. The inset records, taken at constant amplification, show the size of the dorsal root responses. (From Coombs, Eccles and Fatt, 1955*c*.)

response of the sensory nerves was recorded from the dorsal roots, as shown in the inset records of fig. 7.4 (the form of these dorsal root responses shows that all the sensory fibres stimulated were of about the same conduction velocity – all group Ia fibres, in fact). The ventral roots were cut so as to avoid antidromic stimulation of the motoneurons. Each record was obtained by superimposition of faint traces from about forty separate responses to stimuli of the same intensity.

The form of these responses, which are known as *excitatory post-synaptic potentials* (EPSPs), is much the same as that of the end-plate potential in a curarized frog sartorius muscle fibre: a fairly rapid rising phase followed by a slower decay which follows an approximately exponential time course. Notice however that the size of the response is

Fig. 7.5. Initiation of an action potential by the EPSP in a cat gastrocnemius motoneuron. The stimulus intensity to the afferent nerve was increased in the order *a* to *d*, with the result that the EPSP is of sufficient size to produce an action potential in *b* to *d*, and does so progressively earlier in these cases. (From Coombs, Curtis and Eccles, 1957*b*.)

proportional to the stimulus intensity, and therefore to the number of presynaptic fibres which are active. This property is known as *spatial summation*. If the EPSP is large enough, a propagated action potential is set up (fig. 7.5). If a second EPSP is produced a short time later, the total response is greater; thus two successive EPSPs may be able to produce an action potential whereas either alone could not do so. This phenomenon is called *temporal summation.*

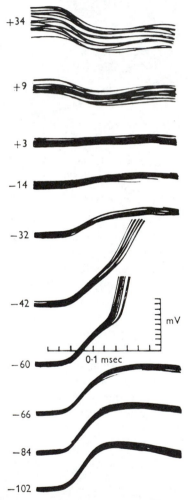

Fig. 7.6. Effect of membrane potential on the size of the EPSP in a cat motoneuron. The membrane potential was set at the values shown to the left of each set of records by passing current through one barrel of a double-barrelled intracellular electrode; the other barrel was used to record the membrane potential. The traces at −42 and −60 mV show the initiation of action potentials. (From Coombs, Eccles and Fatt, 1955c.)

If the cell membrane is progressively depolarized, the EPSP decreases in size and eventually becomes reversed in sign; the reversal potential in cat motoneurons is about 0 mV (fig. 7.6). With small depolarizations an EPSP which is submaximal at the normal membrane potential may cross the threshold for production of an action potential, but larger maintained depolarizations make the cell electrically inexcitable. Hyperpolarization of motoneurons reduces the time constant of the falling phase of the EPSP and does not change the size of the EPSP. More orthodox behaviour can be seen in sympathetic ganglion cells (fig. 7.7), where the size of the EPSP is linearly related to membrane potential for hyperpolarizations as well as for depolarizations.

The ionic mechanism of the EPSP in motoneurons has not been investigated with the precision that has been applied to the ionic basis of the EPP. The reason for this is mainly that it is technically very difficult to change the external ionic environment of motoneurons. However the results described in the previous paragraph indicate that the excitatory transmitter substance causes a simultaneous and non-regenerative

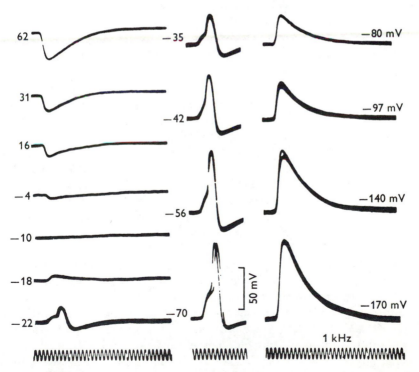

Fig. 7.7. Effect of membrane potential on the EPSPs of a frog sympathetic ganglion cell. (From Nishi and Koketsu, 1960.)

increase in the permeability of the postsynaptic membrane to more than one ion.

We have no idea what the transmitter substance released from the group Ia sensory fibres involved in the monosynaptic reflex is. However, we can be sure that motoneuronal EPSPs are produced by chemical transmission, since they reverse in sign when the membrane potential is made positive. Further evidence is provided by the presence of pre-synaptic vesicles in the terminal boutons, and the apparently quantal nature of the transmission process (Katz and Miledi, 1963).

Initiation of the action potential

As is shown in fig. 7.5, an EPSP of sufficient size elicits an action potential in the motoneuron. The form of this action potential is rather more complicated than that in the axon at some distance from the site at which it is initiated; as recorded in the soma, the action potential consists of three components. These are (1) the M spike, representing activity in the *m*yelinated region of the axon and recorded by a micro-electrode placed in the soma as a small depolarization of 1 to 3 mV; (2) the IS spike, representing activity in the *i*nitial *s*egment of the axon, which is not surrounded by a myelin sheath, and recorded in the soma as a depolarization of 30 to 40 mV; and (3) the SD spike, representing activity in the *s*oma and *d*endrites, and therefore recorded as a full-sized action potential of 80 to 100 mV.

This analysis (due to Coombs, Eccles and Fatt, 1955*a* and Coombs, Curtis and Eccles, 1957*a*, *b*) was derived mainly by examination of the effects of the soma membrane potential on the form of an antidromic action potential produced by stimulation of the ventral roots. The results of such an experiment are shown in fig. 7.8*a*. At the resting potential (−80 mV in this case) only the IS spike is seen. It is reasonable to assume that the size of the IS spike *in the initial segment* is about 100 mV, but the electrotonic currents set up by this activity only pro-duce a depolarization of about 40 mV in the soma, where it is recorded; since this is still below the threshold for the regenerative response of the soma and dendrites, no SD spike arises. However, if the soma is depolarized below about −78 mV, the electrotonic currents produced by activity in the initial segment are now sufficient to excite the soma, so an SD spike arises from the IS spike, with decreasing latency at lower membrane potentials (compare the traces obtained at −77 and −63 mV in fig. 7.8*a*). Many motoneurons are sufficiently excitable to produce SD spikes in the absence of any depolarization prior to the IS spike. If the soma is hyperpolarized beyond about −82 mV (a procedure which must also hyperpolarize the initial segment to some extent), the IS spike disappears, and we are left with a very small deflection, the M spike

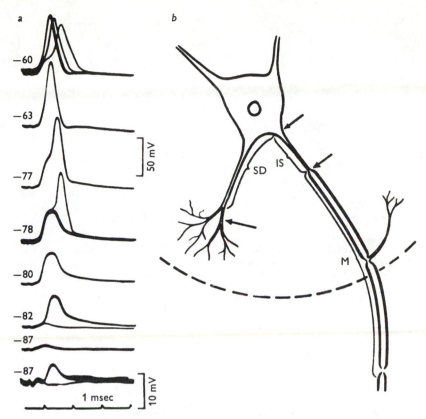

Fig. 7.8. Analysis of action potentials recorded in the motoneuronal soma following antidromic stimulation. *a*: Effect of soma membrane potential on the response. *b*: Diagram of a motoneuron showing the probable sites of the SD, IS and M spikes. Further details in the text. (From Eccles, 1957.)

(shown in the traces obtained at −87 mV in fig. 7.8*a*), which seems to be due to the electrotonic currents produced in the soma by the action potential in the myelinated region of the axon.

The evidence for the localization of the SD, IS and M spikes must, of course, be indirect, since spinal motoneurons are embedded in a mass of nervous tissue so that it is impossible to observe the position of the recording electrode directly. Nevertheless, the analysis outlined above seems very convincing, and it is difficult to see how else the observed potentials could arise.

We now have to consider the nature of the action potentials initiated orthodromically by excitatory synaptic action. An initial IS component can be seen at the beginning of the SD spike produced by an EPSP of sufficient size (fig. 7.9*b*). This implies that the initial segment is the site

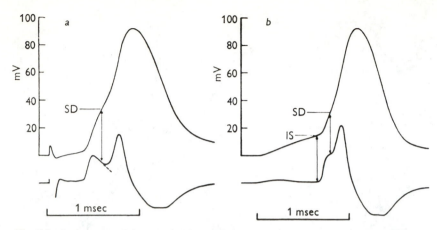

Fig. 7.9. Action potentials recorded intracellularly from cat motoneurons after (*a*) antidromic and (*b*) orthodromic (monosynaptic) stimulation. The lower traces are electrically differentiated records of the upper traces, i.e. they show the rate of change of membrane potential with time. (From Coombs, Curtis and Eccles, 1957*b*.)

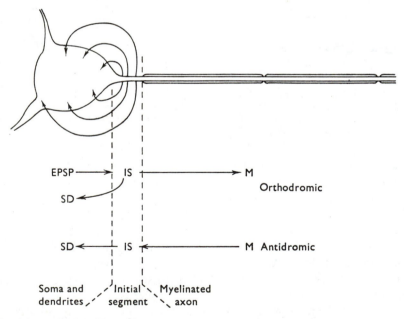

Fig. 7.10. Spike initiation in a vertebrate spinal motoneuron. The arrows in the diagram show how the initial segment is depolarized by the local circuit currents set up by excitatory synaptic action in the soma and dendrites. Also shown are the sequences of events in orthodromic and antidromic activation.

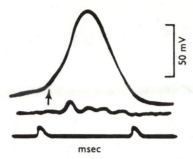

msec

Fig. 7.11. Evidence that the action potential in the axon of a motoneuron is initiated by the IS spike and not by the SD spike. The upper trace shows the action potential recorded intracellularly from a posterior biceps–semitendinosus motoneuron soma following monosynaptic activation; the lower trace shows the response recorded from a small bundle of fibres in the ventral root. The arrow marks the time at which these action potentials in the ventral root must have been initiated centrally. (From Coombs, Curtis and Eccles, 1957*b*.)

of impulse initiation in orthodromic excitation (fig. 7.10). The reason for this is that the threshold is very much lower in the initial segment, so that it can be preferentially excited by the depolarization produced by the EPSP. The IS spike then itself excites both the soma and the myelinated axon. Confirmation of this view is provided by the experiments of Coombs, Curtis and Eccles (1957*b*) in which the M spike of a single neuron (produced in response to orthodromic stimulation) was recorded from a thin bundle of fibres (a 'filament') in the ventral root while the soma was hyperpolarized. It was found that blockage of the SD spike alone did not affect production of an action potential in the axon, whereas blockage of the IS spike always eliminated the axonal response. Furthermore, the action potential in the ventral root filament followed the IS spike by about the time that one would expect from the conduction velocity of the motor fibres, whereas if the SD spike initiated the peripheral action potential, the conduction velocity would have to be impossibly fast (fig. 7.11).

INHIBITION IN MAMMALIAN SPINAL MOTONEURONS

If the contraction of a particular limb muscle is to be effective in producing movement, it is essential that those muscles which oppose this action, the *antagonists*, should be relaxed. In the monosynaptic stretch reflex systems of mammals, this is brought about by *inhibition* of the motoneurons of the antagonistic muscles, through a system known as the direct inhibitory pathway. Fig. 7.12 shows the arrangement of neurons in this pathway. Group Ia afferent fibres from the stretch receptors in a particular muscle (an extensor in fig. 7.12) synapse in the

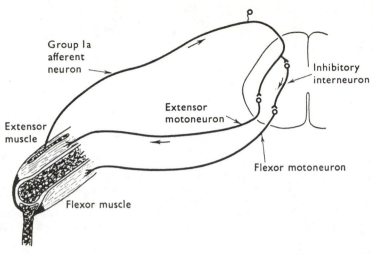

Fig. 7.12. The direct inhibitory pathway. This diagram is much simplified. Many afferent, inhibitory and efferent neurons are involved at each stage; each inhibitory interneuron is innervated by several afferents, and itself innervates several motoneurons.

ventral horn with motoneurons innervating that muscle and, to a lesser extent, with its synergists (i.e. other muscles acting in a similar way). These afferents also synapse, in a region called the intermediate nucleus, with small interneurons which themselves innervate the motoneurons of antagonistic muscles. It is these interneurons which exert the inhibitory action on the motoneurons of antagonistic muscles. This inhibitory action can thus be examined by inserting a microelectrode into a motoneuron and stimulating the group Ia afferents from an antagonistic muscle. Fig. 7.13 shows the results of such an experiment: the records were taken from a biceps-semitendinosus (flexor) motoneuron in response to group Ia afferent volleys from the quadriceps muscle (an extensor). The responses consist of small hyperpolarizing potentials known as *inhibitory postsynaptic potentials*, or IPSPs.

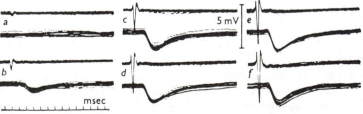

Fig. 7.13. Inhibitory responses in a cat motoneuron (biceps–semitendinosus) to afferent volleys from the antagonistic muscle (quadriceps). Stimulus intensity increases from *a* to *f*. The upper trace shows the afferent volley recorded from the dorsal root. (From Coombs, Eccles and Fatt, 1955*d*.)

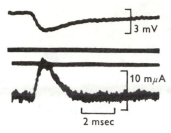

Fig. 7.14. Time course of the inhibitory transmitter action in motoneurons. The upper record shows an IPSP, the lower record shows the current flow when the membrane potential was clamped at the resting level. (From Araki and Terzuolo, 1962.)

Apart from the fact that it is normally hyperpolarizing rather than depolarizing, the shape of the IPSP is similar to the shapes of the EPSP and the EPP. Hence we might expect it to be produced by a similar mechanism, i.e. by a brief change in the ionic conductance of the membrane, which causes the initial hyperpolarization, followed by a passive decay of the charge on the membrane capacitance. This idea can be verified by using a voltage clamp technique (Araki and Terzuolo, 1962), as is shown in fig. 7.14. Observation of the effects of the membrane potential on the size of the IPSP provides a further test of the hypothesis. It is found (Coombs, Eccles and Fatt, 1955*b*) that the IPSP increases on depolarization, but decreases and then reverses in sign on hyperpolarization (fig. 7.15).

The question now arises, to which ions does the postsynaptic membrane alter its permeability during the action of the inhibitory transmitter substance? As we have seen, this type of problem is most readily solved by observing the effects of altering the ionic concentration gradients across the membrane. This was done by Coombs, Eccles and Fatt (1955*b*); since it is very difficult to change the external environment of spinal motoneurons, they altered the ionic concentrations inside the cell by injecting various ions through one barrel of a double micro-electrode, the other barrel being used to measure membrane potentials. The results obtained after chloride injection were the most clear-cut: the reversal potential always moved to a more depolarized level, so that the IPSP at the resting potential became a depolarizing response of increased size (fig. 7.16). This result implies that the potential at the peak of the IPSP is at least partially determined by the Nernst equation for chloride ions, i.e. the inhibitory transmitter substance causes an increase in the permeability of the postsynaptic membrane to chloride ions.

It is probable that the IPSP is also dependent upon an increase in the potassium conductance of the postsynaptic membrane, although the

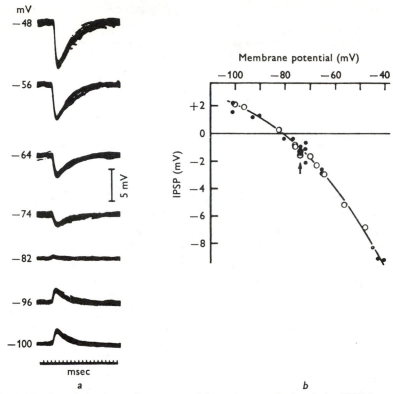

Fig. 7.15. The effect of membrane potential on the magnitude of the IPSP in a cat motoneuron. *a*: A series of records of the IPSP at different membrane potentials. *b*: These results (and others from the same cell) are expressed in graphical form; the arrow indicates the resting potential. (From Coombs, Eccles and Fatt, 1955*b*.)

evidence for this is not so convincing as in the case of chloride. The equilibrium potential for potassium ions, based on measurement of the after-hyperpolarization following an SD spike, is about −90 mV, i.e. about 20 mV more negative than the resting potential. We cannot be sure just what the equilibrium potential for chloride ions is, but it seems

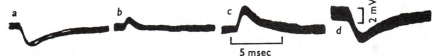

Fig. 7.16. The effect of increasing the internal chloride concentration on the IPSP of a cat motoneuron. The records were obtained by inserting a micropipette electrode filled with 3 M KCl into the cell. Record *a* was obtained immediately after insertion, records *b* and *c* at successively later times. Notice the change in the IPSP following diffusion of chloride out of the electrode. Record *d* was obtained immediately after *c*, but with the membrane potential set at a lower level (−27 mV, instead of −59 mV). (From Coombs, Eccles and Fatt, 1955*b*.)

probable that it is almost the same as, or perhaps slightly less than, the resting potential. If this argument is correct (see Eccles, Eccles and Ito, 1964), then we can only account for the fact that the IPSP is normally a hyperpolarization by assuming that there is an increase in potassium conductance as well as in chloride conductance.

Interactions between inhibitory and excitatory postsynaptic potentials

We have seen that an EPSP which is sufficiently large will cause the production of a propagated action potential in the axon of a motoneuron. If an IPSP is induced so as to coincide approximately with the EPSP, the depolarization produced by the EPSP is reduced, so that the membrane potential does not cross the threshold and no action potential ensues (fig. 7.17). When the EPSP is too small to elicit an action potential we can see the effects of interaction with the IPSP more clearly (fig. 7.18). If the EPSP follows the IPSP by more than about 2 msec, it is not reduced in size (as measured from foot to peak) but the membrane potential at its peak is more negative than usual (*b–f*, fig.

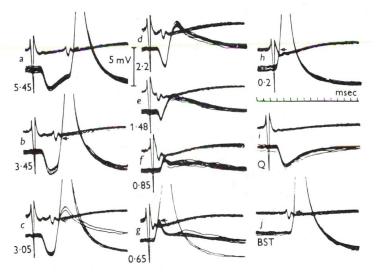

Fig. 7.17. Interactions between inhibitory and excitatory responses in a cat biceps–semitendinosus motoneuron. The lower records show motoneuron membrane potentials, the upper records monitor group Ia afferent responses in the sixth lumbar dorsal root. The excitatory response alone, produced by stimulating the biceps–semitendinosus nerve, is shown in *j* (the peak of the resulting action potential is not shown in these records). The inhibitory response alone, produced by stimulating the quadriceps nerve, is shown in *i*. *a* to *h* show responses to stimulation of both nerves, the excitatory stimulus following the inhibitory stimulus after the times (in msec) shown by each record. The conduction time in the inhibitory pathway is about 1 msec longer than that in the excitatory pathway, so that the EPSP occurs before the IPSP in records *f* to *h*. (From Coombs, Eccles and Fatt, 1955*d*.)

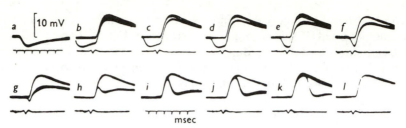

Fig. 7.18. Interaction between inhibitory and excitatory responses in a cat motoneuron. The experimental procedure was similar to that in fig. 7.19, except that the excitatory stimulus was insufficient to produce an action potential in the motoneuron. Trace *a* shows the IPSP alone and trace *l* shows the EPSP alone. Traces *b* to *k* show superimposed records of (i) the EPSP alone and (ii) EPSP plus IPSP at various intervals between the two. Further details in the text. (From Curtis and Eccles, 1959.)

7.18). At much shorter intervals, when the EPSP and IPSP are practically coincident (*g–i*), the EPSP is reduced in size; this is because the increase in potassium and chloride conductance reduces the effects of the excitatory transmitter. If the IPSP occurs after the peak of the EPSP (*j* and *k*) it does not, of course, affect the peak depolarization.

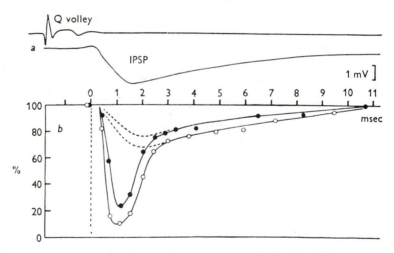

Fig. 7.19. The effect of an inhibitory volley on the size of the monosynaptically excited compound action potential recorded in the ventral root. *a* shows the inhibitory (quadriceps) volley recorded in the dorsal root (upper trace), and also the IPSP (lower trace) that this produces in the biceps–semitendinosus motoneuron. The time scale of these records is the same as the abscissa of the graph *b*. The ordinate of *b* shows the size of the compound action potential in the ventral root, with the response in the absence of inhibition taken as 100 per cent. The two curves in *b* were obtained from different sizes of afferent volley. The broken lines indicate the time-courses of that part of the inhibitory process which can be attributed directly to the hyperpolarization of the IPSPs. (From Araki, Eccles and Ito, 1960.)

Thus there are two effects of inhibition: one, due to the hyperpolarization of the membrane, which lasts for the whole of the IPSP, and the other, due to an antagonism between the inhibitory and excitatory synaptic currents, which lasts only for the duration of the conductance change which occurs during the rising phase of the IPSP. This analysis accords very well with the effect of an inhibitory volley on the monosynaptic response recorded from a ventral root. It is found (fig. 7.19) that the compound action potential in the ventral root is very much reduced in size when the excitatory volley occurs up to 2 msec after the inhibitory volley, but thereafter the degree of inhibition is such as one might expect from the hyperpolarization constituting the IPSP.

Round and flat vesicles

In studying the fine structure of the cerebral cortex by electron microscopy, Gray (1959) was able to discern two structural types of synapse. One of them, called Type 1, showed more marked postsynaptic thickenings and wider synaptic clefts than the other, called Type 2 (see fig. 7.1). It soon became evident that, in those cases where their function was known, Type 1 synapses were excitatory and Type 2 were inhibitory. Then Uchizono (1965) made a remarkable discovery. In aldehyde-fixed sections of the cerebellar cortex, he found that the synaptic vesicles in known excitatory synapses were spherical or 'round' whereas those of inhibitory synapses were ellipsoidal or 'flat'. A similar distinction between round and flat vesicles can be seen in nerve terminals in the spinal cord; presumably they also correspond to excitatory and inhibitory endings.

The functional basis of the difference between round and flat vesicles is obscure. It is quite probable that 'flat' vesicles are in fact 'round' in life, and simply become 'flat' as a result of the fixation process. But the very fact that this happens indicates that there is some difference between the two types before fixation. Whatever the reason for the difference, it does provide another useful tool for the neuroanatomist.

Presynaptic inhibition

The amount of transmitter released from a nerve terminal is related to the size of the action potential in that terminal. This is evident from studies on neuromuscular transmission (e.g. fig. 6.19b suggests that a reduction in the size of the presynaptic action potential by 15 mV would reduce the amount of transmitter released to one tenth of its original value), and is particularly clear in experiments on squid giant synapses, which we shall consider later. If the nerve terminal is slightly depolarized, the size of the action potential will be correspondingly reduced, partly because it will arise from a less negative level of membrane potential, and partly because the depolarization will cause slight increases in the potassium conductance and sodium inactivation of the nerve terminal membrane. Consequently, a depolarization of the nerve terminal by a few millivolts will considerably reduce the amount of transmitter released, and will therefore also reduce the size of the postsynaptic response produced by activity in that terminal.

Now if there were some mechanism in the nervous system whereby such a depolarization of a presynaptic terminal could be brought about by the activity of a second neuron, then this second neuron would be capable of reducing the responses in the postsynaptic cell elicited by the action of the first neuron. Such a mechanism would thus be inhibitory, and since it would act presynaptically, we could call the phenomenon *presynaptic inhibition,* to contrast it with the postsynaptic inhibition mechanism which we have already examined.

There is some good evidence that presynaptic inhibition does in fact occur in the mammalian spinal cord (Frank and Fuortes, 1957; Eccles, Eccles and Magni, 1961). Consider the experiment whose results are shown in fig. 7.20. The EPSPs shown in records *a* to *d* were obtained from a gastrocnemius motoneuron in response to a single stimulus

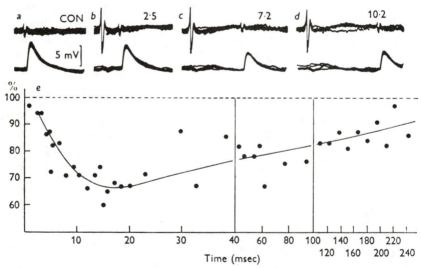

Fig. 7.20. Presynaptic inhibition in the spinal cord of the cat. The lower traces in *a* to *d* show EPSPs recorded from a gastrocnemius motoneuron, produced by stimulation of the gastrocnemius–soleus nerve. The upper traces show extracellular records from one of the dorsal roots. In *b* to *d*, the excitatory volley was preceded by a volley in the posterior biceps–semitendinosus nerve; the figures above each trace give the time intervals between the two volleys. The graph *e* shows the depression of the EPSP (expressed as a percentage of its size in the absence of inhibition) at various times after the inhibitory volley. (From Eccles, Eccles and Magni, 1961.)

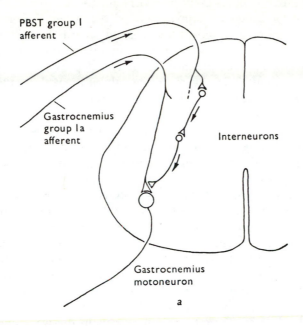

PBST group I
afferent

Gastrocnemius
group Ia
afferent

Interneurons

Gastrocnemius
motoneuron

a

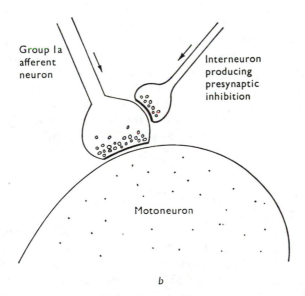

Group Ia
afferent
neuron

Interneuron
producing
presynaptic
inhibition

Motoneuron

b

Fig. 7.21. Diagram to show the suggested anatomical basis of presynaptic inhibition: *a*, neuronal connections; *b*, synaptic structure.

exciting the group Ia fibres in the gastrocnemius–soleus nerve. When the stimulus was preceded by stimulation of the group I fibres in the posterior biceps–semitendinosus nerve, as in records *b* to *d*, the EPSP was depressed in size, although there was no evidence of any inhibitory synaptic action on the motoneuron itself. This depression, or inhibitory action, began about 2.5 msec after the inhibitory volley entered the cord, reached a maximum after about 15 msec, and lasted for over 200 msec (fig. 7.20*c*).

This experiment shows that it is possible for the monosynaptic EPSP to be depressed by neuronal action without there being any IPSP in the motoneuron, and therefore suggests that the inhibitory action takes place presynaptically. But it does not tell us anything about how the inhibitory action is brought about. It has been known for some time (e.g. Barron and Matthews, 1938) that dorsal root volleys are followed by a depolarization (the *dorsal root potential*) which spreads electrotonically along the same or adjacent dorsal roots. Eccles, Magni and Willis (1962) showed that dorsal root afferent fibres subjected to presynaptic inhibition are in fact depolarized, and that the depolarization follows a time course which is apparently identical with that of the inhibitory action. Thus it seems that presynaptic inhibition is brought about by depolarization of the afferent nerve terminals in the way suggested at the beginning of this section.

A central delay of 2.5 msec is much longer than would be expected in a pathway with a single synapse only, so it is probable that the presynaptic inhibitory pathway involves one or two interneuronal stages (fig. 7.21*a*). This could account for the long duration of presynaptic inhibition, if each interneuron fires may times when it is excited. Fig. 7.21*b* shows, in diagrammatic form, the suggested anatomical basis of presynaptic inhibition; 'serial synapses' of this type have in fact been seen in electron micrographs of the spinal cord (Gray, 1962).

MOLLUSCAN SYNAPSES

Fig. 7.22 shows the arrangement of pre- and postsynaptic axons in the stellate ganglion of the squid. The great interest of this preparation is that the presynaptic terminal is large enough for an intracellular microelectrode to be inserted into it (Bullock and Hagiwara, 1957; Hagiwara and Tasaki, 1958). Under normal conditions, an action potential in the presynaptic fibre produces an action potential in the postsynaptic fibre, with a delay of about 0.4 msec at 21 °C. However, if the preparation is fatigued by repetitive stimulation, or synaptic block is induced by magnesium ions or by hyperpolarization of the postsynaptic fibre, an EPSP can be recorded from the postsynaptic fibre (fig. 7.23). Like the

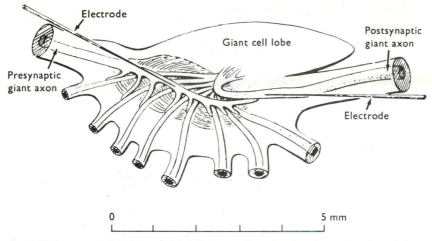

Fig. 7.22. The anatomy of the squid stellate ganglion, showing the positioning of electrodes for recording pre- and postsynaptic activity in the giant axons. (From Bullock and Hagiwara, 1957.)

EPP in frog skeletal muscle and the EPSP in sympathetic ganglion cells, this EPSP increases in size when the postsynaptic membrane is hyperpolarized; extrapolation of these results indicates that the reversal potential occurs at a membrane potential of about zero.

Miledi (1973) has used the squid giant synapse to test the idea that calcium ions are involved in transmitter release. Because of the large size of the presynaptic axon he was able to inject calcium ions into it via an intracellular micropipette. This procedure caused a depolarization of the postsynaptic axon, indicating that raising the presynaptic intracellular calcium ion concentration causes release of the transmitter substance.

The ganglia of many molluscs, including the opisthobranch *Aplysia* (the 'sea hare') in particular, contain nerve cell bodies which may be up

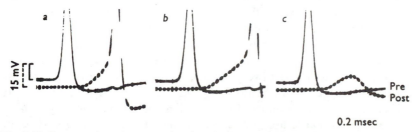

Fig. 7.23. Presynaptic action potentials (continuous records) and postsynaptic responses (records interrupted at 0.2 msec intervals) in a squid giant fibre synapse. The records were obtained when transmission started to fail as a result of prolonged repetitive stimulation. (From Hagiwara and Tasaki, 1958.)

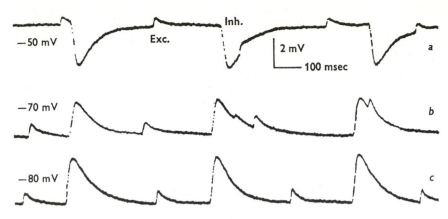

Fig. 7.24. EPSPs and IPSPs recorded from the soma of an *Aplysia* neuron. At the resting potential (*b*), both these responses were depolarizing, but depolarization of the cell membrane to −50 mV (*a*) converts the IPSP to a hyperpolarizing response. (From Tauc, 1959.)

to 800 μm in diameter. Intracellular recording from these cells shows the presence of EPSPs and IPSPs, as is shown in fig. 7.24 (Tauc, 1958). The time course of these potentials is rather longer than one would expect from the time constant of the cell membrane; this is at least partially accounted for by the discovery that the synapses occur on the axons rather than on the soma (Tauc, 1962). The cells frequently show spontaneous activity, which may be modified by light, carbon dioxide or mechanical deformation, as well as by normal synaptic activity.

ELECTRICALLY TRANSMITTING SYNAPSES

An excitatory electrically transmitting synapse is one in which the post-synaptic cell is directly excited by the electrotonic currents accompanying an action potential in the presynaptic axon. If such intercellular excitation is to be possible, the electrical and geometrical characteristics of the junction must be organized in a certain way. Consider fig. 7.25, which represents a junction between two cells, each of which is electrically excitable. The local circuit currents produced by inward movement of sodium ions at *A* in the presynaptic cell will be completed by currents flowing out of the synaptic cleft at *B* and out of the postsynaptic cell at *C*. Electrical transmission can only occur if the current density at *C* is large enough to cause a depolarization which is sufficient to initiate an action potential there. There are three features of electrically transmitting synapses which are designed to secure such an adequate current density in the postsynaptic cell. Firstly, the synaptic cleft is very narrow

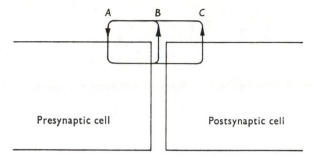

Fig. 7.25. Current flow produced at a synapse by a presynaptic action potential. See text.

or non-existent, so that very little current leaves the system at *B* in fig. 7.27. In chemically transmitting synapses, on the other hand, the synaptic cleft is essential in order to allow the flow of postsynaptic currents consequent upon the postsynaptic membrane permeability change produced by the chemical transmitter. Secondly, the flow of current into the postsynaptic cell will be increased if the resistance of the pre- and postsynaptic membranes is low. Finally, the size of the postsynaptic cell must not be much greater than that of the presynaptic cell, otherwise the postsynaptic current density will be insufficient to cause excitation. (Kuffler and Nicholls, 1976, put this point well in drawing an analogy with heat flow: a hot knitting needle cannot heat up a cannonball.) Hence if transmission is to occur between a small axon and a large postsynaptic cell (as at the neuromuscular junction, for example), then it must be chemical in nature.

Electron microscopy of electrically transmitting synapses shows regions where the intercellular space between the two cells is much narrower than usual, being about 2 nm instead of 20 nm. These regions are known as 'gap junctions' or 'nexuses'. Freeze-fractured gap junctions show lattices of particles which are partly embedded in the membranes and partly traverse the gap between them. Similar junctions in epithelial cells appear to be the sites at which current flow and solute transfer between adjacent cells can occur (Loewenstein, 1976). Structural studies on the gap junctions of mouse liver cells suggest that the particles are pores which connect the cytoplasm of the two cells together (Goodenough, 1976), as is shown in fig. 7.26.

For a time it was thought that electrical transmission occurred via another type of membrane contact between cells, known as the 'tight junction' or 'zonula occludens'. Here the synaptic cleft disappears completely so that the membranes of the two cells are in contact or perhaps partially fused. Their function appears to be to limit diffusion in the extracellular space between different body compartments.

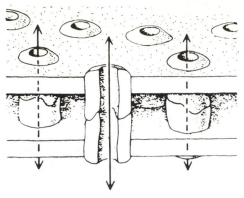

Fig. 7.26. Model of a gap junction between two cells. Each cell-to-cell channel is envisaged as formed from two abutting protein channels traversing the cell membranes. (From Loewenstein, 1976.)

The advantages of electrical transmission are that it does not involve the complex apparatus of the chemical transmitter mechanism and that the synaptic delay can be almost negligible.

Septal synapses in invertebrate giant nerve fibres

Many annelids and crustaceans possess giant fibres in the ventral nerve cord which are multicellular units, each being divided by a transverse septum in each segment. The giant fibres of the earthworm, mentioned in chapter 4, are of this type. The physiology of the transmission process across the septa has been investigated by Watanabe and Grundfest (1961) in the lateral giant fibres of the crayfish. They found that currents passed through a microelectrode into one of the component cells of a fibre would readily cross the septum to cause a potential change in its neighbour. Nevertheless, the resistance of the septum is not negligible, so that its presence must slightly reduce the conduction velocity in the whole fibre. There is some evidence that the septa of annelid giant fibres have a much lower resistance than does the rest of the cell membrane (Kao, 1960).

Electrical interconnections between neurons

A number of cases are now known in which there is some electrical interaction between neurons that frequently fire in synchrony. One example of this is in the two giant cells which occur in the segmental ganglia of leeches (Hagiwara and Morita, 1962). In fig. 7.27, the upper trace shows the response of one of the cells to depolarizing current applied via an intracellular microelectrode, and the lower trace shows the electrical changes which occur simultaneously in the other cell. The

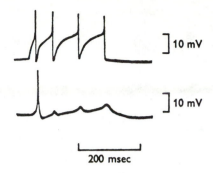

Fig. 7.27. Electrical interaction between two neurons in a leech segmental ganglion. Current is passed through the membrane of one of the cells, whose membrane potential is recorded on the upper trace. The lower trace shows the response of the other cell. (From Hagiwara and Morita, 1962.)

resistance of the electrical interconnections between the two cells is about twice that of the cell membrane of each individual cell.

Similar electrotonic interconnections occur between pacemaker cells in the lobster heart ganglion (Hagiwara, Watanabe and Saito, 1959), between supermedullary neurons in the puffer fish (Bennett, 1960), between some adjacent neurons in *Aplysia* (Tauc, 1959), between spinal electromotoneurons in mormyrid electric fish (Bennett, Aljure, Nakajima and Pappas, 1963), the motoneurons of toadfish sound-producing muscles (Bennett, 1966) and elsewhere (Bennett, 1972). Bennett points out that such electrotonic coupling is much used in synchronizing the activity of neurons innervating effector organs which show marked pulsatile activity. It is essential, for example, that all the cells of the lobster heart, the mormyrid electric organ or the toadfish sound producing muscle should fire at the same time. In many cases, the impedance of the junction increases at higher frequencies, so that brief action potentials have less effect on an adjacent cell than do slow subthreshold potentials.

The giant motor synapses of the crayfish

The flexor muscles in the abdomen of the crayfish are innervated by motor fibres of large diameter which can be excited by impulses in the lateral giant fibres of the nerve cord. This system provides the physiological basis of the 'tail-flick' escape response of the animal. Furshpan and Potter (1959) investigated the transmission process at this synapse using microelectrodes inserted into the pre- and postsynaptic axons, as is shown in fig. 7.28. Fig. 7.29 shows the presynaptic and postsynaptic responses to orthodromic stimulation. Notice particularly that the postsynaptic response begins almost simultaneously with the presynaptic

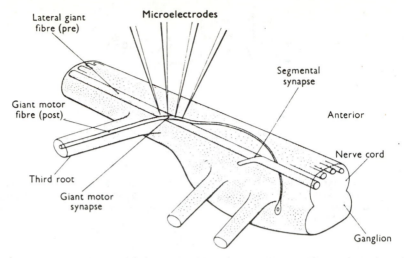

Fig. 7.28. Part of the abdominal nerve cord of a crayfish, to show the position of the electrically-transmitting synapse between the lateral giant fibre and the large motor axon in the third root. (From Furshpan and Potter, 1959.)

action potential; this suggests that the transmission process is electrical in nature. However antidromic stimulation of the motor fibre produces only a very small response (less than 1 mV) in the lateral giant fibre, and so there can be no transmission of an impulse across the synapse in the reverse direction.

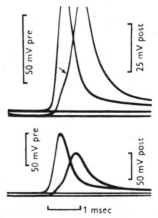

Fig. 7.29. Pre- and postsynaptic action potentials at the crayfish giant motor synapse. Two pairs of records are shown, at different amplifications. The upper trace of each pair is the presynaptic response. The kink in the postsynaptic response (at the arrow) shows the level at which the postsynaptic response crosses the threshold. Notice the negligible latency between the times of onset of the two responses. (From Furshpan and Potter, 1959.)

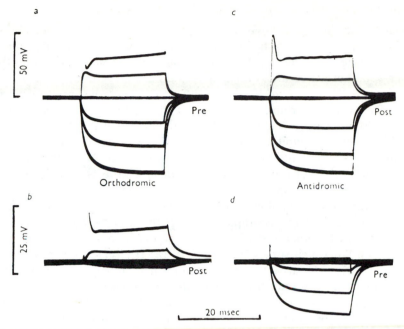

Fig. 7.30. The rectifying properties of the crayfish giant motor synapse. Current is passed through the membrane of one of the fibres, and the membrane potential in both fibres is recorded. *a* and *b*: Responses in the pre- and post-fibres to current through the pre-fibre membrane; notice that only depolarizing currents cross the synapse. *c* and *d*: Responses in the pre- and post-fibres to current through the post-fibre membrane; notice that only hyperpolarizing currents cross the synapse. (From Furshpan and Potter, 1959.)

In further experiments, depolarizing and hyperpolarizing currents were applied through microelectrodes inserted into the axons on either side of the synapse. The results, shown in fig. 7.30, were very interesting. Depolarizing currents passed into the presynaptic axon produced corresponding changes in the postsynaptic axon, whereas hyperpolarizing currents did not; but depolarizing currents passed into the post-synaptic axon did not produce changes in the presynaptic axon, whereas hyperpolarizations did. This indicates that the synaptic membrane is a rectifier: current can only flow from the pre-fibre to the post-fibre, and not in the opposite direction. The rectifying properties of the junction are essential to the animal, since the motor fibre can also be excited (by chemical transmission) by other neurons; if there were no rectification at the junction, such excitation would always produce excitation of the lateral giant fibres and the crayfish would be continually subject to irrelevant escape responses.

An inhibitory synapse operating by electrical transmission

Many fishes possess a large pair of interneurons, the Mauthner fibres, whose cell bodies lie in the brain and whose axons pass down the spinal cord. Using microelectrodes situated near the axon hillock of a Mauthner cell, Furukawa and Furshpan (1963) have recorded an *external hyperpolarizing potential* in response to various nervous inputs, especially following antidromic stimulation of the contralateral axons. Firing of the Mauthner cell is inhibited during the course of this potential, and it is therefore inhibitory in function. The external hyperpolarizing potential is apparently produced by the terminals of a number of fine nerve fibres that surround the axon hillock, and it seems probable that it is due to local circuit currents set up by activity in these terminals.

The function of Mauthner fibres is to elicit the startle response of the fish. This consists of a very rapid flexion of the body, brought about by a massive synchronous contraction of the muscles on that side, followed by a tail-flip which drives the fish forward (Eaton, Bombardieri and Meyer, 1977). Clearly it is most important that only one of the two Mauthner fibres should fire at one time, otherwise the fish's muscles would contract on both sides at once. Hence the need for effective and immediate inhibition of one cell by the other when it fires. The functional value of the electrical inhibition lies therefore in the very high speed with which it takes place.

8 The pharmacology of synapses

In the previous two chapters we have seen that the evidence is very greatly in favour of the view that synaptic transmission in most cases occurs by release of a chemical transmitter substance from the presynaptic terminals. In this chapter we shall look rather briefly into the properties of some of these transmitter substances and some of the drugs which affect their action.

The elucidation of the nature of the transmitter involved at any particular synapse is not an easy task. Paton (1958) has suggested that the following criteria should be satisfied before a transmitter role is accepted for a substance:

1. The presynaptic neuron should contain the substance and be able to synthesize it.
2. The substance should be released on stimulation of the presynaptic axons.
3. Application of the substance to the postsynaptic cell should reproduce the effects of normal transmission.
4. The action of the substance on the postsynaptic cell should be affected by competitive blocking agents in the same way that synaptic transmission is.

Paton originally included a fifth criterion in this list, that an enzyme capable of destroying the substance should exist in the vicinity of the synapse (such as, for example, acetylcholinesterase at the neuromuscular junction). Since then it has become evident that there may be other methods of limiting the duration of transmitter action, such as diffusion or desensitization (Gerschenfeld, 1966) or reabsorption (Brown, 1965).

The application of these criteria can be illustrated by reference to the hypothesis that acetylcholine is the transmitter substance at vertebrate skeletal neuromuscular junctions, as follows:

Criterion 1. The direct demonstration that motor nerves can synthesize acetylcholine is very difficult, because of the relative sparsity of nerve endings in muscular tissue. However, synthesis of acetylcholine has been clearly demonstrated in sympathetic ganglia (Birks and MacIntosh, 1961).

Criterion 2. Release of acetylcholine from motor nerves on stimulation was shown by Dale, Feldburg and Vogt, as described on p. 110.

Criterion 3. The ionophoretic experiments of del Castillo and Katz and others (p. 122) show that acetylcholine has the same effect on the postsynaptic membrane as does the transmitter substance. Takeuchi (1963) has shown that the reversal potential of the responses to iono-phoretically applied acetylcholine is the same as that of the EPP and is similarly affected by changes in the external sodium and potassium ion concentrations.

Criterion 4. We have seen that curare blocks neuromuscular trans-mission by reducing the size of the EPP. Similarly, curare antagonizes the action of externally applied acetylcholine.

Neurons which release acetylcholine from their terminals are known as cholinergic. Not all neurons are cholinergic; many, in vertebrates, release noradrenaline (when they are known as adrenergic), and in many cases the transmitter substance is not known.

THE ACTION OF DRUGS ON NEUROMUSCULAR TRANSMISSION

From the evidence presented in chapter 6 and summarized above, we can safely assume that the transmitter substance at the vertebrate skeletal neuromuscular junction is acetylcholine. In order for acetyl-choline to act effectively as the neuromuscular transmitter it has to be synthesized in the motor nerve, be released from it on stimulation, react with the receptor sites on the postsynaptic membrane so as to increase the ionic permeability of the membrane, and finally it must be broken down by acetylcholinesterase. Some of the choline resulting from the hydrolysis is recycled by being taken up into the presynaptic terminal to be made into acetylcholine again. The different drugs which affect the transmission process may act at any of the stages in this sequence of events.

The drug hemicholinium-3 prevents the synthesis of acetylcholine within the motor nerve, so that neuromuscular transmission is depressed after prolonged stimulation and only recovers very slowly (Desmedt, 1958). This post-tetanic depression is very similar to that which occurs in myasthenia gravis, and Desmedt has suggested that the essential feature of this disease is a failure of acetylcholine synthesis in the motor nerves.

Botulinum toxin, obtained from the bacterium *Clostridium botulinum*, prevents the release of acetylcholine from the presynaptic terminals. This toxin has the distinction of being the most poisonous substance known; 250 g of botulinum toxin A would be sufficient to poison the whole of the population of the world! Black widow spider venom has an almost opposite effect: it produces a massive discharge of acetylcholine quanta and a corresponding reduction in the numbers of vesicles present in the terminals (Longenecker, Hurlbut, Mauro and Clark, 1970). Neuromuscular block ensues as a result of exhaustion of the acetylcholine content of the nerve terminals.

Drugs which act on the postsynaptic membrane in much the same way as acetylcholine does are called *agonists*, whereas those which reduce or prevent the action of acetylcholine are called *antagonists*. Molecules of each of these types are thought to act by combination with the receptor sites at which acetylcholine itself acts. The formulae of acetylcholine and some of its agonists are shown in table 8.1. Many agonists can cause neuromuscular block by desensitization; in some cases (e.g. decamethonium) this action is more marked than the depolarization. Succinylcholine has been used as an immobilizing agent for large wild

TABLE 8.1. *Agonists of acetylcholine*

Acetylcholine	CH_3C $O-CH_2.CH_2-N^+-CH_3$ with CH_3 above and CH_3 below the N, and $=O$ below the C
Choline	$HO-CH_2CH_2N^+(CH_3)_3$
Carbachol	$H_2N.CO.O.CH_2CH_2N^+(CH_3)_3$
Succinylcholine	$CH_2CO.O.CH_2CH_2N^+(CH_3)_3$ $CH_2CO.O.CH_2CH_2N^+(CH_3)_3$
Decamethonium	$(CH_3)_3N^+ -(CH_2)_{10}-N^+(CH_3)_3$
Murexine	imidazole ring $-CH=CH.CO.O.CH_2CH_2N^+(CH_3)_3$
Nicotine	pyridine–pyrrolidine ring with N^+, CH_3, H

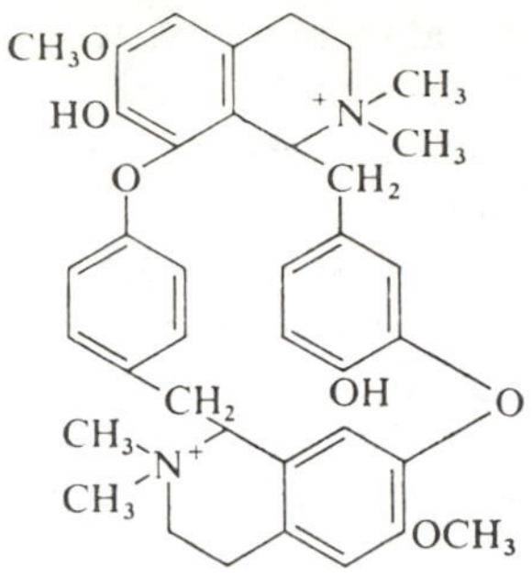

Fig. 8.1. D-tubocurarine, an acetylcholine antagonist at nicotinic synapses.

mammals, since the blocking effect is only temporary as the compound is gradually destroyed by the nonspecific cholinesterase in the blood.

It has been known since the work of Claude Bernard in 1857 that *curare*, the substance used by South American Indians to poison the tips of their arrows, acts at the neuromuscular junction. It is an antagonist to acetylcholine. Crude curare is a mixture of substances obtained from various plants; there are three varieties, known as pot-curare, tube-curare and calabash-curare. One of the active constituents of tube curare is the alkaloid D-tubocurarine (fig. 8.1), which is the most widely used of the various curare extracts. Some of the calabash-curare alkaloids are extremely effective neuromuscular blocking agents; C-alkaloid E, for instance, effectively blocks transmission in frog sartorius muscle as 1/200th the concentration of D-tubocurarine required to produce the same effect. *β-erythroidine* is an alkaloid derived from the American leguminous plant *Erythrina*. Unlike most neuromuscular blocking agents it is active when taken orally. Various snake venoms are also powerful antagonists of acetylcholine; we have seen in chapter 6 how α-bungarotoxin has proved to be very useful in this respect.

Finally we must consider the drugs which inhibit the breakdown of acetylcholine by acetylcholinesterase. The most well-known of these is *eserine* (physostigmine), an alkaloid which can be extracted from the Calabar bean. Compounds with similar structure and activity are miotene, neostigmine (prostigmine) and edrophonium (table 8.2). In the presence of these compounds, the EPP is greatly prolonged and may initiate a series of action potentials in the muscle fibre. The effect of these compounds can be removed by washing the preparation in Ringer solution, when the inhibitor–enzyme complex breaks up; edrophonium seems to be particularly loosely bound. A number of organophosphorus compounds, originally developed for use as war gases and now much used as insecticides, inactivate acetylcholinesterase. Examples of these

TABLE 8.2. *Some acetylcholinesterase inhibitors*

Eserine	$CH_3.NH.CO.O-$ [bicyclic indoline structure with CH_3 at ring junction and two $N-CH_3$ groups]
Miotene	$CH_3.NH.CO.O-$ [benzene ring] $-CH(CH_3)N(CH_3)_2$
Neostigmine	$(CH_3)_2N.CO.O-$ [benzene ring] $-N^+(CH_3)_3$
Edrophonium	$HO-$ [benzene ring] $-N^+(CH_3)_2$ with C_2H_5
DFP (diisopropyl-phosphorofluoridate)	$\begin{array}{c} H_3C \\ \\ H_3C \end{array} CH-\underset{\underset{O}{\downarrow}}{\overset{F}{P}}-CH \begin{array}{c} CH_3 \\ \\ CH_3 \end{array}$
TEPP (tetraethyl pyrophosphate)	$\begin{array}{c} C_2H_5O \\ \\ C_2H_5O \end{array} \underset{\underset{O}{\downarrow}}{P}-O-\underset{\underset{O}{\downarrow}}{P} \begin{array}{c} OC_2H_5 \\ \\ OC_2H_5 \end{array}$

compounds are diisopropylphosphorofluoridate (DFP) and tetraethyl-pyrophosphate (TEPP). The complexes between DFP or TEPP and acetylcholinesterase are very stable, so that the effects of the inhibitor cannot be removed by washing.

OTHER CHOLINERGIC SYNAPSES

Besides being the transmitter substance at voluntary motor nerve endings in vertebrates, acetylcholine is also released at the endings of preganglionic sympathetic fibres, and pre- and postganglionic endings of parasympathetic fibres (fig. 8.2). The pharmacology of the junctions in autonomic ganglia is generally similar to that of the skeletal neuromuscular junction; in particular, nicotine mimics the actions of acetylcholine

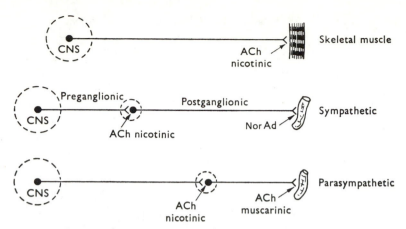

Fig. 8.2. Diagram to show the arrangement and pharmacology of peripheral efferent neurons in mammals. The diagram is oversimplified in that the postganglionic sympathetic fibres to the sweat glands are cholinergic. CNS, central nervous system; ACh, acetylcholine; NorAd, noradrenaline.

and curare is a competitive antagonist. The pharmacology of parasympathetic postganglionic endings is rather different, in that the roles of nicotine and curare are here taken by *muscarine* and *atropine* respectively. Thus the actions of acetylcholine which occur at voluntary motor nerve endings and autonomic ganglia are known as nicotinic actions, whereas those which occur at parasympathetic postganglionic endings are known as muscarinic actions.

Muscarine (fig. 8.3) is obtained from *Amanita muscaria*, a mushroom sometimes eaten in the past for its dionysiac properties (Graves, 1958). Pilocarpine and arecoline (the active principle in betel nuts) have similar actions, although pilocarpine also has weak nicotinic properties.

The deadly nightshade (*Atropa belladonna*) contains the alkaloid L-hyoscyamine; this readily racemizes to form DL-hyoscyamine, which is known as *atropine* (fig. 8.4). Atropine is an antagonist to acetylcholine

$$
\begin{array}{c}
CH_3 \\
|\\
CH \\
|\\
CHOH \\
O \quad | \\
CH_2 \\
|\\
CH \\
|\\
CH_2N^+(CH_3)_3
\end{array}
$$

Fig. 8.3. Muscarine, an agonist of acetylcholine at postganglionic parasympathetic synapses.

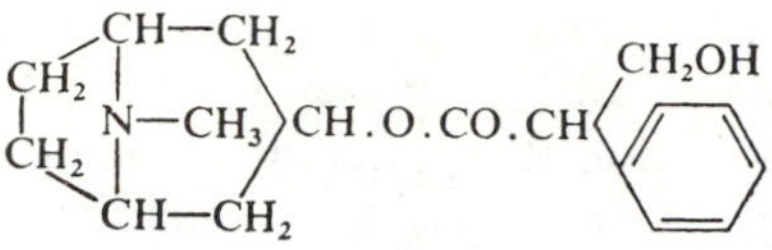

Fig. 8.4. Atropine, an antagonist of acetylcholine at muscarinic synapses.

(and its agonists) at synapses where its action is muscarinic. It has been used since classical times as a cosmetic, since it causes dilation of the pupils by blocking the synapses of parasympathetic fibres on the iris sphincter muscle.

The investigation of the transmitter substances produced by neurons in the central nervous system is a much more difficult problem than the investigation of peripheral synapses. Some of the reasons for this are that synapses of different types may occur very close to each other, that it may be difficult to stimulate a large number of synapses of the same type simultaneously, and that the access of pharmacological agents to the synapses may be restricted by diffusion barriers. However we can be sure of the identity of the transmitter substance at one type of central synapse in vertebrates, that between motor axon collaterals and Renshaw cells in the spinal cord.

Renshaw cells can be easily activated by antidromic stimulation of motor nerves. The response of a Renshaw cell to a single antidromic volley produced by stimulation of a lumbar ventral root, as recorded by an extracellular microelectrode in the vicinity of the cell, is shown in fig. 8.5*a*. Eccles, Eccles and Fatt (1956) investigated the action of various

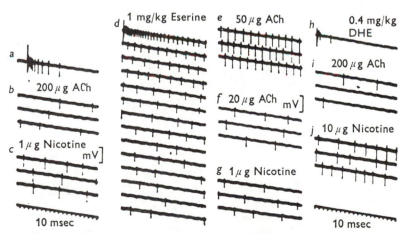

Fig. 8.5. Responses of a Renshaw cell, as recorded by an extracellular microelectrode, to antidromic motor nerve volleys and injections of acetylcholine (ACh), nicotine, eserine and dihydro-β-erythroidine (DHE). Explanation in text. (From Eccles, Eccles and Fatt, 1956.)

drugs on this system by injecting them into the arterial blood supply of this part of the spinal cord. *b* and *c* (fig. 8.5) show the spontaneous activity following injection of acetylcholine and nicotine. After injection of eserine (*d* to *g*), the response to stimulation of the ventral root is enormously prolonged (*d*) and the response to acetylcholine is greatly increased (*e* and *f*). The response to nicotine (*g*) is unchanged, as one would expect, since it is not destroyed by acetylcholinesterase. After treatment with the cholinergic blocking agent dihydro-β-erythroidine, the response to a ventral root stimulus is much reduced (*h*) and the responses to injected acetylcholine and nicotine (*i* and *j*) are also reduced. No effects were seen after injection of drugs, other than acetylcholine, which contain a quaternary nitrogen atom, and which are therefore ionized in solution (such as succinylcholine, D-tubocarine and neostimine); however, Curtis and R. M. Eccles (1958*a*, *b*) were later able to show that responses to these drugs could be obtained if they were ionophoretically ejected from micropipettes in the immediate vicinity of the cells. This suggests that there must be some diffusion barrier closely surrounding the cells. The micropipettes used by Curtis and his colleagues have five barrels (fig. 8.6); the central barrel is filled with 4 M sodium chloride and used as an extracellular recording electrode, and the outer barrels are each filled with a different drug, so that the actions and interactions of four drugs on any one cell can be examined.

It is clear that the evidence for cholinergic transmission at these Renshaw cell synapses satisfies the third and fourth of Paton's criteria.

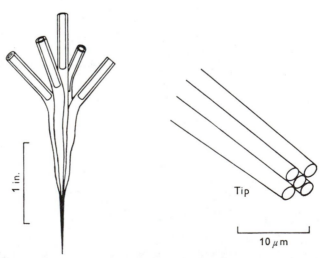

1 in.

Tip

10 μm

Fig. 8.6. Multi-barrelled electrodes used for microionophoresis experiments on single neurons in the central nervous system. (From Curtis, 1965.)

Since we are sure that the motoneuron releases acetylcholine at the neuromuscular junction, we may take the first criterion as satisfied also. The collection of the transmitter substance from the vicinity of the motoneuron collateral endings after stimulation (to satisfy the second of Paton's criteria) is obviously a much more difficult task, but Kuno and Rudomin (1966) have shown that antidromic stimulation of motor nerves in the cat causes release of acetylcholine in the spinal cord. The synapses of motoneurons on skeletal muscles and Renshaw cells are interesting in that they provide evidence in favour of the idea, described by Eccles as 'Dale's principle', that any single neuron always releases the same transmitter substance at its different endings.

The nicotinic action at the motoneuron collateral synapses is not the only action of acetylcholine on Renshaw cells, as has been shown by Curtis and Ryall (1964). They found that the excitatory action of acetyl-β-methylcholine was unaffected by the nicotinic blocking agent dihydro-β-erythroidine, whereas it was blocked by atropine (fig. 8.7). Thus there are some muscarinic receptors on the Renshaw cell, as well as the nicotinic receptors activated by motoneuron collaterals.

Using the ionophoretic application technique, Krnjević and Phillis (1963 a, b) have been able to demonstrate the presence of acetylcholine-sensitive cells with muscarinic receptors in the cerebral cortex. In an investigation of the sensitivity of brain stem neurons to acetylcholine, Bradley and Wolstencroft (1965) showed that some of the sensitive neurons were excited by acetylcholine, whereas others were inhibited (fig. 8.8). The receptors on the inhibited neurons were entirely muscarinic, whereas those on excited neurons had both muscarinic and nicotinic properties.

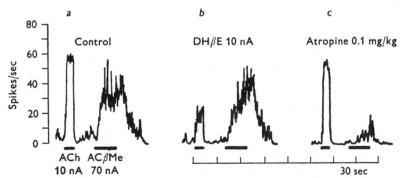

Fig. 8.7. Evidence for muscarinic acetylcholine receptors on Renshaw cells. The records show the firing frequency of the cells in response to ionophoretic application of acetylcholine (ACh) and acetyl-β-methylcholine (ACβMe). *a*: Control responses. *b*: Responses during simultaneous application of dihydro-β-erythroidine, which blocks nicotinic receptors. *c*: Responses after injection of atropine, which blocks muscarinic receptors. (From Curtis and Ryall, 1964.)

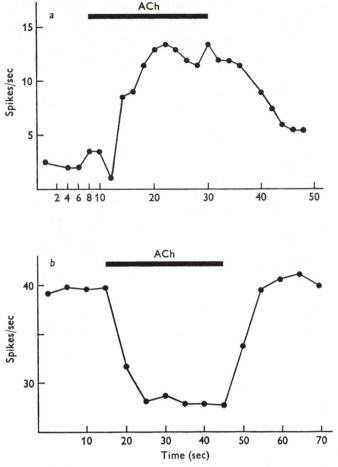

Fig. 8.8. Effect of ionophoretic application of acetylcholine on the rate of firing of two neurons in the brain stem of a cat. *a*: Response of a neuron excited by acetylcholine. *b*: Response of a neuron inhibited by acetylcholine. (From Bradley and Wolstencroft, 1965.)

Tauc and Gerschenfeld (1961, 1962) have provided convincing evidence that acetylcholine acts both as an excitatory and as an inhibitory transmitter substance in the cells of *Aplysia* ganglia. Application of acetylcholine to cells of one type (D-cells) causes depolarization (fig. 8.9), whereas application to the other type (H-cells) causes a hyperpolarization with a reversal potential equal to that of the IPSP. These actions, and also the EPSPs in D-cells and the IPSPs in H-cells, are antagonized by curare and atropine and potentiated by eserine. The excitatory input to H-cells is not cholinergic. Tauc and Gerschenfeld suggested that interneurons which would simultaneously induce excita-

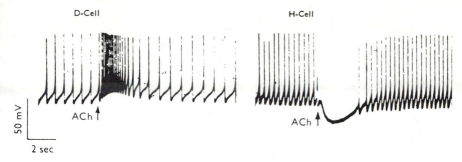

Fig. 8.9. The effect of brief electrophoretic applications of acetylcholine on the membrane potential and firing frequency of *Aplysia* neurons. The D-cell is depolarized by acetylcholine, resulting in an increase in the firing frequency. The H-cell is hyperpolarized, resulting a temporary cessation of activity. (From Tauc and Gerschenfeld, 1961.)

tory and inhibitory actions in different postsynaptic neurons might exist, and Strumwasser (1962) has indeed shown this to be so. It has since been shown that some of the D-cells ('D_{inhi}-cells') are inhibited by a non-cholinergic input, and it is particularly interesting that in these cells the IPSP does not involve an increase in chloride conductance, whereas the IPSP of H-cells does (Gerschenfeld and Chiarandini, 1965).

A number of invertebrate peripheral synapses appear to be cholinergic. The neuromuscular transmitter in annelids and holothurians appears to be acetylcholine; indeed, as we have seen (p. 110), contraction of the dorsal longitudinal muscle of the medicinal leech is frequently used as a bioassay for acetylcholine. Molluscan hearts are inhibited by acetylcholine.

CATECHOLAMINES

The catecholamines are an important group of physiologically active compounds which contain the catechol group, a benzene ring with two adjacent hydroxyl groups. The hormone *adrenaline* (epinephrine) is released by the adrenal glands of mammals and serves, in Cannon's well-known phrase, to prepare the animal for fight or fright; the rate and amplitude of the heart-beat increase, constriction of the blood-vessels (except in the muscles) occurs, the spleen contracts, the pupils dilate, intestinal movements are inhibited, etc. Adrenaline is synthesized from the amino acid tyrosine via a series of other catecholamines, dopa, dopamine and noradrenaline, as is shown in fig. 8.10.

As mentioned earlier, Elliott in 1904 suggested that adrenaline was released from postganglionic sympathetic nerves on stimulation. In later years, a number of discrepancies between the action of adrenaline and sympathetic stimulation were discovered, so the sympathetic transmitter

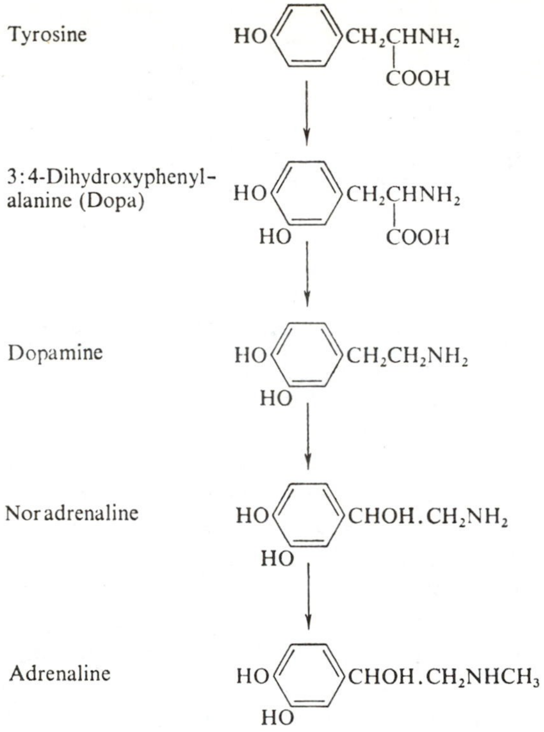

Tyrosine $\qquad$ HO—⟨benzene⟩—CH_2CHNH_2 | $COOH$

3:4-Dihydroxyphenyl-
alanine (Dopa) $\qquad$ HO—⟨benzene⟩—CH_2CHNH_2 | $COOH$, HO

Dopamine $\qquad$ HO—⟨benzene⟩—$CH_2CH_2NH_2$, HO

Noradrenaline $\qquad$ HO—⟨benzene⟩—$CHOH.CH_2NH_2$, HO

Adrenaline $\qquad$ HO—⟨benzene⟩—$CHOH.CH_2NHCH_3$, HO

Fig. 8.10. The synthesis of adrenaline from tyrosine.

substance became known by the noncommittal name 'sympathin'. More recently, von Euler (see von Euler, 1955) has established that in most cases the sympathetic transmitter is not adrenaline but noradrenaline. The results of one of von Euler's experiments, on an extract of bovine spleen nerves, is shown in fig. 8.11. It was found that 0.08 g of extract was equivalent to about 0.6 μg of noradrenaline in its action on cat blood pressure, and that 0.4 g of extract was equivalent to about 5 μg of noradrenaline in its action on the rectal caecum of the fowl; thus in the two reactions 1 g of extract is equivalent to 7.5 and 12.5 μg of noradrenaline respectively. However, the adrenaline equivalents for 1 g of extract work out as 17.5 μg for the cat blood pressure assay but only 0.25 μg for the fowl rectal caecum assay. Since the figures for adrenaline are so different, we must conclude that most of the active agent of the extract is not adrenaline. Chromatographic and other methods of chemical analysis have been used to confirm this conclusion.

Neurons in which noradrenaline is the transmitter substance are fairly widespread in the mammalian central nervous system. Our evidence for this is based largely on a useful technique in which freeze-dried sections

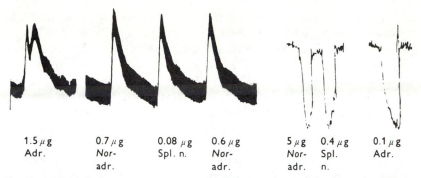

1.5 μg   0.7 μg   0.08 μg   0.6 μg   5 μg   0.4 μg   0.1 μg
Adr.   Nor-   Spl. n.   Nor-   Nor-   Spl.   Adr.
        adr.   adr.   adr.   n.

Fig. 8.11. Experiment to show that an extract of beef splenic nerves contains predominantly noradrenaline rather than adrenaline. The records show the effects of particular quantities of adrenaline (Adr.), noradrenaline (Nor-adr.) and the splenic nerve extract (Spl. n.) of the arterial pressure of a cat (left) and on the length of the isolated rectal caecum of a hen (right). (From von Euler, 1955.)

of brain tissue are exposed to hot formaldehyde vapour; the catecholamines are then converted to substances which fluoresce strongly under ultraviolet light. Neurons in which dopamine appears to be the transmitter are also present.

Adrenergic receptors

Ahlquist (1948) compared the activity of adrenaline, noradrenaline and isoprenaline (which is similar in structure to adrenaline but for the substitution of an isopropyl group for the N-methyl group) on various tissues. He found that for some tissues (some blood vessels, the pregnant uterus, nictitating membrane, ureter) the order of potency was adrenaline > noradrenaline > isoprenaline; these tissues were described as having α-receptors. For other tissues (other blood vessels, non-pregnant uterus, heart) the order was isoprenaline > adrenaline > noradrenaline; these were described as having β-receptors. Some tissues (e.g. intestinal muscle) appear to have both α and β receptors. There are different competitive blocking agents (antagonists of noradrenaline) at the two types of receptors: α receptors are blocked by phentolamine whereas β receptors are blocked by pronethalol, for example.

In many cases the action of noradrenaline on β receptors is followed by an increase in the intracellular concentration of cyclic adenosine monophosphate (cyclic AMP). The enzyme which produces cyclic AMP from ATP, adenyl cyclase, is membrane bound. It is therefore an attractive idea to suggest that the β receptors are closely linked to adenyl cyclase molecules so that the conformational change in a β receptor, brought about by its combination with noradrenaline, activates the associated adenyl cyclase molecule so that the intracellular concentration of cyclic AMP rises (see, for example, Beam and Greengard, 1976). It is not yet always clear how the cyclic AMP acts in the effector cell.

Combination of noradrenaline with α receptors usually results in a change in membrane permeability (Jenkinson, 1973). Thus in intestinal smooth muscle the potassium conductance is raised by noradrenaline, so that the membrane potential is made more negative and contractile activity inhibited. In the smooth muscle of most blood vessels, however, noradrenaline. is excitatory (leading to vasoconstriction) as a result of increases in membrane permeability to sodium and calcium ions.

Adrenergic vesicles

The adrenergic neurons of the sympathetic nervous system have non-myelinated axons which terminate in various effector organs. The terminals branch, and have a large number of swollen portions called *varicosities* which are packed with mitochondria and vesicles. Most of the vesicles have electron-dense cores and are hence known as 'granular vesicles'. They are of two sizes, large (diameter up to 100 nm) and small (diameter 40 to 50 nm). The small ones are predominant in the varicosities, whereas the large ones are also found in the cell body and axon. Both types contain noradrenaline and ATP. The proteins chromogranin A and dopamine β hydroxylase (the enzyme which converts dopamine to noradrenaline) are found in the large vesicles and may be also present in the small ones (see Smith, 1972, and Burnstock and Costa, 1975). In some electron micrographs there are also some 'agranular vesicles' present; it seems likely that these are derived from small granular vesicles by loss of some of their contents.

The fate of released noradrenaline

We have seen that in vertebrate skeletal muscles acetylcholine combines with receptors for a short time but is then rapidly hydrolyzed by acetylcholinesterase. The primary activity of released catecholamines is similarly a brief combination with postsynaptic receptors, but thereafter there are several possibilities.

Adrenaline and noradrenaline are enzymically inactivated in two main ways: by oxidative deamination (by the mitochondrial enzyme monoamine oxidase) or by methylation of a hydroxyl group (by the cytoplasmic enzyme catechol-O-methyl transferase). Both these enzymes are intracellular and hence can only act on the catecholamines after they have been taken up by cells. This situation is in marked contrast with that at cholinergic synapses, where the extracellular localization of acetylcholinesterase in the synaptic cleft enables it to inactivate the transmitter substance immediately and effectively.

Clearly then, enzymic inactivation cannot be the prime method of terminating the effects of noradrenaline on the postsynaptic membrane. In recent years it has become evident that many tissues, and especially adrenergic nerve endings, are able to take up catecholamines from the extracellular fluid. Thus Whitby, Axelrod and Weil-Malherbe (1961) found that tritium-labelled noradrenaline was rapidly concentrated in adrenergic nerve endings. Iversen (1963, 1967) found that the adrenergic nerve terminals in a perfused rat heart were capable of clearing the entire extracellular space of the heart in about ten seconds. He considers that this uptake system would therefore remove noradrenaline from the immediate vicinity of the terminals in a matter of milliseconds, thus constituting a most effective method of inactivating the transmitter (Iversen, 1971). It is also an economical method, since the noradrenaline is accumulated intracellularly in the synaptic vesicles, from which it can be released again in further synaptic transmission.

(It is possible to fool the uptake process by feeding it with an analogue of noradrenaline or one of its metabolites. The analogue is then taken up into the nerve terminal, stored in the synaptic vesicles and then released into the synaptic cleft when a presynaptic nerve impulse arrives. Such compounds are known as *false transmitters*. A useful example is 5-hydroxydopamine; when taken up into the synaptic vesicles it makes them stain heavily with osmium tetroxide, so that adrenergic terminals become readily distinguishable by electron microscopy (Tranzer and Thoenen, 1967). A number of analogues of noradrenaline have been used in medicine to relieve high blood pressure, and it may be that some of them act by becoming false transmitters. If this were so, activity of adrenergic neurons could result in the release of substances which were less effective than noradrenaline in causing vasoconstriction.)

Noradrenaline is also taken up by a variety of non-neural cells in the body. Thus noradrenaline may be removed from the synaptic region by diffusion to the blood stream or by uptake into the postsynaptic cell. In each case it is usually enzymically inactivated by catechol-O-methyl transferase or monoamine oxidase.

Finally, some of the released noradrenaline molecules may combine with α receptors on the presynaptic membrane, where they serve to reduce further transmitter release (see Burnstock and Costa, 1975). There is thus a negative feedback system which appears to act so as to prevent excessive catecholamine release. In some cases there is a parallel system whereby the postsynaptic cell may release prostaglandin E, which has a similar inhibiting effect on the presynaptic terminal.

Drugs active at adrenergic synapses

In view of the variety of processes occurring at adrenergic synapses (summarized in fig. 8.12), it is not surprising that there is a large number of drugs which affect transmission in some way. Such drugs may affect the synthesis, release or re-uptake of catecholamines, may inhibit their intracellular enzymic breakdown, or may act as agonists or antagonists at α or β receptors. Most of these drugs are active because they resemble the natural catecholamines in some way, and it is therefore not surprising that they may be active in more than one way. Hence there are often difficulties in deciding how a particular drug produces its effect. For example, the drug propanolol is an effective β receptor blocking agent, and is useful in reducing the blood pressure. But this hypotensive effect is gradual in onset and is produced by both optical

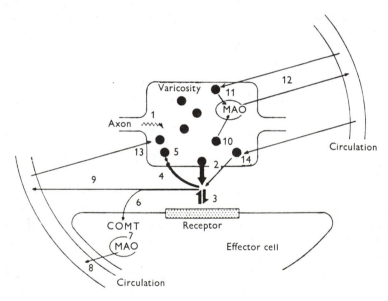

Fig. 8.12. Sites of action of drugs at an adrenergic synapse. 1, Propagation of action potential into presynaptic terminal; 2, release of noradrenaline; 3, interaction of noradrenaline with α or β receptors; 4, neuronal uptake of released noradrenaline; 5, uptake of axoplasmic noradrenaline into storage vesicles; 6, uptake of released noradrenaline into extraneuronal cells; 7, catabolism of noradrenaline in extraneuronal cells; 8, loss of noradrenaline catabolites to the circulation; 9, overflow of unchanged noradrenaline to the circulation; 10, spontaneous loss of stored noradrenaline to intraneuronal monoamine oxidase; 11, reserpine-induced loss of stored noradrenaline; 12, loss of deaminated catabolites to the circulation; 13, uptake of circulating catecholamines into adrenergic terminal; 14, displacement of stored noradrenaline by exogenous sympathomimetic amines; these may themselves be released as false transmitters. MAO, monoamine oxidase. COMT, catechol O-methyl transferase. (From Iversen and Callingham, 1971.)

isomers of propanolol, whereas β blockade is immediate and stereo-specific for the *laevo* form (see Laverty, 1973). The precise mechanism whereby propanolol reduces blood pressure is not clear; it is possible that it acts via adrenergic neurons in the central nervous system.

A further complexity is that the drug may be metabolically altered, the product being pharmacologically active. Thus α-methyldopa is readily taken up by noradrenergic nerve endings and converted into α-methyldopamine and α-methylnoradrenaline. It has been a matter of much discussion as to whether these conversions are important in the hypotensive action of α-methyldopa.

A very large number of substances show sympathomimetic activity, although in not all cases we can be sure that the substance is acting directly as an agonist of noradrenaline at the receptor sites. Here, besides isoprenaline, we may merely list the amines tyramine, amphetamine, ephedrine octopamine and α-methylnoradrenaline; there are many others.

Antagonists, or blocking agents, include yohimbine, phentolamine and chlorpromazine at α receptors, and dichloroisoprenaline and pro-panolol at β receptors. Catecholamine breakdown is reduced by pyrogallol (which inhibits catechol-O-methyl transferase) and iproniazid (which inhibits monoamine oxidase), among others.

Noradrenaline uptake into nerve terminals is prevented by phenoxybenzamine, chlorpromazine and ephedrine, and the uptake from the terminal cytoplasm into vesicles is blocked by reserpine. Uptake into non-neural tissues is blocked by metanephrine and phenoxybenzamine.

CENTRAL INHIBITORY SYNAPSES IN VERTEBRATES

Many convulsant drugs act by depressing the inhibitory processes in the spinal cord. Bradley, Easton and Eccles (1953) showed that intravenous injection of strychnine greatly reduced the inhibitory effect of group Ia volleys from antagonist muscles on the size of the efferent volley in a monosynaptic reflex (fig. 8.13). Further experiments showed that the activity of the inhibitory neurons was unaffected by this treatment, and that local application of strychnine by microelectrophoresis (Curtis, 1959) caused a reduction in the size of the IPSP without otherwise affecting the membrane potential of the motoneuron. Since the onset of strychnine action is quite rapid, Bradley *et al.* suggested that strychnine acts as a competitive inhibitor, as curare does at the neuromuscular junction.

Tetanus toxin also inhibits spinal postsynaptic inhibition, possibly by preventing release of the inhibitory transmitter substance, as botulinum toxin does at cholinergic synapses. These actions of strychnine and

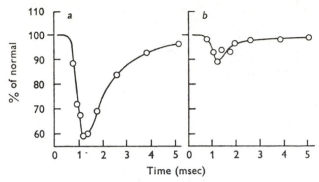

Fig. 8.13. The effect of strychnine on direct inhibition of the monosynaptic reflex. The ordinates show the size of the compound action potential in the ventral root following stimulation of group Ia afferents in the biceps–semitendinosus nerve. At zero time an inhibitory volley in the quadriceps nerve was initiated. *a* shows the normal response, *b* shows the response after injection of strychnine. (From Bradley, Easton and Eccles, 1953.)

tetanus toxin occur at all the postsynaptic inhibitory synapses in the spinal cord that have been investigated, and it would therefore seem probable that the transmitter substance is the same in each case.

The most probable candidate for the role of spinal postsynaptic inhibitory transmitter is the amino acid glycine. This suggestion was first made by Aprison and Werman (1965) as a result of the demonstration that the local concentration of glycine in the spinal cord is much higher in those regions where inhibitory interneurons occur. Later work has shown that electrophoretic application of glycine results in the hyperpolarization of spinal neurons, and that strychnine interferes with this action (Werman and Aprison, 1968; Curtis, 1969). Thus glycine appears to meet the first, third and fourth of Paton's criteria. The inhibitory action of glycine ceases very rapidly after the cessation of the electrophoretic current, suggesting that some efficient inactivation process is present; this might occur by means of an uptake mechanism similar to that which occurs for noradrenaline at sympathetic nerve endings.

The pharmacology of presynaptic inhibition in vertebrates is quite different from that of postsynaptic inhibition; this is perhaps not surprising since the action of the presynaptic transmitter substance is to produce a depolarization rather than a hyperpolarization. Presynaptic inhibition is not depressed by strychnine (except in the Mauthner cells of fishes), whereas it is depressed by picrotoxin, which has no effect on postsynaptic inhibition. Barbiturates and chloralose, which are general anaesthetics, potentiate presynaptic inhibition and the associated depolarization of dorsal root afferents, possibly by interfering with the inactivation of the transmitter substance.

GAMMA-AMINOBUTYRIC ACID

Kuffler and Edwards (1958) showed that the action of the inhibitory transmitter substance on crustacean stretch receptor cells could be imitated by γ-aminobutyric acid (GABA). Both synaptic inhibition and GABA produce a hyperpolarization or depolarization according as to where the membrane potential is set, and the reversal potential is the same for each. Changes in external potassium or chloride concentrations produce similar changes in the two responses. Both responses are depressed by picrotoxin. A similar imitation of synaptic inhibition by GABA is seen in the peripheral inhibition of crustacean and insect muscle fibres. Kravitz, Kuffler and Potter (1963) showed that the inhibitory fibres from the legs of lobsters contain appreciable quantities of GABA while the excitatory fibres do not. And finally Otsuka, Iversen, Hall and Kravitz (1966) found that GABA appeared in a perfusate of the lobster claw opener muscle when the inhibitory axon was stimulated. Thus, at crustacean peripheral inhibitory junctions, GABA fulfils all of Paton's criteria for the identification of a transmitter substance.

GABA also exerts a depressant action on mammalian central nervous activity, and there has been some discussion about whether or not it is an inhibitory transmitter substance. It now seems likely that it is (Krnjević, 1974), at least in the cerebral cortex and at the synapses between the Purkinje cells of the cerebellum and the neurons which they inhibit.

EXCITATORY AMINO ACIDS

Curtis, Phillis and Watkins (1960) found that glutamic, aspartic and cysteic acids cause excitation of spinal interneurons and motoneurons when applied ionophoretically. Later studies have extended these results to a wide variety of neurons and a considerable number of other acidic amino acids.

L-glutamic acid has an excitatory effect on arthropod muscle fibres, producing depolarization and contracture. Takeuchi and Takeuchi (1964) have shown that responsiveness to ionophoretically applied L-glutamic acid is localized at the neuromuscular junctions in crayfish muscle, that desensitization occurs with prolonged application and that this desensitization affects synaptic responses as well as the responses to applied L-glutamic acid. They conclude that the receptors responsive to L-glutamic acid are identical with those activated by the excitatory transmitter substance.

5-HYDROXYTRYPTAMINE

5-Hydroxytryptamine (5-HT, serotonin) is widely distributed in the tissues of the mammalian body, including the brain. It exerts an excitatory action on smooth muscles and sensory nerves; these actions are antagonized by lysergic acid diethylamide (LSD). Because of its occurrence in the brain, and because of the hallucinogenic properties of LSD, it has been suggested that 5-HT may act as a transmitter substance in the vertebrate central nervous system. Some brain cells are excited by applied 5-HT, others are inhibited. Whether or not 5-HT is a neurotransmitter in the vertebrate brain remains an open question (Krnjević, 1974).

Molluscan central nervous systems contain 5-HT, and certain cells in the cerebral and suboesophageal ganglia of snails are excited by it (Kerkut and Walker, 1962; Kerkut and Cottrell, 1963). Molluscan hearts are accelerated by 5-HT, and the maintained contractions of some molluscan muscles are relaxed by it. In arthropods, 5-HT and tryptamine block insect neuromuscular transmission (Hill and Usherwood, 1961) and LSD causes spiders to spin unusual webs.

ADENOSINE TRIPHOSPHATE

Adenosine and its derivatives, including adenosine-5'-triphosphate (ATP) have an inhibitory, or sometimes an excitatory, action on various smooth muscle preparations. ATP is also released on stimulating the intestinal wall electrically. Burnstock (1972) concludes that ATP or something very like it is a neurotransmitter in these tissues, and uses the term 'purinergic' to describe the nerves thought to release it.

9 The integrated activity of neurons

The essential feature of a nervous system, or a part of a nervous system, is that its activity is coordinated and adaptively related to the external influences affecting the system. This feature of nervous activity is known as *integration*. As Bullock and Horridge (1965) put it, 'integration is a process or a set of processes resulting in an output not identical to input but still some function of the input'. We can study integrative processes at a variety of different levels. For instance, we can see how the behaviour of an animal is affected by the interaction of external stimuli with internal activity, how the activity of its motor nerves is coordinated so as to produce locomotion, or how sensory input affects motor activity in a simple reflex. These problems are outside the scope of this book. In this chapter we shall be concerned with integration at the lowest level, that of single neurons, in an attempt to draw together some of the material which has been presented in earlier chapters.

ELECTRICALLY EXCITED AND NON-ELECTRICALLY EXCITED RESPONSES

In the previous chapters, we have seen that the electrical activity of excitable membranes is of two distinct types. First, there are those systems in which the ionic conductances of the membrane are determined by the potential across it. Activity of this kind can be described as *electrically excited*; it occurs *par excellence* in the all-or-nothing propagated action potentials of nerve axons, but may also be a graded phenomenon, as in the subthreshold local responses of axons and (as we shall see in chapter 13) in the electrically excited responses of arthropod muscle fibres. Secondly, there are those systems in which the primary response involves changes in the ionic conductance of the membrane which are largely independent of the membrane potential, being produced by the action of some external agent. Such *non-electrically excited* responses occur at postsynaptic membranes in response to transmitter substances and other drugs and in sensory endings in response to various sensory stimuli.

An example may help to make this distinction clear. The end-plate potential of frog sartorius muscle fibres is a non-electrically excited response. The changes in the sodium and potassium conductances of the membrane are synchronous and largely independent of the membrane potential; their intensity and time course are determined by the concentration of acetylcholine at the external surface of the membrane. However the propagated action potential of the muscle fibre, which is produced by the depolarization constituting the end-plate potential, is an electrically excited response. The changes in sodium and potassium conductance in this case are not synchronous and are dependent upon the membrane potential. The response of a sartorius muscle fibre to stimulation of its motor nerve, recorded by means of an intracellular electrode inserted into the end-plate region (as in trace *N* of fig. 6.7), contains both electrically excited and non-electrically excited components.

The distinction between these two types of response has been much emphasized by Grundfest (e.g. 1957, 1966), except that Grundfest refers to electrically excitable and electrically inexcitable *membranes* rather than electrically excited and non-electrically excited *responses*. It would be equally instructive to refer to two types of *channels*. For example, the ionic channels connected to the acetylcholine receptors at the neuromuscular junction are non-electrically excited, whereas the sodium and potassium channels of nerve axons are electrically excited.

FUNCTIONAL AND ANATOMICAL REGIONS OF NEURONS

From a functional point of view, a neuron can be divided into four sections:*

1. *The input region.* In interneurons and motoneurons this is the post-synaptic region; its activity is usually induced by the chemical transmitter substance released by presynaptic neurons. If transmission at the synapse is electrical, the pre- and postsynaptic membranes are probably connected by gap junctions. If the neuron is a primary sensory neuron, then the membrane of the input region is responsive to changes in a particular modality of its environment. The primary response of the input region membrane is non-electrically excited; i.e. it is essentially a graded electrical change, depolarization or hyperpolarization, brought about by changes in ionic conductance which are independent of the membrane potential. In addition to this primary response, the input region may show some degree of electrical excitability.

---

* The analysis given here is very similar to that of Grundfest (1957), differing only in the conceptual separation of the impulse initiation site from the conductile region.

2. *The impulse initiation site.* This is a region of the neuron which is electrically excitable, with a relatively low threshold, situated near the input region. It responds to depolarizations of sufficient intensity (produced either spontaneously or, as in most cases, by activity at the input region) by producing an all-or-nothing depolarization (the action potential) dependent upon a specific, regenerative increase in the sodium conductance of the membrane.

3. *The conductile region.* This is the axon, whose function is to conduct the all-or-nothing activity which arises at the impulse initiation site to the terminal region. The physiological properties of the axon are not fundamentally distinct from those of the impulse initiation site; the responses are electrically excited, by virtue of the regenerative relation between membrane potential and sodium conductance.

4. *The output region.* This region includes all those parts of the neuron whose activity affects other neurons or effector cells. In most cases this activity is secretory, by release of a transmitter substance. As we have seen however, there are some instances in which the postsynaptic cell is directly affected by the electrical currents in the presynaptic cell; in these cases the specialization of the output region seems to be merely one of geometrical arrangement and the pressure of gap junctions.

These functional divisions can to some extent be related to the structural elements of neurons. In vertebrate motoneurons, for instance, the input region is quite clearly the soma and the dendrites arising from it;

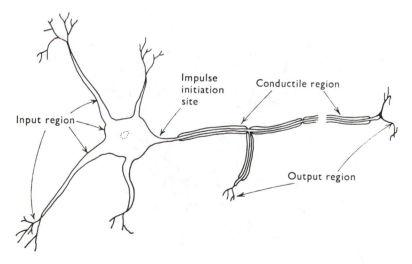

Fig. 9.1. Diagram of a vertebrate spinal motoneuron to show its various functional regions.

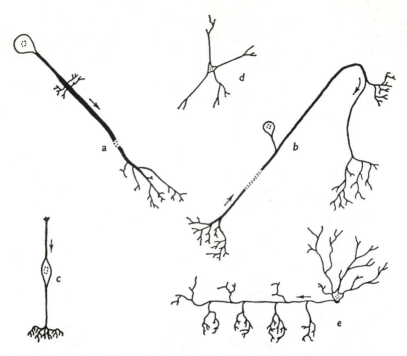

Fig. 9.2. The variety of form in neurons. The diagrams, which are rather schematic, represent: *a*, an arthropod motoneuron; *b*, a mammalian spinal sensory neuron; *c*, a bipolar neuron from the vertebrate retina; *d*, a neuron from the nerve net of a coelenterate; and *e*, a basket cell from the mammalian cerebellum.

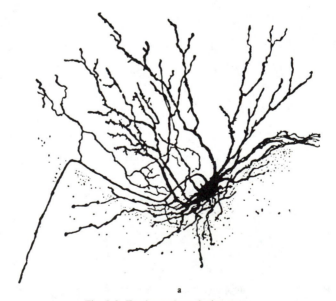

a

Fig. 9.3. For legend see facing page.

there is a single impulse initiation site, the initial segment; the conductile region consists of the main axon and its collateral branch; and there are two output regions, the axon terminals at the neuromuscular junctions and on Renshaw cells (fig. 9.1). But the vertebrate motoneuron is by no means the only type of neuron. Fig. 9.2 shows a number of different types of neuron in schematic form, and fig. 9.3 shows three different neurons drawn by the great neurohistologist

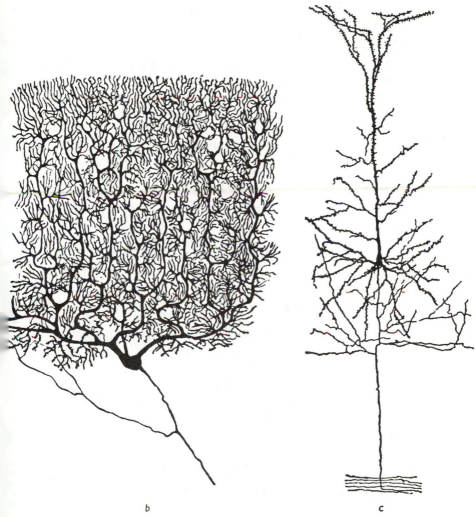

b   *c*

Fig. 9.3. Three neurons from the mammalian central nervous system, drawn from silver-stained preparations. *a*: Motoneuron from the spinal cord of a cat foetus. *b*: Purkinje cell from the human cerebellum. *c*: Pyramidal cell from the cerebral cortex of a rabbit. (From Ramon y Cajal, 1909.)

Ramon y Cajal from preparations stained by the Golgi method (in which a small proportion of the neurons present in a tissue take up silver or osmium dichromate).

Morphologically, neurons can be divided into soma, dendrites, axon and axon terminals. The soma contains the nucleus and its surrounding cytoplasm. The functions of the soma are mainly concerned with the long-term maintenance processes in the cell. As a consequence, its position is largely irrelevant to the electrical activity of the cell and shows considerable variation, as is shown in fig. 9.2. Dendrites are branches of the neuron which are thought to coincide with the input region of the cell. Synapses are found on the dendrites of interneurons and motoneurons; they may also be found on the soma if this lies between the dendrites and the axon, as in most vertebrate neurons. Axons can frequently be distinguished from dendrites by their length and by the absence of synapses. In some cases, however (such as in fig. 9.2*c*, *d*), no particular branch of the cell can be shown to be axonic in character. Axons may branch considerably before the terminals are reached; in these cases the shorter branches are known as collaterals.

INPUT–OUTPUT RELATIONS OF NEURONS

We may define the *input* to a neuron as those extrinsic factors which affect its activity. Such inputs may be unitary in nature, as when the input is derived from action potentials in presynaptic fibres, or graded, as in sensory neurons (when the input is a sensory stimulus) and in neurons which interact with each other via electrotonic interconnections. In this section we shall restrict ourselves to consideration of unitary inputs. Let us regard the *output* of a neuron as the action potentials set up in its axon or axons. Then the input–output relation is the ratio between the number of action potentials in the presynaptic fibres during any given time and the number of impulses in the postsynaptic axon that arise as a result of these presynaptic impulses.

It is convenient to divide input–output relations, as they have been defined here, into three classes: (i) 'one-to-one' relations, where a single impulse in the presynaptic cell leads to a single impulse in the postsynaptic cell, (ii) 'many-to-one' relations, where a number of presynaptic impulses are needed to produce one postsynaptic impulse, and (iii) 'one-to-many' relations, where a single presynaptic impulse can lead to many postsynaptic impulses. It is important to realize that any one neuron may have more than one of these types of relation with its different presynaptic neurons; for example, the giant motor fibres innervating the abdominal flexor muscles of the crayfish have a one-to-one relation (involving electrical transmission) with the lateral giant fibres of the nerve cord and a many-to-one relation (involving chemical transmission) with small fibres in the cord.

One-to-one relations are found only in situations where no integration of the information carried in the presynaptic fibre is required. Such a situation is one to which, provided the presynaptic and postsynaptic fibres are of similar size, electrical transmission is ideally suited. Thus we find a one-to-one relation involving electrical transmission at the segmental synapses of the giant fibres in the nerve cords of annelids and arthropods, and at the giant motor synapses of crayfish. The other principle site at which one-to-one relations occur is at the neuromuscular junction in vertebrate 'twitch' fibres; here, as we have seen, transmission has to be chemical because of the disparity in size of the presynaptic and postsynaptic cells.

The great majority of synapses in the central nervous systems of higher animals appear to be of the many-to-one type. Their essential

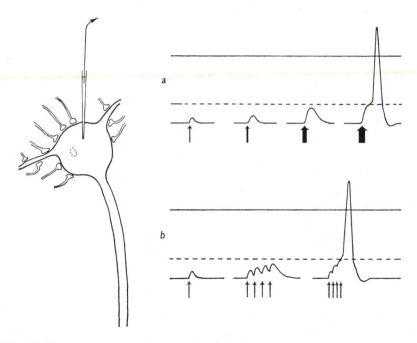

Fig. 9.4. Diagram to show spatial summation (*a*) and temporal summation (*b*). The curves show the responses in the postsynaptic cell to stimulation (at the arrows) of presynaptic fibres. The continuous horizontal lines indicate the zero potential and the broken lines indicate the threshold for action potential production in the postsynaptic cell. In *a*, stimulation of more and more fibres (indicated by the thickness of the arrows) produces progressively larger EPSPs, the final one being large enough to produce an action potential. In *b*, EPSPs of the same size (produced by activity in the same group of presynaptic fibres) summate if the time interval between one EPSP and the next is short enough.

feature is that a single impulse in an excitatory presynaptic axon produces an EPSP which is too small to lead to excitation of the postsynaptic axon. Postsynaptic action potentials can only be produced by summation of the effects of many presynaptic action potentials: a situation which gives a great deal of scope for integrative action. The effect of spatial summation is that the postsynaptic cell will only 'fire' (i.e. produce an action potential in its axon) if a sufficient number of presynaptic fibres fire simultaneously (fig. 9.4a). The effect of temporal summation is that the postsynaptic cell will only fire if a number of presynaptic impulses arrive in a short space of time (fig. 9.4b). The effects of inhibitory presynaptic impulses are also summated both spatially and temporally with each other and with the effects of excitatory impulses.

The existence of all these properties must lead us to the conclusion that the input region of a neuron showing many-to-one relations is the primary unit basis of decision making in the central nervous system. It is not surprising that a system containing very large numbers of such units is capable of extremely complex behaviour.

One-to-many relations occur when a superthreshold EPSP of long duration is produced by a single presynaptic impulse. An example is in the Renshaw cells of the spinal cord; a single impulse in the motoneuron collateral produces a burst of impulses in the Renshaw cell axon.

SPONTANEOUS ACTIVITY

As well as the various types of induced activity that we have discussed, many neurons show spontaneous activity that arises in the cells themselves. Cells which produce action potentials spontaneously at a more or less constant frequency are called pacemakers; they are frequently used to drive effector organs (such as hearts or electric organs) which must be activated rhythmically and continuously.

A characteristic feature of spontaneous activity is that the action potentials, at the site at which they are initiated, are each preceded by a progressive depolarization known as a 'pacemaker potential'. Huxley (1959) has shown that behaviour of this type is to be expected from the Hodgkin–Huxley equations if slight modifications are made in some of the parameters. Qualitatively, what happens during the pacemaker potential is probably as follows. At the end of the preceding action potential the potassium conductance is high and the sodium channel is inactivated. As the potassium conductance and the sodium inactivation process fall, the membrane potential becomes less negative. This increases the sodium conductance, and depolarization therefore continues until the threshold is crossed and the process becomes regenerative, producing another action potential.

10 The electrical properties of glial cells

We have seen in chapter 4 that the axons of peripheral nerve fibres are surrounded by accessory cells known as Schwann cells. Neurons in central nervous systems are similarly associated with accessory cells known as glia, glial cells or neuroglia. There are many more glial cells than there are neurons in the mammalian central nervous system, but the total volumes of the two types are approximately equal.

In the vertebrate brain there are two main types of glial cells: *astrocytes*, which have extensive cytoplasmic processes and a rather 'watery' cytoplasm, and *oligodendrocytes*, which are more compact in form and possess a more 'granular' cytoplasm. Oligodendrocytes are largely restricted to the white matter, where they are responsible for myelin formation. In a third type, the *microglia*, the cells are motile and phagocytic; it is a matter of definition as to whether they should be regarded as glial cells or not.

The morphological relations of astrocytes with capillaries, neurons and the extracellular space are shown in fig. 10.1. Two points are of

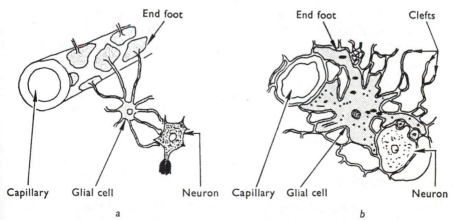

Fig. 10.1. The relations of glial cells to the extracellular spaces in the vertebrate brain. *a*: As determined by light microscopy. *b*: As determined by electron microscopy. The size of the intercellular clefts is greatly exaggerated in *b*. (From Kuffler, 1967, by permission of the Royal Society.)

particular interest here. Firstly, the capillaries in the brain are largely surrounded by processes of astrocytes that are swollen to form 'end-feet'. These astrocytes are also in contact with neurons which do not themselves form any direct link between the capillaries. Secondly, the extracellular space is limited to clefts about 150 to 200 Å wide.

The morphological picture has led to considerable speculation about how nutrient substances, waste products and ions pass through the tissue. Broadly speaking, there are three main hypotheses:

(i) The extracellular clefts are insufficient to allow diffusion of substances through them, and therefore the passage of these substances through the nervous tissue must be via the glial cells.

(ii) The extracellular clefts do allow diffusion of substances through them.

(iii) The apparent narrowness of the extracellular clefts is an artefact produced by the fixation procedure for electron microscopy. Van Harreveld and his colleagues have found that electron micrographs of material frozen very rapidly within seconds of circulatory arrest show extracellular spaces which are much larger than the clefts seen in material prepared by conventional methods (van Harreveld, Crowell and Malhotra, 1965).

The functions of glial cells have been and are still subject to much discussion. Oligodendrocytes are definitely concerned with myelin formation, and most glial cells are probably useful in providing a structural matrix for the neurons of the central nervous system. If the nervous tissue is damaged, glial cells undergo considerable changes in connection with degeneration and regeneration. It is widely believed that glial cells act as nutritive reservoirs for neurons, but Kuffler and Nicholls (1966) state that 'in our opinion, no convincing demonstration has yet been made to show that substances are in fact exchanged between neurons and glia across the narrow intercellular clefts.' Ten years later they were still of the same opinion (Kuffler and Nicholls, 1976).

Most of these questions, concerned as they are with metabolic and long-term processes, are outside the scope of this book. For the rest of this chapter we shall examine some of the electrical properties of neuroglia as determined by the work of Kuffler and his colleagues.

Resting potentials

The resting potentials of glial cells appear to be larger than those of neurons. In the leech ventral nerve cord, glial cells have resting potentials in the region of -75 mV, whereas those of the neurons are about -50 mV. In the optic nerve of *Necturus* (the mud-puppy, related to

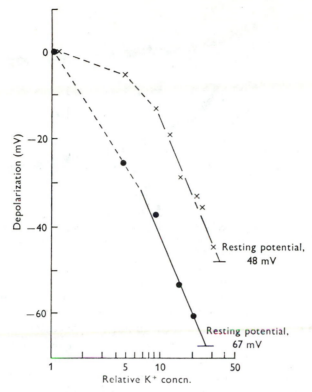

Fig. 10.2. The differential sensitivity of leech glial cells (circles) and neurons (crosses) to changes in potassium ion concentration in the medium. The abscissa shows the ratio of the potassium ion concentration in the experimental solution to that in the normal Ringer. (From Kuffler, 1967, by permission of the Royal Society.)

salamanders), glial cells have resting potentials in the region of -90 mV, whereas the usual range for vertebrate central neurons is -70 to -80 mV.

The membrane potentials of glial cells are sensitive to the external potassium ion concentration, and follow the Nernst equation for potassium ions closely over a wide range. In particular, the membrane potential follows the Nernst equation at low external potassium ion concentrations much more closely than do the membrane potentials of neurons and muscle cells, which are rather insensitive to changes in potassium ion concentration near the resting potential. These facts are illustrated schematically in fig. 10.2.

Membrane potential responses to nervous activity

The membranes of glial cells are not electrically excitable, their resistance being almost independent of membrane potential. However they

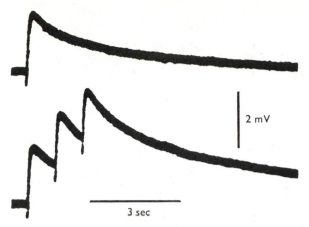

2 mV

3 sec

Fig. 10.3. Responses of glial cells in the optic nerve of *Necturus* to nerve impulses. The upper trace shows the response to a single volley of nerve impulses, the lower trace to three volleys. (From Orkand, Nicholls and Kuffler, 1966.)

do show changes in membrane potential following nervous activity. As is shown in fig. 10.3, these responses have a much slower time course than do nerve action potentials, and in fact the membrane potential does not begin to rise until after the local circuit currents associated with the nerve action potential (recorded as a downward deflection in fig. 10.3) have died away (Orkand, Nicholls and Kuffler, 1966).

In view of the sensitivity of the glial membrane potential to the external potassium ion concentration, a possible explanation of these glial potentials is that they are caused by an increase in the potassium ion concentration in the intercellular clefts, produced by the passage of action potentials in the nerve axons. This hypothesis was subjected to an elegant experimental test by measuring the size of the glial responses to a train of impulses in the optic nerve in solutions of different potassium ion concentrations. The results of this experiment, shown in fig. 10.4, require a little explanation. The resting potential of the glial cell in Ringer solution containing 3 mM potassium was -89 mV; on stimulating the optic nerve a depolarization of 12.1 mV occurred. If this depolarization were due solely to an increase in potassium ion concentration in the intercellular clefts, then it is possible to calculate the extent of this increase from the equation

$$\Delta E = 59 \log_{10}\left(\frac{[K]_0 + \Delta[K]}{[K]_0}\right), \tag{10.1}$$

where ΔE is the change in membrane potential (in mV), $[K]_0$ is the potassium ion concentration in the Ringer solution, and $\Delta[K]$ is the increase in potassium ion concentration in the intercellular clefts. In the experiment shown in fig. 10.4, for the middle point, we have

$$12.1 = 59 \log_{10}\left(\frac{3 + \Delta[\mathrm{K}]}{3}\right),$$

therefore $\Delta[\mathrm{K}] = 1.8\,\mathrm{mM}$.

The experiment is now repeated at external potassium concentrations of 1.5 and 4.5 mM; with a $\Delta[\mathrm{K}]$ of 1.8 mM, the stimulus should produce depolarizations (calculated from 10.1) of 16.6 and 8.5 mV respectively, as shown by the broken lines in fig. 10.4. In fact, as shown by the open circles in fig. 10.4, the actual depolarizations were 17.5 and 8.8 mV. These figures are sufficiently close to the predicted values for the hypothesis to be accepted.

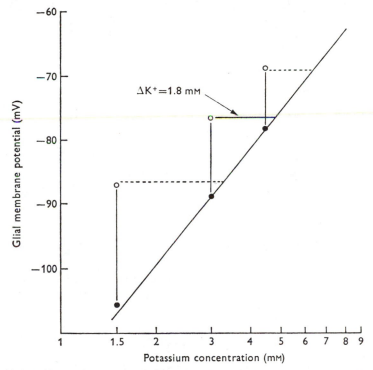

Fig. 10.4. Evidence that the depolarization of glial cells by nerve impulses in *Necturus* optic nerve is produced by liberation of potassium ions from the nerve fibres. Solid circles show glial resting potentials at three different potassium ion concentrations; the sloping line through them is that predicted from the Nernst equation. Open circles show the peak depolarization produced by a single volley in the optic nerve fibres. The result in normal Ringer (middle point) suggests an increase in potassium ion concentration in the inter-cellular clefts of 1.8 mM: the dotted lines show the expected levels of response if the same amount of potassium were released in the solutions with lower and higher potassium ion concentrations. (From Orkand, Nicholls and Kuffler, 1966.)

The significance of these observations for the functioning of nervous tissues is not yet entirely clear. One possibility is that the glia may act as a 'spatial buffer' for extracellular potassium, transporting it, by means of the currents set up during depolarization, from regions of high concentration to those of low concentration. Another possibility is that the depolarizations produced by nervous activity may act as some sort of signal for metabolic changes in glial cells. The observations do, however, indicate that ions can diffuse fairly freely through the intercellular clefts, and it is therefore not necessary to invoke the glia as essential pathways for the transport of small molecules between neurons and other cells.

11 The organization of muscle cells

In this and the next two chapters we shall examine some of the properties of muscle cells. The function of muscle cells is to shorten and develop tension. Thus the end product of cellular activity can be measured with considerable precision, by mechanical measurement of the change in length or tension or both. Such a contraction* must obviously involve the consumption of energy, some of which may appear as heat. In this chapter and the next we shall mainly be concerned with the properties of rapidly contracting vertebrate skeletal muscles, such as frog sartorius and the rabbit psoas. Some of the special properties of other muscles will be examined in chapter 13.

The sartorius muscle of the frog consists of a number of fibres about 70 μm in diameter and up to 3 cm long. Each fibre is bounded by a cell membrane and is therefore, for our purposes, a single cell, although it contains a number of nuclei. The fibres are bound together in the muscle by connective tissue sheaths. At each end, they are connected to collagen fibres which form the tendons connecting the muscle to the skeleton. The sartorius muscle is attached at one end to the pelvic girdle, and at the other end to the tibia at the knee joint. Contraction of the muscle moves the leg forward and flexes it at the knee. For physiological experiments, the muscle can be removed intact from the animal by cutting the tendons, or the bones to which they are attached, which can then be connected to the recording apparatus.

ISOMETRIC CONTRACTIONS

We have seen in chapter 6 that the fibres of the frog sartorius are electrically excitable. Thus the muscle can be stimulated by direct application of electric shocks, as well as via its motor nerve. For many experiments it is more convenient to use such direct stimulation, often

---

* In muscle physiology, the word 'contraction' is usually used to denote mechanical activity of the muscle, whether or not this includes shortening. If we hold the length of a muscle constant, it may still 'contract' in this sense, the contraction being seen as tension development.

via a number of electrodes spaced along the length of the muscle so that all parts of the muscle are activated simultaneously.

If both ends of a muscle are rigidly fixed so that it cannot shorten, then when it is stimulated it contracts *isometrically*. In this case the contraction is measured as a change in tension, so that one end of the muscle has to be attached to a tension-recording device. A simple method of recording tension is provided by an isometric lever, which is a lever pivoted on a torsion spring, so that its deflection is proportional to the force applied to it. The deflection can be recorded by a revolving smoked drum on which the tip of the lever writes or, in more sophisticated designs, by a beam of light and a photoelectric cell. Of course, a deflection of the lever can only be produced by some movement of the end to which the muscle is attached. This is a feature of any tension-recording device, so that it is never possible to record a completely isometric contraction; the muscle must always shorten to some extent. A good device for recording isometric tension is therefore one which will permit very little shortening of the muscle, i.e. one which is very stiff. One such device is a triode valve (RCA type 5734) whose anode projects through a stiff spring diaphragm. Movement of the anode then alters the current flowing through the valve, so that, using a suitable electric circuit, the force applied to the anode peg can be recorded as a voltage change on an oscilloscope (fig. 11.1*b*). An additional advantage of this arrangement is that it has negligible inertia, and so can follow very rapid changes in tension; this is not so for the lever shown in fig. 11.1*a*. Another useful method is to attach the muscle to a steel bar of suitable stiffness which has semiconductor strain gauges bonded onto it. The resistance of the strain gauges then varies with muscle tension.

The tension produced by a muscle is a force, and is therefore measured in newtons or, more usually, in grams weight. Fig. 11.2 shows

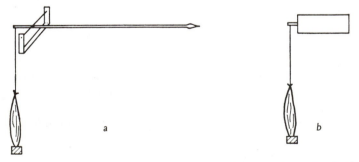

Fig. 11.1. Methods of measuring the tension produced by a muscle. *a*: Simple isometric lever, pivoted on a torsion wire, for use with a smoked drum. *b*: Transducer valve with a movable anode, giving an electrical output proportional to tension.

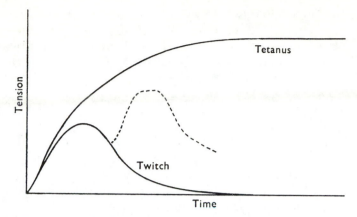

Fig. 11.2. Isometric tension changes in response to stimulation. A single stimulus produces a twitch; the dotted curve shows the time course of a second twitch induced by another stimulus before the relaxation of the first twitch is complete. On repetitive stimulation at a fairly high frequency, twitches of this type fuse to form a tetanus.

the time course of the tension development in isometric contractions. A single stimulus produces a rapid increase in tension which then decays; this is known as a *twitch*. The duration of the twitch varies from muscle to muscle, and decreases with increasing temperature. For a frog sartorius at 0 °C, a typical value for the time between the beginning of the contraction and its peak value is about 200 msec, tension falling to zero again with 800 msec. If a second stimulus is applied before the tension in the first twitch has fallen to zero, the peak tension in the second twitch is higher than that in the first; this effect is known as

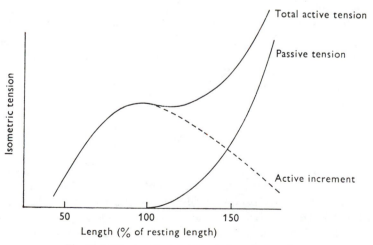

Fig. 11.3. The length–tension relation of a muscle.

mechanical summation. Repetitive stimulation at a low frequency thus results in a 'bumpy' tension trace. As the frequency of stimulation is increased, a point is reached at which the bumpiness is lost and the tension rises smoothly to reach a steady level. The muscle is then in *tetanus*, and the minimum frequency at which this occurs is known as the *fusion frequency*.

A resting muscle is resistant to stretching beyond a certain length, so that it is possible to determine a passive length–tension curve, as is shown in fig. 11.3. The full isometric tetanus tension of the stimulated muscle is also dependent on length, as is shown in the 'total active tension' curve in fig. 11.3. The difference between the two curves is known as the 'active increment' curve; notice that this passes through a maximum at a length near to the maximum length in the body, falling away at longer or shorter lengths.

BIOCHEMICAL ASPECTS OF CONTRACTION

In order to understand what is known about the structural and physiological aspects of muscular contraction, it is necessary to know something of the biochemistry of the process. A very large amount of work has been done on this subject in recent years; we shall merely examine a few of the main results of this work in the following paragraphs.

The nature of the energy source

The energy for muscular contraction is ultimately derived from the chemical energy released by the oxidation of food substances. This is a rather trivial statement, and it is much more pertinent to enquire what the *immediate* source of the energy for contraction is. What may be called 'the modern period' in the study of muscle biochemistry began with the work of Fletcher and Hopkins in 1907. They showed that muscles can continue to contract in the absence of oxygen, and that they produce lactic acid under these conditions. It was later shown by Meyerhof and others that this lactic acid is formed by the breakdown of glycogen, which itself is formed by polymerization of glucose derived from carbohydrates in the food. The next step was the discovery of 'phosphagen', later shown to be creatine phosphate, and the demonstration that it was broken down during contraction. Finally, the discovery of adenosine triphosphate (ATP) led to understanding of the basic scheme of energy production and utilization in cells which is shown in fig. 11.4, and which is now familiar to all students of biochemistry.

If this scheme is correct, then, as A. V. Hill (1950*a*) pointed out in his 'challenge to biochemists', it should be possible to show that ATP is

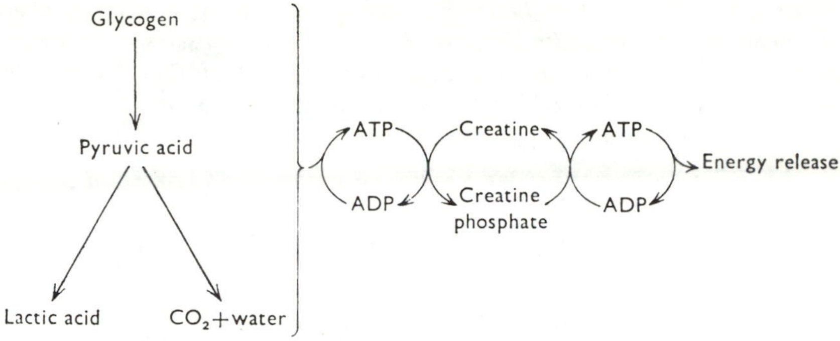

Fig. 11.4. Diagram to show, very schematically, the respiratory energy production in an animal cell. Oxygen is needed to convert pyruvic acid to carbon dioxide and water. Many invertebrates use arginine phosphate instead of creatine phosphate for storage of high energy phosphate.

broken down during the contraction of living muscle cells. The general technique used is to stimulate a muscle, very rapidly freeze it so as to prevent any further biochemical changes, and then determine how much of the substance one is interested in is present. A similar determination is performed on an unstimulated muscle, usually the equivalent muscle on the opposite leg of the same animal. Using these techniques, it was shown by a number of workers (see Carlson, 1963) that contraction leads to the breakdown of creatine phosphate in muscles poisoned with iodoacetic acid (which prevents glycolysis, and hence prevents resynthesis of creatine phosphate). This could mean either that creatine phosphate is the immediate source of energy for contraction, or that creatine phosphate reacts with adenosine diphosphate (ADP) to give ATP and creatine, ATP being the immediate energy source.

The substance 1-fluoro-2,4-dinitrobenzene (FDNB) blocks the action of the enzyme creatine phosphotransferase, so that ADP cannot be rephosphorylated to ATP by the breakdown of creatine phosphate. Using this substance, Davies and his colleagues were able to show that breakdown of ATP occurs during contractions of frog rectus abdominis (Cain, Infante and Davies, 1962) and sartorius (Infante and Davies, 1962) muscles. The results of their experiments on sartorius muscles are given in table 11.1. The average amount of ATP lost during the rising phase of each twitch was 0.22 μmoles per gram of muscle. Is this sufficient to account for the energy used in the contraction? In these experiments, the muscle shortened against a light load, and the average value for the work done in each contraction was 17.4 gm cm/gm; this is equivalent to 1.71×10^{-3} J/gm. The heat of hydrolysis of the

TABLE 11.1 *Changes in the ATP content of frog sartorius muscles after treatment with fluorodinitrobenzene. The stimulated muscles were frozen in liquid propane at the peak of a single isotonic twitch at 0 °C. The control muscles were unstimulated. ATP contents are given in μmoles per gram of muscle.* (From Infante and Davies, 1962.)

Muscle pair	Control ATP	Twitch ATP	Difference in ATP
1	2.55	2.39	−0.16
2	3.38	3.25	−0.13
3	2.55	2.53	−0.02
4	2.82	2.58	−0.24
5	3.25	3.12	−0.13
6	2.93	2.57	−0.36
7	3.38	3.20	−0.18
8	2.16	2.01	−0.15
9	3.16	2.98	−0.18
10	2.51	2.40	−0.11
11	1.90	2.09	+0.19
12	2.90	2.40	−0.50
13	3.56	2.64	−0.92
Mean ± S.E.	2.85 ± 0.14	2.63 ± 0.10	−0.22 ± 0.07

terminal phosphate bond in the conversion of ATP to ADP is probably about 34 kJ/mole (Wilkie, 1976). Hence the dephosphorylation of 0.22 μmoles of ATP should make available about 7.5×10^{-3} J. The amount of ATP broken down is therefore more than enough to account for the work done in a twitch; the excess energy appears as heat.

The contractile proteins and their interaction with ATP

The contractile machinery of striated muscle cells consists of a small number of different proteins which are aggregated together in filaments (we shall examine the structure of these filaments later). The two major components of this system are the proteins *actin* and *myosin*, which interact with each other to produce the contraction. A number of minor constituents serve either to modify the interaction between actin and myosin or in determining the structural organization of the system.

Myosin is a rather complex protein with a molecular weight of about 470 000. One of its most important properties is that it is an ATPase, i.e. it will enzymatically hydrolyze ATP to form ADP and inorganic phosphate. This reaction is activated by calcium ions, but inhibited by magnesium ions. Treatment with the proteolytic enzyme trypsin splits the myosin molecule into two sections, known as light meromyosin and heavy meromyosin; of these, only heavy meromyosin acts as an ATPase. Examination of the meromyosins by electron microscopy shows that heavy meromyosin has a more or less globular 'head' with a short 'tail', whereas light meromyosin is a rod-like molecule. Heavy meromyosin can be further split into two subfragments, a globular S_1 portion and a rod-like S_2 portion, by digestion by papain; the ATPase activity is confined to the globular subfragment. Light meromyosin molecules will aggregate to form filaments under suitable conditions, but neither heavy meromyosin nor its two subfragments will.

The myosin molecule consists of a number of different protein chains. These can be dissociated from one another by treatment with suitable reagents such as disulphide-link reducing agents, alkaline conditions and sodium dodecyl sulphate. The mixture of the resulting complexes (of protein chains to which dodecyl sulphate ions are attached) can then be separated by electrophoresis in a column of polyacrylamide gel; the negatively charged complexes move towards the anode. The gel acts as a three-dimensional net which impedes the passage of large complexes more than that of small ones. This enables the complexes to be separated from one another, and the distance which a particular complex moves is fairly precisely related to its molecular weight: the lighter the protein chain, the further its dodecyl sulphate complex moves.

This technique has been used by a number of workers to investigate the complex nature of the myosin molecule (see, for example, Lowey and Risby, 1971; also Offer, 1974, and Mannherz and Goody, 1976). It is found that the molecule consists of two heavy chains (each of molecular weight about 200 000) and four light chains of three different types (fig. 11.5). Two of the light chains have a molecular weight of about 18 000 and are detachable from the whole molecule by treatment with DTNB (5,5'-dithiobis-(2-nitrobenzoate), a substance which breaks disulphide linkages in proteins) and are hence known as the DTNB light chains. The other two light chains are separable from the whole molecule by treatment with alkali; they are of two types, known as the A-1 and A-2 light chains, with molecular weights of 21 000 and 17 000 respectively. There are about twice as many A-1 chains as there are A-2

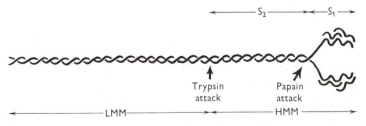

Fig. 11.5. Schematic diagram of the myosin molecule. Binding sites for actin and ATP occur in the S_1 heads.

chains in extracts from rabbit psoas muscles; perhaps some myosin molecules have A-1 chains and others A-2 chains. Removal of the DTNB light chains does not affect the ATPase activity of the myosin, but removal of the A-1 and A-2 light chains abolishes it.

Isolated *actin* exists in two forms: G-actin, a more or less globular molecule of molecular weight about 42 000, and F-actin, a fibrous protein which is a polymer of G-actin. The structure of F-actin is described later (p. 231). Neither form has any ATPase activity.

If solutions of actin and myosin are mixed, a great increase in viscosity occurs, due to the formation of a complex called *actomyosin*. Actomyosin is an ATPase but, unlike myosin ATPase, it is activated by magnesium ions. 'Pure' actomyosin (a mixture of purified actin and purified myosin) will split ATP in the absence of calcium ions, but 'natural' actomyosin (an actomyosin-like complex which can be extracted from minced muscle with strong salt solutions) can only split ATP if there is a low concentration of calcium ions present. In the absence of calcium ions, addition of ATP to a solution of natural actomyosin results in a decrease in viscosity, suggesting that the actin-myosin complex becomes dissociated.

These properties of actomyosin solutions can be paralleled by the properties of glycerol-extracted muscle fibres, which are prepared by soaking a muscle in cold 50 per cent glycerol for a period of weeks. Such treatment removes most of the sarcoplasmic material from the fibres, leaving only the contractile structures. In the absence of ATP the fibres do not contract, but cannot be extended without exerting considerable force; they are said to be in *rigor*. This stiffness is caused by the formation of cross-linkages between the actin and myosin, and corresponds to the formation of actomyosin when solutions of actin and myosin are mixed. On addition of ATP in the presence of magnesium ions the fibres become readily extensible. This 'plasticizing' action of ATP is due to the breakage of the cross-links, and corresponds to the dissociation of actomyosin in solution by ATP. Finally, if calcium ions are added to the glycerinated fibres in the presence of magnesium and ATP, ATP is split and the fibre contracts; this corresponds to the calcium-activated splitting of ATP by 'natural' actomyosin.

The responsiveness of 'natural' actomyosin, and of glycerinated muscle fibres, to calcium ions depends upon the presence of two accessory proteins, *tropomyosin* and *troponin* (see Ebashi and Endo, 1968; Ebashi, Endo and Ohtsuki, 1969). Tropomyosin is a fibrous protein of molecular weight about 70 000, which is not in itself very reactive, except that it binds to F-actin and to troponin. Troponin (molecular weight 76 000) consists of three subunits with different properties which we shall examine later (p. 262). Addition of tropomyosin and troponin to purified actomyosin systems inhibits their ATPase activity in the absence of calcium, but does not do this if calcium ions are present. Thus troponin and tropomyosin serve to sensitize the actomyosin ATPase to the presence of calcium ions; calcium ions appear to combine with troponin so as to inhibit the inhibition of actomyosin ATPase by the troponin-tropomyosin system. We shall see later that this effect is of prime importance in the control of contractile activity in the living muscle cell.

A number of other proteins are present in the contractile machinery of muscle cells. α-*actinin* binds to actin, and is probably localized at the Z line (see below) in the intact myofibril; other proteins may also be present at the Z line. β-*actinin* appears to be rather like actin in structure, and may be concerned with determining the length of the F-actin polymers; in its absence they may be much longer. The function of *C-protein*, which binds to myosin, is not known (Craig and Offer, 1976). And there is at least one special protein found at the M line (see below).

THE STRUCTURE OF THE MYOFIBRIL

Examination of fixed and stained frog sartorius muscle fibres by light microscopy shows that each fibre contains a large number of thin longitudinal elements called *myofibrils*. These myofibrils have characteristic banding patterns, and the bands on adjacent myofibrils are aligned so that the whole fibre appears striated. Besides the myofibrils, the fibres contain many mitochondria (sometimes called 'sarcosomes') and a few nuclei. There is also an internal membrane system consisting of the endoplasmic reticulum (or sarcoplasmic reticulum) and tubular invaginations of the cell membrane; this cannot normally be seen by light microscopy. The remaining cytoplasm of the fibre is called the sarcoplasm. The whole fibre is bounded by a sarcolemma, which consists of the cell membrane plus some extracellular material.

The striation pattern of myofibrils, as seen by conventional light microscopy after fixing and staining, or by phase-contrast, polarized light or interference microscopy, is shown in fig. 11.6*a*. The two main bands are the dark, strongly birefringent A band and the lighter, less birefringent I band. These bands alternate along the length of the myofibril. In the middle of each I band is a dark line, the Z line. In the middle of the A band is a lighter region, the H zone, which is sometimes bisected by a darker line, the M line.* The unit of length between two Z lines is called the *sarcomere*.

* The origins of these letters used in the description of the striation pattern are seen in the original names: *I*sotropisch, *A*nisotropisch, *Z*wischenscheibe, *H*ensen's disc, and *M*ittlemembran. These names are now no longer used.

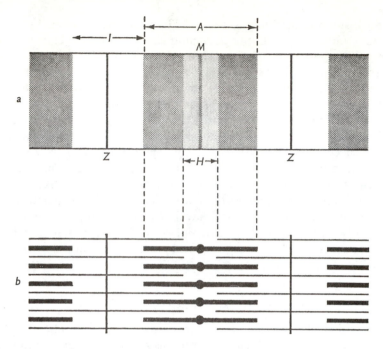

Fig. 11.6. The striation pattern of a myofibril (*a*), and its structural basis; interdigitating arrays of thick and thin filaments (*b*).

Techniques used for the determination of myofibrillar fine structure

Two principal techniques are used in determining the finer details of myofibrillar structure: electron microscopy and X-ray diffraction. We have met a number of examples of the use of electron microscopy in previous chapters, but it may be as well to consider at this point some of the difficulties encountered in making accurate measurements with the technique. The study of the structure of cells and tissues usually involves the preparation of thin sections of the material. The tissue is first fixed in a suitable fixative such as a solution of osmium tetroxide or glutaraldehyde. It is then dehydrated in alcohol and embedded in a substance such as methacrylate ('Perspex') or various epoxy resins (such as 'Araldite') so as to form a relatively hard block. This block is then sectioned on a special microtome to provide sections of the order of 0.02 to 0.05 μm thick.

The electron microscope can only produce an image of an object if parts of the object are 'electron dense', i.e. if they are opaque to electrons. In practice, this means that the tissue has to be 'stained' by depositing heavy metal atoms upon it; some suitable stains are osmium tetroxide, potassium permanganate, lead hydroxide and uranyl acetate. In the final electron micrograph, therefore, one observes the distribution of heavy metal atoms on a thin section of tissue which has been fixed, dehydrated and embedded. Hence it is as well to check that the structures seen after any particular treatment are compatible with those seen after other procedures (use of a different fixative, for example) or deduced from the use of other techniques, such as light microscopy or X-ray diffraction.

The study of muscle structure may involve a number of accurate measurements of size. Unless special precautions are taken, a piece of tissue sectioned for electron microscopy undergoes an appreciable amount of distortion in the preparatory procedures. Shrinkage is particularly likely to occur during fixation and dehydration. The pressure of the microtome knife during sectioning may cause considerable compression in the direction in which the cut is made.

Alternative methods are available for the examination of small particles (such as viruses, large molecules and some subcellular components) by electron microscopy. In these methods, the particles are deposited on a thin supporting film, and then a suitable electron dense material is deposited over them. In the 'shadow-casting' technique, a heavy metal is vaporized by heating it in a vacuum above and to one side of the specimen, so that a thin film of heavy metal atoms is produced, with a 'shadow' at the side of each particle. In the 'negative staining' technique, an electron dense stain, such as sodium phosphotungstate, is deposited over the surface of the preparation, with the result that it is much more thinly distributed over the tops of the particles.

X-rays are a type of electromagnetic radiation; they differ from visible light in that their wavelength is extremely short, being about 1 Å. The resolution of an optical microscope is ultimately limited by the wavelength of the light which it uses. Thus if we could make a microscope which used X-rays instead of visible light, we should be able to examine very small structures indeed. This cannot be done, however, since it is not possible to focus X-rays. But the light in an optical microscope passes through two stages between the object and its image; first it is scattered by the object, then it is focused by the lenses of the microscope. X-rays are also scattered by objects. The essence of the X-ray diffraction technique is that the pattern produced by this scattering is interpreted by mathematical means so as to indicate what a focused image would look like. In practice, it is only possible to do this if we possess certain information about the X-rays and if the scattering pattern shows a sufficient degree of regularity. A regular pattern is produced only if there are corresponding regularities in the object, such as are produced by the regular packing of atoms and molecules in crystals or fibres. It would be inappropriate here to go any further into the theory of X-ray diffraction methods; the interested reader should consult, for example, the book by Wilson (1966).

The arrangement for an examination of a muscle fibre by the X-ray diffraction technique is shown schematically in fig. 11.7. If we were to draw a line on the photographic plate parallel to the fibre axis and passing through the point at which the undiffracted X-ray beam hits the plate, such a line would lie along the *meridian* of the pattern. A similar line passing through the undiffracted beam but at right angles to the fibre axis would lie along the *equator* of the pattern. Spots or lines produced on the plate by diffracted X-rays are called 'reflections'. Reflections lying on the meridian are produced by regularly repeating structures spaced along the fibre axis. Reflections lying on the equator are produced by regularly repeating structures in the transverse plane of the fibre. Structures which repeat in directions other than axially or radially (as, for example, in helically arranged units) produce 'off-meridional reflections', which may be arranged in a series of lines, parallel to the equator, known as 'layer lines'.

Each set of repeating structures produces a series of reflections on the photographic plate at regular distances from the line of the original X-ray beam, and their intensity usually decreases at increasing distances from this line. The first member of a series of this type (i.e. that nearest to the centre of the pattern) is known as a first order reflection, the second is known as a second order reflection, and so on. The distance between reflections of successive orders (which is equal to the distance between the first order reflection and the centre of the pattern) is inversely related to the distance between the repeating units in the fibre.

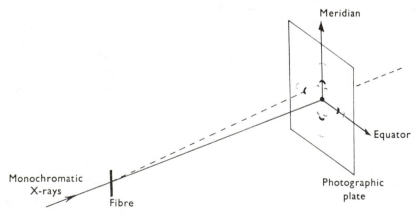

Fig. 11.7. Schematic diagram to show the arrangement for making low-angle X-ray diffraction observation on fibres. The dotted line shows one of the many diffracted rays, producing a reflection where it hits the photographic plate.

Hence low-angle patterns (patterns in which the emerging rays do not diverge much from their original direction) give information about structures repeating at relatively long distances, such as might be seen with the electron microscope, and wide-angle patterns give information about repeating structures which are closer together.

It will be evident that, although the X-ray diffraction technique can give rather precise measurements of the distance between repeating structures, it cannot give any direct information as to what these structures are. On the other hand, electron microscopy can enable us to identify structures, but the precision of size measurement is usually much less. Thus the low-angle X-ray diffraction technique can be used to provide precise information about the distances between the regularly repeating structures seen with the electron microscope.

The striation pattern

The structural basis of the myofibrillar striation pattern was obscure until the advent of the techniques of electron microscopy and thin sectioning, and their application to striated muscle by Hanson and Huxley (Huxley and Hanson, 1954; Hanson and Huxley, 1955; H. E. Huxley, 1957). It was then found that the myofibrils are composed of two interdigitating sets of filaments, about 50 and 110 Å in diameter. The thin filaments are attached to the Z lines and extend through the I bands into the A bands. The position of the thick filaments is coincident with that of the A band. The H zone is that region of the A band between the ends of the two sets of thin filaments, and the M line is caused by cross-links between the thick filaments in the middle of the sarcomere. Because of their positions, the thick filaments are sometimes called A filaments and the thin filaments are called I filaments. This arrangement can be seen in the electron micrographs shown in plates 11.1 and 11.2, and, diagrammatically, in fig. 11.6b.

As we have seen, the major part of the myofibrillar material consists of two proteins, *actin* and *myosin*, and the interaction between these two seems to be the chemical basis of muscular contraction. Consequently, it was desirable to determine the localization of actin and myosin in the myofibrils, and to see if their distribution was related to that of the thick and thin filaments discovered by electron microscopy. The question was investigated by Hanson and H. E. Huxley (1953, 1955) using isolated myofibrils from a muscle which had been previously extracted with glycerol; glycerol extraction removes most of the sarcoplasmic material from rabbit psoas muscles, leaving only the contractile structures. The appearance of such a myofibril, viewed by phase-contrast microscopy, is shown in fig. 11.8a. After treatment with a 0.6 M solution of potassium chloride containing some pyrophosphate and a

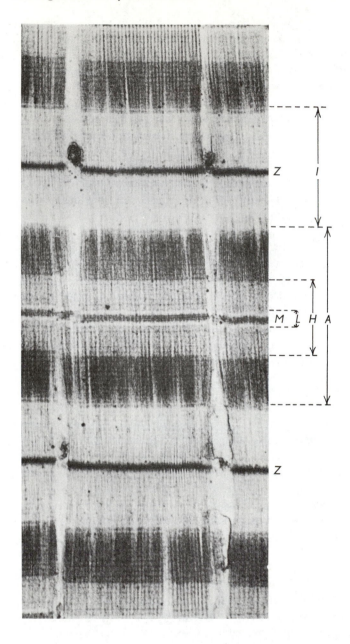

Plate 11.1. Thick longitudinal section of a frog sartorius muscle fibre, showing the striation pattern as seen by electron microscopy. Magnification 28 000 times. (Photograph kindly supplied by Dr H. E. Huxley.)

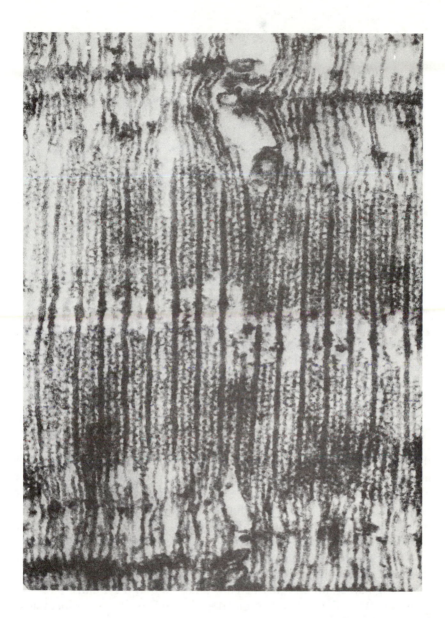

Plate 11.2. Thin longitudinal section of a glycerol-extracted rabbit psoas muscle fibre. Notice, particularly, the cross-bridges between the thick and thin filaments. (Photograph kindly supplied by Dr H. E. Huxley.)

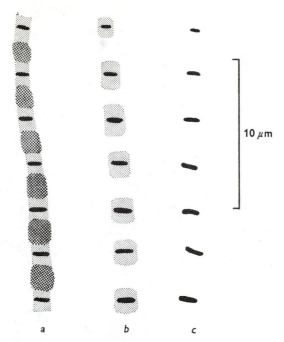

Fig. 11.8. Diagram showing the appearance of a stretched myofibril from glycerol-extracted rabbit muscle, as viewed by phase-contrast microscopy. *a*: Before treatment; *b*: after extraction of myosin; *c*: after extraction of actin. (Drawn from a photograph in Hanson and Huxley, 1955.)

little magnesium chloride, the dark material of the *A* bands disappeared, as is shown in fig. 11.8*b*. This solution had previously been used to extract myosin from minced muscle, and it was therefore concluded that the *A* filaments are composed of myosin. The myosin-extracted fibre was then treated with a 0.6 M potassium iodide solution, which was known to extract actin from muscle. This removed the substance of the *I* bands (fig. 11.8*c*), showing that the *I* filaments are composed of actin. It would seem that the *Z* lines, which were not affected by the extraction of myosin and actin, are composed of some other substance.

A very interesting and rather puzzling feature was seen in myosin-extracted fibrils (fig. 11.8*b*). If such a fibril was stretched, it would return to its original length on release. This indicates that there must be some material left in the *A* band connecting the ends of the actin filaments. Hanson and Huxley suggested that there was a set of filaments crossing the *H* zone and connected to the ends of the filaments, and called these hypothetical structures 'S filaments'. But no such filaments can be seen by electron microscopy, either in intact or myosin-extracted fibrils. It is clear that there must be something present to give continuity to the myosin-extracted fibril, but what this is is unknown.

Transverse sections through the *A* band in the region of overlap show that the filaments are arranged in a hexagonal array so that each myosin filament is surrounded by six actin filaments, and each actin filament is surrounded by three myosin filaments (fig. 11.9). Hence there are twice as many actin filaments as there are myosin filaments. Cross-sections through the *H* zone show only myosin filaments, and cross-sections through the *I* band show only actin filaments. Previous to these electron microscope studies, H. E. Huxley (1953) had obtained evidence for a hexagonal array of two types of filaments from equatorial measurements of low-angle X-ray diffraction patterns. Taking his results in conjunction with the electron micrographs later obtained, it was deduced that the centre-to-centre distance between each myosin filament in the array was about 450 Å at rest length (this distance increases at shorter lengths and becomes less if the muscle is stretched). In electron micrographs the distance usually appears as about 200 to 300 Å, but this value is undoubtedly affected by the shrinkage which occurs during fixation and embedding.

High magnification electron micrographs of glycerol extracted muscles (plate 11.2) show that the thick filaments are covered with

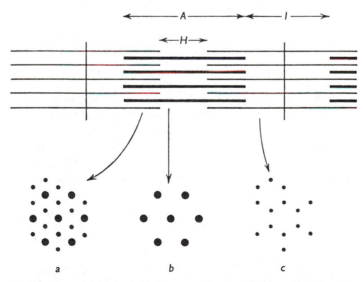

Fig. 11.9. The array of thick and thin filaments in a vertebrate muscle fibre, as seen by electron microscopy in transverse sections. *a*: Section through the *A* band outside the *H* zone. *b*: Section through the *H* zone. *c*: Schematic section through the *I* band; in actual sections the filament array in the *I* band is not so regular as is indicated here.

projections (H. E. Huxley, 1957). In the overlap region, these projections may be joined to the thin filaments, when they are known as
cross-bridges. There are no projections in the very middle of the
filaments; this produces a light region, the '*L* zone' or 'pseudo *H* zone',
about 0.15 μm long, in the middle of the sarcomere (plate 11.1).

The ultrastructure of the filaments

H. E. Huxley (1963) managed to isolate myosin filaments by homogenizing portions of glycerinated fibres. The filaments were then
examined by electron microscopy, using the negative staining technique.
The most noticeable feature of these filaments was the presence of fairly
regularly spaced projections on them, which almost certainly correspond to the projections and cross-bridges seen in thin sections of
glycerinated muscle fibres. In the middle of each filament was a section,
0.15 to 0.2 μm long, from which these projections were absent, and
which must correspond to the 'pseudo *H* zone' of intact muscle fibres.

The myosin molecule, as observed by the shadow casting technique, is
a tadpole-like structure consisting of a rod about 15 to 20 Å in diameter
with a terminal thicker region containing two 'heads' (S_1 fragments), in
accordance with the biochemical analysis described earlier. The whole
molecule is about 1500 Å long (H. E. Huxley, 1963). A most interesting
feature of these molecules, discovered by Huxley, is that they are able to
aggregate, under suitable conditions, to form filaments. The 'artificial
fllaments' so formed are of varying lengths, but all otherwise show the
same general structure as the isolated natural filaments, including the
projection-free region in the middle. Huxley suggested that the 'tails' of
the myosin molecules become attached to each other to form a filament
as is shown in fig. 11.10, with the 'heads' projecting from the body of the
filament. Notice particularly that this type of arrangement accounts for
the bare region in the middle, and also that it implies that the polarity of
the myosin molecules is reversed in the two halves of the filament.

In his pioneer investigation on the low-angle X-ray diffraction pattern
of living muscle, H. E. Huxley (1953) observed a series of reflections
corresponding to an axial repeat distance of about 420 Å. Using more
accurate methods, Worthington (1959) observed a strong reflection at
145 Å, with a fainter one at 72 Å, and suggested that these corresponded to the third and sixth orders of a 435 Å repeat distance.

Fig. 11.10. H. E. Huxley's suggestion as to how the myosin molecules aggregate to form an
A filament with a projection-free shaft in the middle and reversed polarity of the
molecules in each half of the sarcomere. (From Huxley, 1971.)

These reflections did not correspond to those obtained from actin (which could also be seen as a separate series), and Worthington therefore suggested that they were produced by myosin. It seems very probable that these reflections were produced by the projections on the myosin filaments. Work by Huxley and Brown (1967) suggests that the projections emerge in pairs from opposite sides of the myosin filaments every 143 Å, there being a rotation of 120° between each successive pair, so that projections oriented in the same direction occur every 429 Å (fig. 11.11).

Huxley and Brown were also able to make some deductions about the orientations of the thick filaments relative to each other. They observed that the width of the meridional reflection at 143 Å was very small,

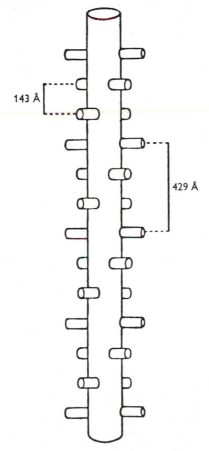

Fig. 11.11. Diagram to show the arrangement of the projections on the myosin filaments. (From Huxley and Brown, 1967.)

corresponding to a diffracting structure several thousand Å wide. This suggests that the thick filaments are arranged in lateral register with a high degree of accuracy. Off-meridional reflections on the myosin layer lines are also small in width, which implies that a definite relationship exists between the rotational orientation of neighbouring filaments. In detail, Huxley and Brown concluded that the filaments are arranged on a 'superlattice' in which each filament is surrounded by three filaments whose orientation is rotated through 120° with respect to its own, and three more whose orientation is at 240°, as is shown in fig. 11.12*a*.

This arrangement has a curious consequence. It implies that there are two populations of actin filaments: half of them are opposite single projections from adjacent myosin filaments every 143 Å, whereas the other half are opposite three projections at once, every 429 Å. This odd situation led Squire (1974) to propose an alternative model of the myosin filament, in which three projections emerge from it every 143 Å, instead of two. As fig. 11.12*b* shows, this arrangement can allow all the actin filaments to receive a cross-bridge every 143 Å.

In principle it is possible to distinguish between the different models of the myosin filament by measuring the number of molecules in each filament. This demands very careful extraction techniques so that all the

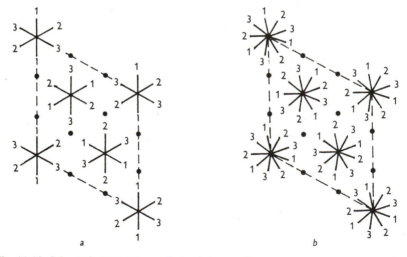

Fig. 11.12. Schematic transverse sections of the myofilament array to show two possible arrangements of the myosin filaments in a superlattice. Actin filaments are represented by black circles. The radii of the 'stars' show projections from the myosin filaments, numbered according to their axial position. Thus all projections at one level are labelled 1, all at the next (143 Å away in a direction perpendicular to the page) are labelled 2, and so on. *a*, Huxley and Brown's suggestion, based on the two-stranded myosin filament model of fig. 11.11; *b*, Squire's suggestion for a three-stranded myosin filament. (From Squire, 1974.)

myosin, and nothing but the myosin, is measured. In a recent analysis, Morimoto and Harrington (1974) conclude that there are four myosin molecules in every 143 Å of myosin filament, whereas Squire's model predicts three and Huxley and Brown's model predicts two. One possibility, compatible with Huxley and Brown's model, is that two myosin molecules are involved in each cross-bridge, but it is difficult to see at present what the point of such an arrangement could be.

The thin filaments of all muscles consist principally of the fibrous form of actin, F-actin. X-ray diffraction patterns of actin were first obtained by Selby and Bear (1956), using dried muscles of the clam *Venus*. They concluded that the actin monomers were arranged either in a net-like structure or in helices (topologically a helix is similar to a cylindrical net), the pitch of the helix model being 350 or 406 Å, corresponding to 13 or 15 G-actin monomers respectively. Using a variety of techniques Hanson and Lowy (1963, 1964) reached the conclusion that F-actin consists of two chains of monomers connected together in a double helical form, as is shown in fig. 11.13. At the time there was some uncertainty about the pitch of the helix and the number of monomers per turn. Huxley and Brown (1967) later concluded that the pitch of the helix is about 2×370 Å (or, less probably, about 2×360 Å) and that

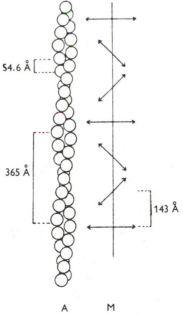

A   M

Fig. 11.13. Arrangement of the subunits in an actin filament (A). Alongside is a representation of the positions of crossbridges on a myosin filament (M). Note that the pitches and subunit repeat of the actin and myosin helices are different. (From H. E. Huxley, 1972.)

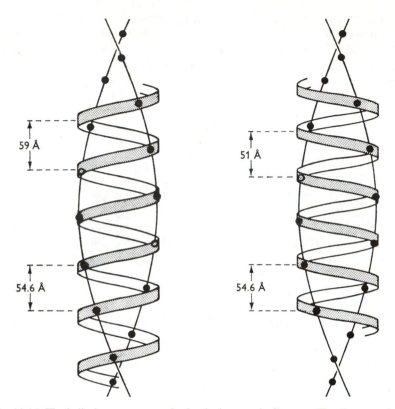

Fig. 11.14. The helical arrangement of subunits in an actin filament. The diagrams show the two primitive helices with pitches of approximately 51 and 59 Å. The double non-primitive helix with a pitch of 2×370 Å is shown in both diagrams. Lateral distances are not to scale.

there is a non-integral number of monomers per turn. Let us look briefly at the evidence for this conclusion.

The principal reflections in the X-ray diffraction pattern from actin are a meridional one at 27.3 Å, and two off-meridional ones at 59.1 and 50.97 Å (the figures are those given by Huxley and Brown). The interpretation of the meridional reflection is straightforward: it represents the axial distance between successive monomers. The '59' and '51' reflections represent the pitches of the two primitive helices present, one of them righthanded and the other lefthanded; fig. 11.14 shows what this means. We can calculate the pitch of the non-primitive helix if we know precisely what the pitches of the two primitive helices are, as follows. In one turn of the non-primitive helix there will be $(n+1)$ turns of the shorter primitive helix, i.e. the one with the same 'handedness' as the non-primitive helix, and $(n-1)$ turns of the longer. Hence, using

Huxley and Brown's figures, the pitch of the non-primitive helix is given by

$$L = 50.97(n + 1)$$

and

$$L = 59.1(n - 1).$$

Therefore

$$n = \frac{50.97 + 59.1}{59.1 - 50.97}$$

$$= 13.54.$$

Therefore

$$L = 50.97(13.54 + 1)$$

$$= 740.$$

So that the length of the non-primitive helix is 2×370 Å. The number of subunits per turn is then (740/54.6), i.e. 13.54, and so the helix is non-integral. An alternative but less probable value for the pitch of the longer primitive helix is 59.4 Å; in this case the pitch of the non-primitive helix would be 2×360 Å and the number of subunits per turn would be 13.2.

In addition to those reflections which can be ascribed to the arrangement of the actin and myosin molecules in the thin and thick filaments, there are other reflections which must arise from other periodicities associated with the filaments. In this category, Huxley and Brown observed particularly strong reflections on the meridian at 385 and 442 Å; they suggest that the meridional reflection observed at about 410 Å by Worthington (1959) was an unresolved combination of these two. Electron micrographs show periodicities in the form of regular fine striations about 400 Å apart in both I and A bands, that in the I band being more prominent. Since the A band periodicity is rather longer than that in the I band, Huxley and Brown conclude that the 442 Å reflection is associated with the A filaments and the 385 Å reflection with the I filaments. If a myofibril is stained with the specific antibody anti-troponin, the periodicity in the I band becomes greatly enhanced (Ohtsuki *et al.*, 1967), hence it seems very likely that the 385 Å reflection represents the distance between points at which troponin is attached via tropomyosin molecules to the actin filament. Also the ratio of actin to tropomyosin to troponin molecules is $7 : 1 : 1$. If each tropomyosin molecule contacts seven actin monomers, then the distance between successive tropomyosin molecules will be $(7 \times 55) = 385$ Å. It has therefore been suggested (e.g. by Ebashi, Endo and Ohtsuki, 1969) that the tropomyosin molecules lie end-to-end along the grooves between the two strings of actin monomers, each tropomyosin with a troponin molecule attached to it, as is shown in fig. 11.15. We shall

Fig. 11.15. A model for the structure of the thin filament, showing the probable position of the tropomyosin and troponin components. (From Ebashi *et al.*, 1969.)

return to this model later in considering how the contractile mechanism is activated (p. 263).

Functionally, both thick and thin filaments are polarized. This is very evident in the thick filament, where all the myosin molecules in one half of the filament are arranged with their heads towards that end of the filament, and all those in the other half are arranged in the other direction (fig. 11.10). The polarity of the actin filaments can be demonstrated by allowing them to be 'decorated' with the heads of myosin molecules (H. E. Huxley, 1963). Negatively stained electron micrographs of such filaments show 'arrowheads' which always point away from the *Z* line.

A more detailed study of the structure of the thin filaments has been carried out by H. E. Huxley and his colleagues using a three-dimensional reconstruction technique developed by DeRosier and Klug (1968) and used by them to determine the structure of the tail of the T4 bacteriophage. The principle of the method is essentially similar to that whereby it is possible to determine the structure of molecules by X-ray analysis of their crystals. Electron micrographs of negatively stained material are transmission images in which the total density through the three-dimensional object is projected onto a two-dimensional plane. From a number of such images obtained with different orientations of the object (or, in the case of a helical object, from images at different levels in the helix) it is possible to reconstruct the appearance of the original three-dimensional object. In order to do this the electron micrograph is first scanned by a microdensitometer which feeds its measurements into a computer. The computer then carries out a Fourier transformation on these measurements (so producing the information that would have been obtained from an optical diffraction pattern of the electron micrograph) and then combines transformations from different orientations to produce maps of the three-dimensional distribution of density.

Moore, Huxley and DeRosier (1970) applied this method to aggregations ('paracrystals') of F-actin, and produced confirmation for the 'double string of beads' model illustrated in fig. 11.13. In their models of

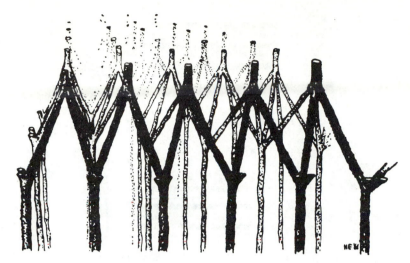

Fig. 11.16. The interconnections of the *I* filaments at the *Z* line. (From Knappeis and Carlsen, 1962.)

whole thin filaments the two strings of actin monomers were somewhat further away from each other and there was some additional material in the grooves between them; clearly this must be tropomyosin, providing confirmation of the model shown in fig. 11.15. Finally, the structure of thin filaments decorated with myosin subfragment 1 was determined. Here each actin monomer was attached to an S_1 subfragment lying at an angle to the filament axis.

The *Z* line of vertebrate muscle has the appearance of a square lattice, when viewed in cross-section. Serial sections show that this is produced by the ends of a series of tetragons formed by connecting filaments ('*Z* filaments') between the *I* filaments on each side of the *Z* line. Each *I* filament is connected by four '*Z* filaments' to four *I* filaments on the other side of the *Z* line, as is shown in fig. 11.16 (Knappeis and Carlsen, 1962; Reedy, 1964).

THE SLIDING FILAMENT THEORY

Prior to 1954, most suggestions as to the mechanism of muscular contraction involved the coiling and contraction of long protein molecules, rather like the shortening of a helical spring. In that year, the *sliding filament theory* was independently formulated by H. E. Huxley and J. Hanson (from phase-contrast observations on glycerinated myofibrils) and by A. F. Huxley and R. Niedergerke (using interference microscopy of living muscle fibres). In each case the authors showed that

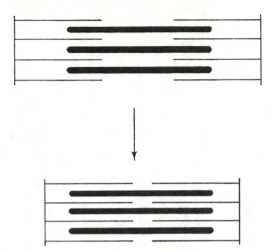

Fig. 11.17. The structural changes in a sarcomere on shortening, according to the sliding filament theory.

the *A* band does not change in length either when the muscle is stretched or when it shortens actively or passively. This observation, interpreted in terms of the interdigitating filament structure described in the previous section, suggests that contraction is brought about by movement of the *I* filaments between the *A* filaments, as is shown in fig. 11.17. The sliding is thought to be caused by a series of cyclic reactions between the projections on the myosin filaments and active sites on the actin filaments; each projection is first attached to the actin filament to form a cross-bridge, then it moves or contracts and finally releases it, moving back to attach to another site further along the actin filament.

Much further evidence for this theory has accumulated since it was first formulated. We shall consider here three lines of this evidence: the measurement of *A* and *I* filament lengths, the relation between sarcomere length and isometric tension, and X-ray diffraction measurements of the repeat distances in resting and contracting muscles at different lengths.

The lengths of the A and I filaments

As we have seen earlier, the preparatory procedures used in electron microscopy frequently result in shrinkage of the tissue involved. Hence it is no easy matter to determine the lengths of the *A* and *I* filaments in living or glycerinated muscles. However, Page and Huxley (1963) investigated this problem in a piece of work that is a most beautiful example of mensurative electron microscopy. They began by showing that the sarcomere length of unrestrained muscles (measured by an

optical diffraction technique) decreased during the fixation and dehydration processes, but was unaffected by the embedding process. These
effects could be eliminated if the muscle was fixed at each end, so this
procedure was followed in the rest of their experiments.

By homogenizing glycerinated fibres in a suitable medium, it was
possible to obtain '*I* segments' consisting of a *Z* line and a set of *I*
filaments. These were examined by negative staining and shadow-
casting; both of these techniques can be used without fixation or
dehydration in alcohol, and so shrinkage from these causes could be
eliminated. Shadowed segments and those stained with sodium
phosphotungstate were 2.05 μm long; those stained with uranyl acetate
were slightly shorter. These results on their own can only give a rather
tentative estimate of the *I* filament length; but the value so obtained
must be 2.05 μm.*

Particular care was used in the electron microscopy of longitudinal
sections of muscles: the blocks were cut at right angles to the longitudinal axis so as not to shorten the filaments by the pressure of the
microtome knife, the effectiveness of this procedure was checked by
measuring the width of the sections and the width of the block, the
magnification of the electron microscope was calibrated at frequent
intervals, and allowances were made for the shrinkage of the photographic prints.

Muscles were fixed in osmium tetroxide when contracting isometrically under the influence of electrical stimulation or solutions
containing a high potassium ion concentration. The filament lengths
were 2.01 to 2.05 μm for the *I* filaments (except at a sarcomere length
of 3.7 μm, when the average *I* filament length was 1.98 μm), and 1.56
to 1.61 μm for the *A* filaments (fig. 11.18). These results indicate that
very little shrinkage of the filaments occurs during osmium fixation of
active muscles, in spite of the fact that appreciable shrinkage occurs
during osmium fixation of resting muscles. Page and Huxley suggested
that the reason for this is that the *A* and *I* filaments are cross-linked
during activity, whereas in the resting condition they are able to shrink
by sliding past each other. At a sarcomere length of 3.7 μm, the
filaments will not be overlapping, therefore no cross-bridges can be
formed, which accounts for the shrinkage of the *I* filaments at this
length.

The conclusion from this set of experiments, illustrated diagrammatically in fig. 11.18, is that the lengths of the filaments do not change
in the resting or active muscles, whatever the length of the sarcomere. In

* The term '*I* filament length' is here used to indicate the total length of *I* substance on
each side of a *Z* line, including the thickness of the *Z* line, not the length of *I* substance on
only one side of a *Z* line.

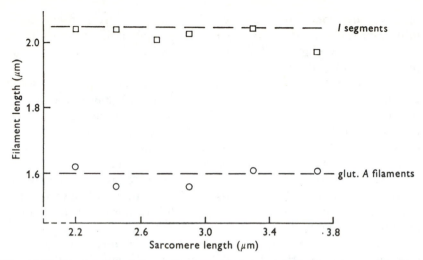

Fig. 11.18. The lengths of the *A* filaments (circles) and *I* filaments (squares) in frog muscles fixed in osmium tetroxide during isometric tetani at various sarcomere lengths. The broken lines show the lengths of isolated *I* segments and the lengths of the *A* filaments in glutaraldehyde-fixed muscles. (From Page, 1964, by permission of the Royal Society.)

frog striated muscles the *I* filaments are 2.05 μm long and the *A* filaments are 1.6 μm long.

The relation between sarcomere length and isometric tension

If the sliding filament theory is correct, and contraction depends on the interaction of the actin and myosin filaments at the cross-bridges, then the isometric tension should be proportional to the degree of overlap of the filaments. In terms of fig. 11.3, the decline in the active increment curve at longer lengths should be due to the reduced degree of overlap (and therefore smaller number of cross-bridges) between the two types of filaments as the muscle is extended (A. F. Huxley and Niedergerke, 1954). The way to test this suggestion is to measure the active increment length–tension curve, with the abscissa as sarcomere length rather than the length of the whole muscle, and compare it with the lengths of the *A* and *I* filaments. However, this turns out to be not so easy as one might think. A. F. Huxley and Peachey (1961) found that the sarcomere length in single frog muscle fibres was not constant along the length of the fibre, being shorter at the ends. At long lengths there may be overlap of the filaments in the sarcomeres at the ends of the fibre, but not in those in the middle, so that the ends contract at the expense of the middle section if the fibre is fixed at each end. In order to overcome this difficulty, it is necessary to hold a small portion of the fibre (in which the

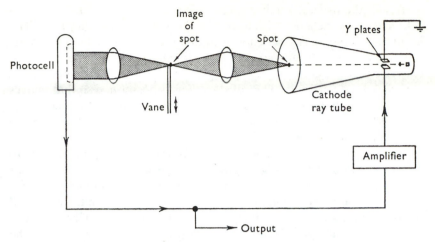

Fig. 11.19. A simple 'spot follower' circuit. Explanation in text.

sarcomere length *is* constant) at a constant length while its tension is measured. This has been done by A. F. Huxley and his colleagues (Gordon, Huxley and Julian, 1966*a, b*), whose work we shall now consider.

Before examining the apparatus used in these experiments, let us consider a useful electronic feedback device known as a 'spot follower' (fig. 11.19). This consists of a cathode ray tube, a photocell, two lenses, and a movable vane placed between the lenses. The object of the system is to ensure that the movement of the spot follows exactly the vertical movement of the edge of the vane, however irregular this may be. This is achieved by feeding the output of the photocell into an amplifier connected to the *Y* plates of the cathode ray tube. The system is arranged so that if the photocell can 'see' more than half the spot, the spot is raised, and if it can see less than half the spot, it is lowered (remember that the spot on the screen moves in the opposite direction to its image on the vane). Thus the movement of the spot follows (inversely) the movement of the edge of the vane. It follows that the output of the photocell, since it is this that directly determines the position of the spot, must be proportional to the position of the edge of the vane.

Now let us examine some of the details of the apparatus used by Gordon, Huxley and Julian (fig. 11.20). A single muscle fibre is used, mounted on a microscope stage, and stimulated electrically. It is connected via its tendons to a tension transducer valve at one end and to the arm of a galvanometer at the other. The galvanometer arm moves according to the force exerted on it by the muscle fibre and the current in its coil (the usual zeroing spring is not present). Two small pieces of gold leaf are attached to the fibre with grease, to act as markers. The position of these markers is chosen so that the sarcomere length (observed through the microscope) does not vary along the length of the fibre between them. Below the microscope is a double beam cathode ray tube, mounted vertically. The substage lens is positioned so that images of the two spots on the cathode ray tube screen can be focused on an edge of each marker. Light from these images is then collected separately onto two photocells.

The output of the right hand photocell is fed to the amplifiers controlling the position of both spots. Thus if both markers move to the right or left by the same amount, the follower system works so that both spots move correspondingly, and there is therefore no change in the output of the left hand photocell. But if the distance between the markers alters, the left hand photocell (and its spot-follower loop) is activated, and the right hand spot (which is focused on the left hand marker) moves independently of the left hand spot. The output of the left hand photocell then indicates the distance between the markers, shown as the box marked 'length' in fig. 11.20.

Besides the spot-follower feedback loops, there is another feedback loop, indicated as the 'length-regulator loop' in the diagram. The output of the left photocell ('length') is fed to the main amplifier, which drives the galvanometer. This circuit is arranged so that the galvanometer moves so as to bring the output of the photocell back to its mid-position, i.e. so as to keep the 'length' (the distance between the markers on the muscle fibre) constant. This, of course, is just what the apparatus is required to do.* The length can be set at different values by altering the 'length input signal', since the input to the main amplifier is the difference between this and the 'length' signal.

A single experiment with this apparatus consists of a series of isometric tetani obtained at different sarcomere lengths; for technical reasons it is not possible to measure isometric tension over the whole range of sarcomere lengths in any one experiment. The results of these experiments are summarized in fig. 11.21. It is evident that the length–tension diagram consists of a series of straight lines connected by short curved regions. There is a 'plateau' of constant tension at sarcomere lengths between 2.05 and 2.2 μm. Above this range, tension falls linearly with increasing length; the projected straight line through most

---

* This account is rather simplified as also is fig. 11.20. The apparatus also has facilities for (i) a tension-regulator loop so that the tension can be held constant while the length changes, (ii) switching between length regulation and tension regulation according to the magnitude of either of these parameters, and (iii) forcing either length or tension to follow an input command signal.

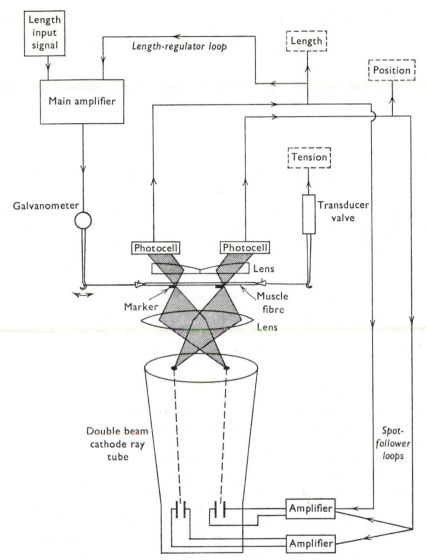

Fig. 11.20. Apparatus used for experiments on the mechanics of portions of a muscle fibre. Simplified; explanation in text. (Based on Gordon, Huxley and Julian, 1966*a*.)

of the points in this region passes through zero at 3.65 μm, but there is in fact a very slight development of tension at this point. Below the plateau, tension falls gradually with decreasing length down to about 1.65 μm, then much more steeply, reaching zero at about 1.3 μm.

According to the sliding filament theory, the isometric tension is directly proportional to the number of cross-bridges that can be formed

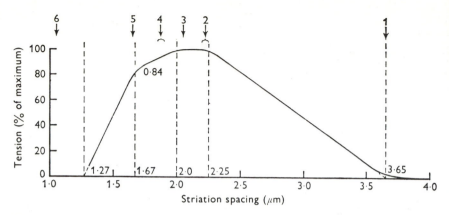

Fig. 11.21. The isometric tension (active increment) of a frog muscle fibre, measured as a percentage of its maximum value, at different sarcomere lengths. The numbers 1 to 6 refer to the myofilament positions shown in fig. 11.22*b*. (From Gordon, Huxley and Julian, 1966*b*.)

between the *A* and *I* filaments, less any internal resistance tending to extend the sarcomeres. Thus, at long lengths, the tension ought to be proportional to the degree of overlap of the *A* and *I* filaments. In order to see whether the length–tension diagram fits with this prediction we need to know the dimensions of the filaments and the position of the cross-bridge-forming projections on the myosin filaments. From the measurements of Page and Huxley (1963), we know that the *A* filaments are 1.6 μm long (symbol *a* in fig. 11.22) and the *I* filaments, including the *Z* line, are 2.05 μm long (*b*). The middle region of the *A* filaments, including the *Z* line, are 2.05 μm long (*b*). The middle region of the *A* filaments, which is bare of projections and therefore cannot form cross-bridges, (*c*) is 0.15 to 0.2 μm long and the thickness of the *Z* line (*z*) is about 0.05 μm.

Now let us see if the length–tension diagram shown in fig. 11.21 can be related to these dimensions, starting at long sarcomere lengths and working through to short ones. Fig. 11.22*b* shows the points at which qualitative changes in the relations between the elements of the sarcomere occur. These stages are numbered 1 to 6, and the corresponding lengths are shown by the arrows in fig. 11.21. Above 3.65 μm (1) there should be no cross-bridges, and therefore no tension development. In fact there is some tension development up to about 3.8 μm; the reason for this is not clear. Between 3.65 μm and 2.25 to 2.2 μm (1 to 2) the number of cross-bridges increases linearly with decrease in length, and therefore the isometric tension should show a similar increase. It does. With further shortening (2 to 3) the number of cross-bridges remains constant and therefore there should be a plateau

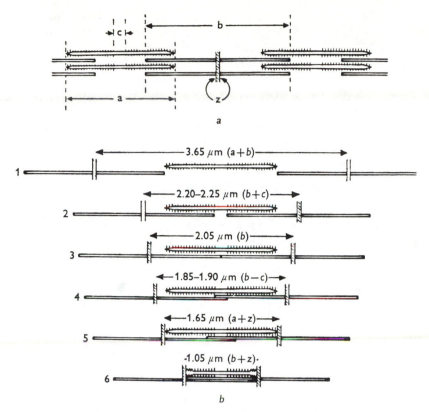

Fig. 11.22. *a*: Myofilament dimensions in frog muscle. *b*: Myofilament arrangements at different lengths; the letters *a*, *b*, *c* and *z* refer to the dimensions given in part *a*. The sarcomere lengths corresponding to the positions labelled 1 to 6 are indicated by the arrows in fig. 11.21. (From Gordon, Huxley and Julian, 1966*b*.)

of constant tension in this region; there is. After stage 3, we might expect there to be some increase in the internal resistance to shortening, as the *I* filaments must now overlap; after stage 4 the *I* filaments from one half of the sarcomere might interfere with the cross-bridge formation between the *A* and *I* filaments in the other half of the sarcomere (it is reasonable to assume that the cross-bridges on one half of an *A* filament can only pull in one direction, an idea that is reinforced by H. E. Huxley's discovery that the two halves of an *A* filament are polarized in opposite directions). These effects would be expected to reduce the isometric tension. This does in fact occur, although no extra reduction corresponding to stage 4 is detectable. At 1.65 μm (5), the *A* filaments hit the *Z* lines, and therefore there should be a considerable increase in the resistance to shortening; it is found that there is a distinct kink in the length–tension curve at almost exactly this point, after which the tension falls much more sharply. The curve reaches zero tension at about 1.3 μm, before stage 6 (when the I filaments would have hit the Z line) is reached.

It would be difficult to find a more precise test of the sliding filament theory than is given by this experiment, and the theory obviously passes the test with flying colours. In addition, the results tell us something about the mode of action of the cross-bridges. The filament array in the myofibrils is a constant-volume system, so the transverse distance between the actin and myosin filaments must decrease as the muscle is stretched. However, the fact that the decrease in isometric tension with increasing sarcomere length above 2.2 μm is linear means that the tension per cross-bridge is independent of length. Therefore the tension produced per cross-bridge is independent of the transverse distance between the filaments. We shall return to this point later (p. 249).

X-ray diffraction measurements

Wide angle X-ray diffraction measurements on muscle show the alpha helix pattern, irrespective of the length of the muscle (see H. E. Huxley, 1960). This alpha helix pattern appears to be mainly derived from myosin, and its constancy at different lengths implies that there is little change in the lengths of the myosin molecules during stretching or contraction. This observation is in accordance with the sliding filament theory, although it does not provide any strong evidence in its favour.

Much more conclusive evidence is provided by low-angle X-ray diffraction measurements of meridional reflections, which indicate the distances between repeating units along the axis of the myofibrillar filaments. H. E. Huxley (1953) showed that such measurements on resting muscles were independent of the length of the muscle. This is just what one would expect from the sliding filament theory; if, on the other hand, the filaments did shorten during shortening of the muscle, we would expect the distance between their repeating units to decrease.

It is obviously desirable that this conclusion should be confirmed for actively contracting muscles. This has been done by two groups of workers, in London (Elliott, Lowy and Millman, 1965) and in Cambridge (H. E. Huxley, Brown and Holmes, 1965). The chief technical difficulty of this type of experiment is that the exposure times for X-ray diffraction experiments may have to be. a matter of hours, whereas it is impossible to excite a vertebrate striated muscle for more than a few seconds at a time without fatiguing it. The problem was overcome by stimulating the muscle at intervals and, using refined X-ray diffraction methods, only passing the X-ray beam through the muscle when the isometric tension was above a certain level. In each case it was found that the axial spacings during isometric contractions were not significantly different from those in the resting muscle and were unaffected by the length of the muscle.

More precise measurements by Huxley and Brown (1967) have shown that there is in fact a slight *increase* in the axial spacings derived from the myosin filaments during an isometric contraction. Their measurements give the value of the principal myosin subunit spacing as 143 ± 0.1 Å in the resting muscle and 144.6 ± 0.1 Å during contraction, an increase of just over 1 per cent. There was no significant change in the actin subunit spacing during contraction.

The sliding filament theory postulates that the contractile force is mediated via the cross-bridges originating on the myosin filaments. This raises the possibility that the cross-bridges could move during the contraction, and there is now some evidence that this is so. Reedy,

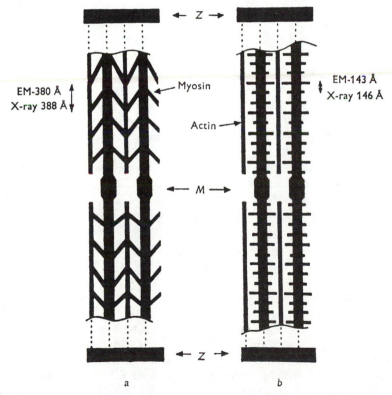

Fig. 11.23. Diagram, derived from X-ray diffraction and electron microscopy measurements, to show the positions of the cross-bridges of glycerinated insect flight muscle in the relaxed state (*b*) and in rigor (*a*). (From Reedy, Holmes and Tregear, 1965.)

Holmes and Tregear (1965) found that glycerol-extracted insect flight muscles produced different X-ray diffraction and electron micrograph patterns according to whether they were in rigor (p. 197) or were relaxed. They concluded that the projections from the myosin filaments in resting muscle stick out at right angles from the myosin filaments, but that in rigor they become attached to the actin filaments and move through an angle of about 45 degrees so as to pull the actin filaments towards the centre of the sarcomere (fig. 11.23).

Changes in the orientation of the cross-bridges in living muscles during isometric contractions are apparent from the X-ray diffraction measurements of Huxley and Brown. This conclusion is based on the observation that, during contraction, the intensity of the off-meridional reflection at 429 Å is only about 30 per cent of that in resting muscle (fig. 11.24), whereas the intensity of the meridional reflection at 143 Å is about 66 per cent of that in resting muscle. This implies that the helically repeating structures on the myosin filament becomes much less regularly arranged than do the axially repeating structures, and therefore that the outer portions of the myosin reflections move (in an asynchronous fashion) with respect to their bases.

Further evidence for cross-bridge movement during contraction comes from X-ray diffraction measurements on insect fibrillar muscle (Tregear and Miller, 1969). The physiological properties of these muscles are dealt with in chapter 13; suffice it to say here that their glycerinated fibres will perform oscillatory work when subjected to sinusoidal length changes at the right frequencies in the presence of ATP and calcium ions. Tregear and Miller measured the fluctuations in the intensity of the 145 Å meridional reflection during such oscillations at different frequencies by means of a rather ingenious technique. A metal plate was placed in the X-ray beam, with a rectangular aperture positioned so as to let through only the diffracted rays comprising the 145 Å meridional reflection. Behind the aperture was a proportional counter to measure the X-rays passing through it, and the output of this counter was recorded on an eight-channel scaler which scanned synchronously with the length oscillation. In this way the results shown in fig. 11.25 were obtained.

At low frequencies, the diffraction intensity of the spot varied in time with the variations in tension, being low when tension was high and high when it was low, as is shown in fig. 11.25. A diminution in the intensity of a meridional reflection must be produced by a decrease in the degree of axial order of structures occurring every 145 Å in the muscle. One way in which this might occur is by axial displacements of the mass of each cross-bridge, such as would be produced by bending. Tregear and Miller calculated that the 30 per cent changes in intensity which they

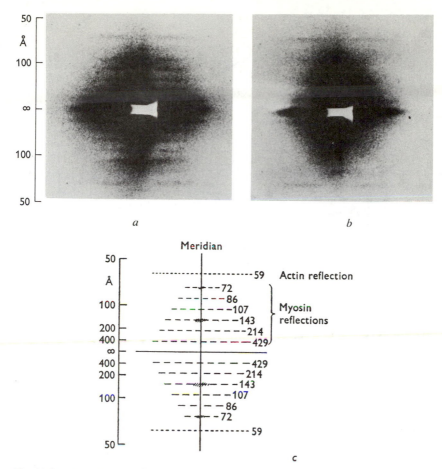

Fig. 11.24. X-ray diffraction patterns of living frog sartorius muscle at rest (*a*) and during isometric contractions (*b*). *c* is a schematic diagram showing the reflections apparent in *a*. Notice the disappearance of the system of layer lines based on 429 Å during contraction. The 72 Å and 143 Å meridional reflections remain (the latter is not evident in this reproduction), as does the 59 Å actin reflection. (*a* and *b* from Huxley, 1972.)

observed would be produced by an axial movement of the ends of all of the cross-bridges by 20 Å, or of 30 per cent of them by 100 Å.

Of course we do not know precisely how the cross-bridges move, or how their movement is brought about. However the X-ray evidence can be combined with the results of other studies to give a rather convincing picture of what happens. H. E. Huxley (1969) suggests that the myosin molecule possesses flexible linkages at two points in its structure: at the link between light meromyosin and heavy meromyosin, and within heavy meromyosin at the link between the rod-like portion and the

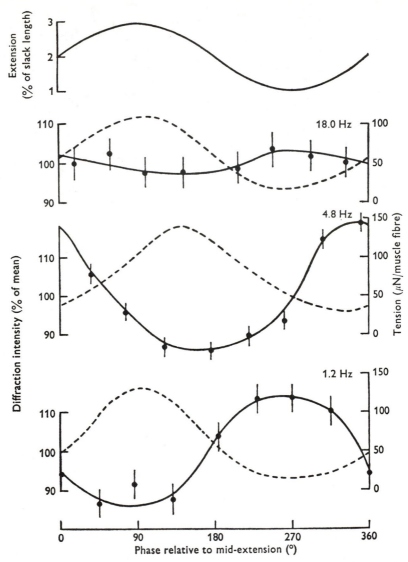

Fig. 11.25. X-ray diffraction measurements on a bundle of glycerol-extracted flight muscle fibres from a giant water bug, *Lethocerus cordofanus*, showing changes in intensity (black circles and full lines) of the 145 Å meridional reflection during sinusoidal oscillations at three different frequences in the presence of ATP and calcium ions. The dashed lines show tension changes. (From Tregear and Miller, 1969.)

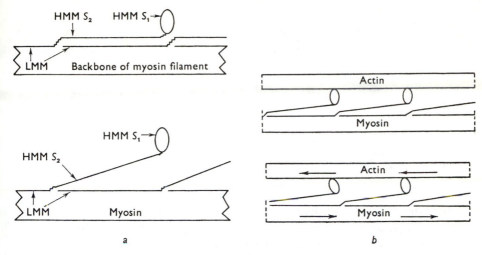

Fig. 11.26. A possible structural basis for cross-bridge action. *a* shows how the cross-bridge could extend different distances from the 'backbone' of the myosin filament. Each myosin molecule is composed of light meromyosin (LMM) and the globular (S_1) and rod-like (S_2) portions of heavy meromyosin (HMM). *b* shows how a rotational movement at the link between HMM S_1 and actin could produce sliding of the filaments past each other. (From Huxley, 1969.)

globular portion. The primary source of movement is a rotation about the link between the globular portion of heavy meromyosin and actin (fig. 11.26). This model explains the appearance of insect flight muscles in rigor (fig. 11.23) and the fact that tension per cross-bridge is independent of the distance between the actin and myosin filaments (p. 244).

BIOCHEMICAL EVENTS DURING THE CROSS-BRIDGE CYCLE

The cycle of cross-bridge activity – attachment, movement, detachment and resetting ready for reattachment – must somehow be related to the splitting of ATP, which provides the energy for the whole process. We may ask how the reactions between the contractile proteins and ATP in test-tubes relate to the situation in the intact myofibril, and, particularly, at what stage in the cross-bridge cycle is the ATP split?

Some progress in answering these questions has been made in recent years (see, for example, Taylor, 1973, and White and Thorson, 1975). Lymn and Taylor (1971) used a rapid mixing technique to investigate the reaction kinetics of ATP hydrolysis by a mixture of actin and heavy meromyosin (HMM). They found that the rate of dissociation of acto-HMM by ATP was much greater than the rate of hydrolysis of ATP. This suggests that dissociation precedes splitting. In accordance with

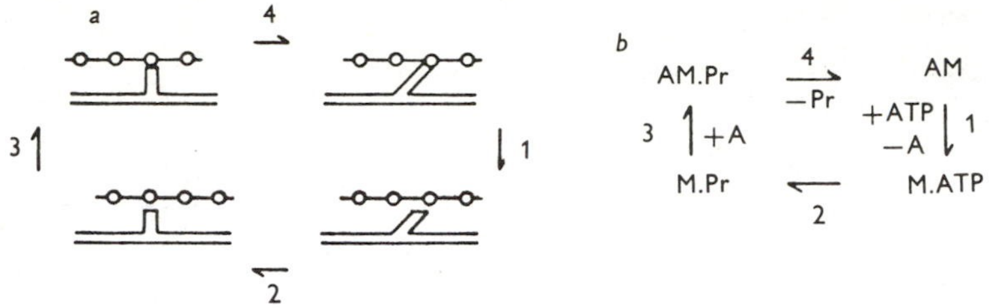

Fig. 11.27. How the cross-bridge cycle (*a*) may be related to the chemical events of actin–myosin interactions and ATP hydrolysis (*b*). A, actin; M, myosin; Pr, reaction products (ADP plus phosphate). (From Lymn and Taylor, 1971.)

this, the initial rate of hydrolysis of ATP was much the same for both HMM and acto-HMM, although the steady-state rate for HMM was much reduced. This suggests that the first molecule of ATP which an HMM molecule meets is readily split, whereas later ones are not, unless actin is present. This situation could arise if actin promotes the dissociation of the products of splitting from myosin.

Experiments and conclusions of this sort led Lymn and Taylor to propose a cycle of biochemical reactions paralleling the cycle of cross-bridge action, as is shown in fig. 11.27. We begin in fig. 11.27 with the cross-bridge at the end of its power stroke, still attached to the actin filament (compare fig. 11.23*a*). Reaction (1) in the cycle is the combination with ATP, which promotes dissociation of the myosin cross-bridge–actin link. Reaction (2) is the splitting of ATP, which may well be associated with a repositioning of the myosin cross-bridge ready for reattachment; the products of hydrolysis (ADP and phosphate) remain attached to the myosin at this stage. Reaction (3) is the attachment of the myosin cross-bridge to a new actin monomer. This promotes reaction (4), which is the release of the products from the actomyosin complex; this stage appears to be the logical one to associate with the power-stroke of the cross-bridge.

THE EXCITATION-CONTRACTION COUPLING PROCESS

The normal stimulus for the contraction of a muscle fibre in a living animal is an impulse in the motor nerve by which it is innervated. The sequence of events following the motor nerve impulse is shown schematically in fig. 11.28. We have examined stages 1 to 4 of this sequence (the excitation processes) in previous chapters, and this chapter has so far been concerned with some of the features of stage 6 (the contraction process). We must now consider how excitation of the muscle fibre membrane initiates contraction of the myofibrils in the interior of the

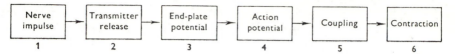

Fig. 11.28. The control sequence leading to contraction in a vertebrate 'twitch' muscle fibre.

fibre; this constitutes stage 5 in fig. 11.28, the excitation–contraction coupling process.

The importance of depolarization of the cell membrane

When muscle fibres are immersed in solutions containing a high concentration of potassium ions, they undergo a relatively prolonged contraction known as a *contracture*. Contractures can also be produced by various drugs, such as acetylcholine, veratridine and others. In 1946, Kuffler showed that many of these substances produce depolarization of the cell membrane; furthermore, if the substance was applied locally the resulting contracture was limited to that part of the muscle fibre where depolarization occurred. The relation between membrane potential and tension in potassium contracture was determined quantitatively by Hodgkin and Horowicz (1960). They found that no contracture occurred when the membrane potential was more negative than about -55 to -50 mV; above this threshold value, tension increased rapidly with increasing depolarization, reaching a maximum above about -40 mV (fig. 11.29).

These results show quite clearly that contracture tension is related to the degree of depolarization of the cell membrane. Can we apply this conclusion to the twitch contraction which follows a muscle fibre action potential? As we have seen in chapter 4, the electrical events constituting a propagated action potential include, in addition to the change in membrane potential, longitudinal currents inside the fibre and transverse currents across its membrane. From time to time it has been suggested that these currents are the operative features in the coupling process. This idea seems to have been conclusively disproved by Sten-Knudsen (1960), who showed that longitudinal currents are only effective in so far as they cause depolarization of the cell membrane.

We must digress a little here to examine a rather more subtle influence of membrane potential on the coupling process, which is seen in certain types of muscle. When a frog rectus abdominis muscle (a 'tonic' muscle) is depolarized by placing it in a solution containing a high potassium ion concentration, the resulting contracture lasts for several minutes. On the other hand, a frog sartorius muscle (a 'twitch' muscle) subjected to the same treatment contracts to maintain a steady tension

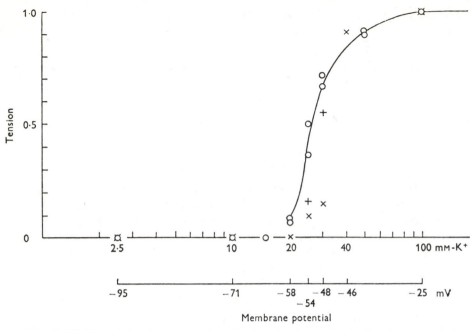

Fig. 11.29. The relation between peak contracture tension and potassium ion concentration or membrane potential in single frog muscle fibres. (From Hodgkin and Horowicz, 1960.)

for a few seconds and then relaxes; this relaxation is not accompanied by any change in membrane potential. A second contracture can only be obtained if the potassium ion concentration is sufficiently reduced for a short time. There is thus some kind of restitution or 'priming' process occurring. Hodgkin and Horowicz showed that the extent of this restitution is inversely related to the potassium ion concentration (and therefore to the membrane potential) in the intervening period between two exposures to a high potassium ion concentration (fig. 11.30). This 'priming' process does not occur in those muscles which show maintained potassium contractures.

The importance of calcium ions

It is clear that depolarization of the cell membrane cannot itself be the *ultimate* trigger for the contraction process. Apart from the difficulty of seeing how this could work, we know that glycerinated fibres will contract even though there is no membrane present. Now glycerol-extracted muscle fibres bathed in a solution containing ATP and magnesium ions are extremely sensitive to the calcium ion concentration; concentrations as low as 10^{-6} M are sufficient to cause some

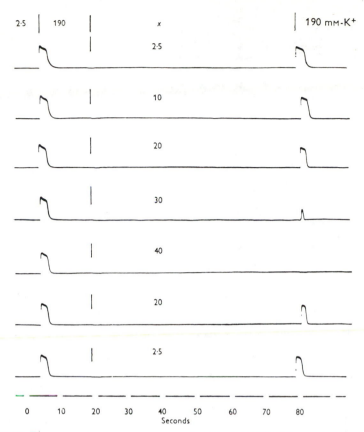

Fig. 11.30. Experiment to show the 'priming' process in a single frog muscle fibre. After a contracture induced by 190 mM K$^+$, the fibre was allowed to recover for 1 minute in x mM K$^+$ before retesting with 190 mM K$^+$. (From Hodgkin and Horowicz, 1960.)

contraction and ATP splitting. Thus we might expect that depolarization of the cell membrane of an intact muscle fibre would cause an increase in the internal calcium ion concentration, so leading to contraction. There is considerable evidence, which we must now consider, that this is so.

A number of investigations have been made on the effect of calcium ions on the potassium contractures of various muscles. Contractures eventually fail in the absence of calcium, and low calcium ion concentrations reduce the contracture tension (Niedergerke, 1956; Frank, 1960; Edman and Schild, 1962). An example of this effect in an insect muscle is shown in fig. 11.31; after treatment with a calcium-free solution containing the chelating agent ethylenediamine tetraacetate, the muscle did not contract on immersion in potassium chloride solution,

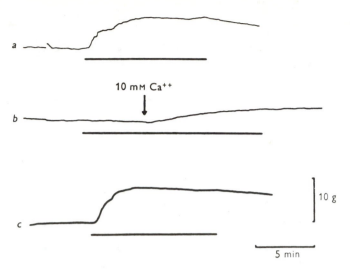

Fig. 11.31. The effect of calcium ion concentration on potassium contracture in a locust leg muscle. The thick horizontal lines under each record show the times of exposure to the high potassium solution. Pretreatment as follows: *a*, Ringer solution (4 mM Ca^{++}); *b*, calcium-free solution containing a chelating agent; *c*, Ringer solution again. (From Aidley, 1965*a*.)

but addition of calcium ions to the depolarized muscle was immediately followed by contraction (Aidley, 1965*a*).

Caldwell and his colleagues have developed a technique whereby it is possible to inject solutions into the interior of the large muscle fibres of the spider crab *Maia*. Injection of potassium, sodium or magnesium ions or ATP did not produce any contraction, but the injection of calcium ions did (Caldwell and Walster, 1963). In further experiments (Portzehl, Caldwell and Rüegg, 1964), calcium ion buffer solutions, prepared by mixing known quantities of calcium chloride and the chelating agent EGTA,* were injected, so that the free calcium ion concentration inside the fibre could be stabilized at a predetermined value. It was found that the threshold calcium ion concentration necessary for contraction was 0.3 to 1.5×10^{-6} M. This is significantly close to the calcium ion concentration necessary to activate the ATPase system of isolated myofibrils (Weber and Herz, 1963).

A very direct demonstration that the calcium ion concentration in the sarcoplasm rises immediately after stimulation has been made by Ashley and Ridgeway (1968, 1970). Their experiments made use of the protein aequorin, isolated from a bioluminescent jellyfish, which emits light in the presence of calcium ions. Solutions of aequorin were injected into

* Ethylene glycol *bis*(β-aminoethyl ether)-N,N'-tetraacetate.

the large muscle fibres of the barnacle *Balanus nubilis*. When such a fibre was stimulated electrically it produced a faint glow of light, indicating the presence of calcium ions in its interior. Relative measurements of the internal calcium ion concentration at different times and under different conditions can be made by measuring the light emission with a photomultiplier tube; the time course of light emission after a stimulus can be called a 'calcium transient'.

Fig. 11.32 gives an example of the results to be got from this preparation, showing the effects of increasing stimulus strength (and therefore increasing depolarization) on the amount of calcium released and the subsequent tension development. Examination of this experiment reveals a number of interesting features. The size of the calcium transient increases with the degree of depolarization, beyond a certain threshold value. The degree of tension development rises with increasing size of the calcium transient. The calcium transient starts after a short latent period and begins to fall soon after the end of the stimulus pulse. But the tension keeps rising until the calcium transient is long past its peak and almost finished; the time course of the calcium transient is

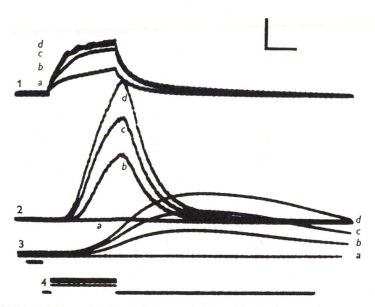

Fig. 11.32. Calcium transients in *Balanus* muscle fibres as measured by the aequorin technique described in the text. Trace 4 shows the stimulus pulse, which is applied via an intracellular silver wire electrode. Trace 1 shows the resulting depolarization, trace 2 the photomultiplier output (the calcium transient) and trace 3 the tension. *a, b, c* and *d* are the results of applying four different current pulses of increasing intensity. The calibration marks are 100 msec (horizontal), 20 mV (for trace 1) and 5 g (for trace 3). (From Ashley and Ridgeway, 1968.)

in fact very similar to the rate of change of tension during the rising phase of the tension change.

Calcium transients following stimulation were first demonstrated by means of a method which is only slightly less direct than that described above. Jöbsis and O'Connor (1966) used the calcium indicator substance murexide, whose absorption spectrum alters when it combines with calcium. Murexide was taken up by fibres of the sartorius muscle of toads after the substance had been injected into the whole animal some time previously. Then, by shining a monochromatic light beam through the muscle onto a photomultiplier tube, changes in calcium ion concentration could be detected after stimulation of the muscle fibre. The results were very similar to those obtained with the aequorin method from barnacle muscle fibres.

The T system

The experiments that have been so far described lead to the suggestion that depolarization of the cell membrane causes an increase in the calcium ion concentration in the interior of the muscle fibre. A possible hypothesis as to how this is done is that depolarization releases calcium from the cell membrane or allows calcium ions to enter the muscle fibre from outside; these calcium ions would then diffuse into the interior of the fibre and activate the contractile system. Evidence compatible with this idea has been obtained from radioactivity measurements on calcium fluxes; both the influx and efflux of calcium ions are increased on stimulation (see Bianchi, 1961). There are, however, considerable objections to the idea. First, it seems that the amount of calcium entering the fibre during a twitch is too little to account for the necessary increase in calcium ion concentration inside the fibre. Bianchi and Shanes (1959) showed that the calcium influx during a twitch of frog sartorius muscle is 0.2 pmoles/cm^2. Sandow (1965) calculates that such an influx would give an internal calcium ion concentration of 8×10^{-8} M (assuming the fibre diameter to be 50 μm), which, as he says, 'is too small by a factor of 10 to cause even threshold effects, and it would be too small by a factor of 100 to elicit maximum activation'. The second objection arises from a calculation by Hill (1948, 1949a). He showed that the time taken for a substance released from the membrane to diffuse into the interior of the fibre would be much too long to account for the speed with which activation becomes maximal after the stimulus.

How then does excitation at the cell surface cause release of calcium inside the fibre? The first step in the solution of this problem was provided by the demonstration by A. F. Huxley and Taylor (1955, 1958) that there is a specific inward-conducting mechanism located (in

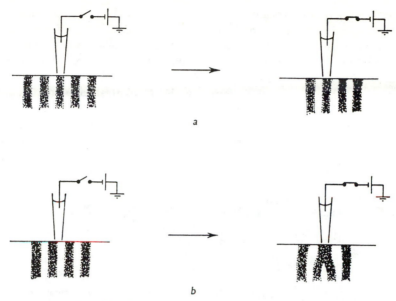

Fig. 11.33. The effect of local depolarizations on an isolated frog muscle fibre. Diagrams on the left show the resting condition, those on the right show the condition during passage of a depolarizing current. *a*, With the electrode opposite the *A* band; *b*, with the electrode opposite the *I* band (including the *Z* line). (Based on Huxley and Taylor, 1958.)

frog sartorius muscles) at the *Z* line. In these experiments, the fibres were viewed by polarized light microscopy (so as to make the striation pattern visible) and stimulated by passing current through an external microelectrode, of diameter 1 to 2 μm, applied to the fibre surface. Hyperpolarizing currents did not produce contraction. Depolarizing currents did produce contraction, but only when the electrode was positioned at certain 'active spots' located at intervals along the *Z* line (fig. 11.33). In these cases the *A* bands adjacent to the *I* bands opposite the electrode were drawn together, and the extent to which this contraction passed into the interior of the fibre was proportional to the strength of the applied current.

At first it was thought that the inward-conducting mechanism was the *Z* line itself, but on repeating the experiments with crab muscle fibres (Huxley and Straub, 1958), it was found that the 'active spots' were localized not at the *Z* line but near the boundary between the *A* and *I* bands. This suggests that there is some transverse structure located at the *Z* lines in frog muscles and at the *A*–*I* boundary in crab muscles.

Such a structure was found by Porter and Palade (1957) in their study, by electron microscopy, of the endoplasmic (or 'sarcoplasmic') reticulum in various vertebrate skeletal muscles. The sarcoplasmic reticulum consists of a network of vesicular elements surrounding the myofibrils (fig. 11.34). At the Z lines in frog muscle, and at the $A-I$ boundaries in most other striated muscles (including crab muscle), are structures known as 'triads', in which a central tubular element is situated between two vesicular elements. These central elements of the triads are in fact tubules which run transversely across the fibre and are known as the transverse tubular system or T *system*. There is no communication between the lumina of the T system and those of the

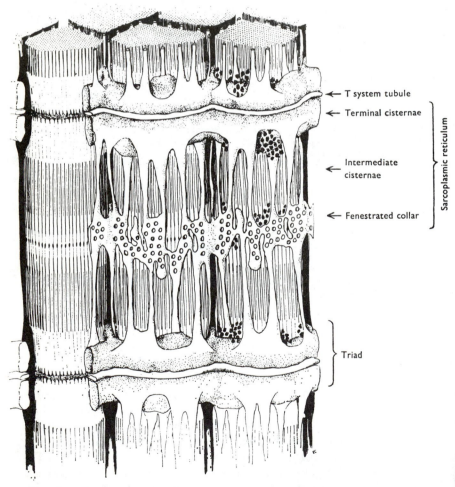

← T system tubule
← Terminal cisternae
← Intermediate cisternae
← Fenestrated collar
Sarcoplasmic reticulum
Triad

Fig. 11.34. The internal membrane systems of a frog sartorius muscle fibre. (From Peachey, 1965.)

sarcoplasmic reticulum vesicles, although their respective membranes are in close contact.

It is clear that the T system tubules are in just the right position to account for the 'inward-conducting mechanism' suggested by the experiments of Huxley and Taylor. Hence it is important to know whether or not the T system is connected to the cell membrane. In material fixed with osmium tetroxide, as in the earlier studies, the T system appears as a row of elongated vesicles, but in material fixed with glutaraldehyde it is clear that the system really is tubular, and in favourable preparations it is possible to see that the tubules are invaginations of the cell membrane (e.g. Franzini-Armstrong and Porter, 1964). Further convincing evidence that this is so was provided by H. E. Huxley (1964), from electron microscopy of muscle fibres which had been soaked in ferritin solutions. Ferritin is an iron-storage protein found in the spleen; because of its high iron content it is electron-dense, and the individual molecules can be seen by electron microscopy. In Huxley's experiments, ferritin appeared in the T system tubules, indicating that they must be in contact with the external medium. No ferritin appeared in the sarcoplasmic reticulum or in the rest of the sarcoplasm.

How does the electrical signal at the cell surface membrane travel down the T tubules into the interior of the fibre? There are two possibilities to be considered: either it is a passive process of electrotonic spread, or the T tubule membranes are electrically excitable and can conduct action potentials. The T tubules are too small for their membrane potentials to be measured directly, and hence it is necessary to use contraction of the myofibrils in different parts of the muscle fibre cross-section as an indicator of the inward spread of activity. Using voltage-clamped fibres, Costantin (1970) found that the inward spread of activity was reduced in fibres treated with tetrodotoxin or a low-sodium solution. This suggests that inward propagation is an electrically excited process with a propagated action potential involving sodium ion flow, just as at the cell surface membrane.

The speed of inward propagation has been measured in a most ingenious way by Gonzales-Serratos (1971). He placed single muscle fibres in a Ringer solution containing gelatine, which was then allowed to cool and set. He then compressed the gelatine block so as to shorten the muscle fibre without bending it. This made the myofibrils wavy. But when the myofibrils were activated after stimulation of the fibre, they shortened and so became straight. Using a high-speed movie camera, it was possible to measure the time at which straightening of the myofibrils occurred at different points in the fibre cross-section, and so to determine the speed of propagation of the activating signal. This was about 7 cm/sec at 20 °C, with a Q_{10} of 2. Thus in a fibre 100 μm in diameter the wave of activation would take about 0.7 msec to propagate from the surface to the centre.

The sarcoplasmic reticulum

If a skeletal muscle is homogenized, the myofibrils in the homogenate do not contract on the addition of ATP, and the rate of ATP splitting is very low. However, if the myofibrils are isolated from the rest of the homogenate by low speed centrifugation and subsequent washing, then they do contract in the presence of ATP and have a high ATPase activity. Marsh (1952) suggested that the ATPase activity of the myofibrils was inhibited by a 'relaxing factor' present in the muscle homogenate. Bendall (1953) showed that this factor prevented the contraction of glycerol-extracted muscle fibres. A crucial observation on the nature of relaxing factor was made by Portzehl (1957). She found that if a relaxing factor extract is centrifuged for some time at a high speed, the relaxing factor activity is confined to the resulting precipitate, which is of course particulate; the supernatant is without relaxing effect. Electron microscopy of this precipitate by Nagai, Makinose and Hasselbach (1960) and others showed that it consists of small vesicles, often ellipsoidal or tubular in shape. It would seem that these vesicles are derived from the sarcoplasmic reticulum.

Several workers have shown that the vesicular fraction is able to accumulate calcium from solutions containing ATP, magnesium ions and a small amount of calcium ions (see Hasselbach, 1964). This calcium uptake is associated with ATP splitting; Hasselbach and Makinose (1963) calculate that two calcium ions are taken up for each ATP molecule split at calcium ion concentrations above 10^{-7} M, this ratio falling to one in the range 10^{-7} to 10^{-8} M. When the calcium ion concentration is below 10^{-8} M, the rate of ATP splitting is very low, and no further accumulation of calcium occurs. These experiments show that the vesicles of the sarcoplasmic reticulum can reduce the calcium ion concentration to below that necessary for contraction, by means of an ATP-driven 'calcium pump' in the vesicular membrane. The action of the pump is perhaps aided by binding of calcium inside the sarcoplasmic reticulum by a membrane-bound protein called 'calsequestrin' (MacLennan and Wong, 1971).

This pumping of calcium ions from the myofibrils into the sarcoplasmic reticulum appears to be the means whereby relaxation is brought about. We have seen that activation of contraction is effected by a sudden increase in myofibrillar calcium ion concentration, and so it seems very likely that this calcium is released from the sarcoplasmic reticulum.

How does the T system signal to the sarcoplasmic reticulum?

The nature of the link between depolarization of the T system tubule and release of calcium ions from the sarcoplasmic reticulum has puzzled workers in this field for some time. Two clues have recently emerged. The detailed structure of the triad has been examined by Franzini-Armstrong (1970, 1975). She finds that the T tubule and its adjacent sarcoplasmic reticulum sac are connected by an array of electron-dense structures called 'feet'. It is tempting to see these feet as possible channels of information transfer.

The other clue comes from investigations of charge transfer in the T tubule. Since the activation of contraction is voltage-dependent, it may be that activation depends upon the movement of charged particles in the potential gradient of the T tubule membrane. The problem has been investigated by Schneider and Chandler (1973), using techniques similar to those used in investigations of the gating currents in nerve axons. They found that asymmetrical displacement currents that appear to be due to the movement of membrane charge can indeed be detected. Charge flow was not evident at membrane potentials more negative than $-80\,\text{mV}$ and saturated above about $-10\,\text{mV}$. The time course of the currents was much slower than that of nerve gating currents, by twenty to a hundred times. Perhaps the most convincing evidence that these charge movements are related to excitation–contraction coupling comes from a comparison with the mechanical 'priming' process discovered by Hodgkin and Horowicz (p. 252). After a maintained depolarization followed by a return to the normal resting potential, it takes some seconds for the membrane charge movements to regain their full size; and the time course of their recovery is closely paralleled by the recovery of the ability to contract (Adrian, Chandler and Rakowski, 1976).

The mode of action of these charged particles remains enigmatic. Their density in the T tubules is about $600/\mu\text{m}^2$, which is similar to the density of the 'feet' seen at the triad. Perhaps, as Chandler and his colleagues have suggested (1976), each charge forms part of a long molecule which extends from the T tubule membrane to the sarcoplasmic reticulum membrane. Movement of the charge (produced by depolarization of the T tubule membrane) might then unplug channels in the sarcoplasmic reticulum membrane, so allowing calcium ions to escape. It is an attractive idea, but still, of course, a speculative one.

The molecular basis of activation

As mentioned earlier, pure actin will react with pure myosin so as to split ATP in the absence of calcium ions. But if tropomyosin and

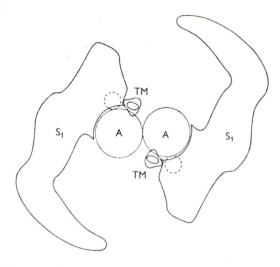

Fig. 11.35. Diagram to show how movement of tropomyosin molecules may affect actin–myosin interactions. A thin filament is seen in cross-section with actin (A) and tropomyosin (TM) molecules. Two myosin S_1 fragments are shown in the position they are thought to occupy in 'decorated' thin filaments. Tropomyosin positions are shown for the muscle at rest (dotted circle) and when active (solid contours). This model suggests that on activation the tropomyosin molecules move into the groove in the actin filament, so enabling S_1 fragments to attach to their binding sites on the actin molecules. The diagram is based on X-ray diffraction studies and on the three-dimensional reconstruction method described in the text. (From Huxley, 1973.)

troponin are also present, the actin–myosin interaction and ATP splitting will only occur in the presence of calcium ions. Hence it seems likely that tropomyosin and troponin are intimately involved in the control of muscular contraction.

Troponin is a complex molecule consisting of three subunits with different properties (see, for example, Mannherz and Goody, 1976). The tropomyosin-binding subunit (troponin T, molecular weight 37 000) links readily to tropomyosin. The calcium-binding subunit (troponin C, molecular weight 18 000) combines with calcium ions, up to four calcium ions being bound per molecule. Various physicochemical techniques indicate that such binding produces a conformational change in the troponin C molecule. The inhibitory subunit (troponin I, molecular weight 21 000) is so called because under the right conditions it inhibits the interaction of actin and myosin. Actin, tropomyosin and the three troponin subunits form a group in which each protein is bound to two of its neighbours, so that a circular linkage is formed thus:

Actin–tropomyosin–troponin T–troponin C–troponin I

It is not hard to believe that the structural arrangements of this system would be altered by a conformational change in one of its members, such as occurs when troponin C binds to calcium.

Direct evidence that the structure of the thin filaments alters during activation has come from X-ray diffraction measurements in which it is found that the relative intensity of the different actin layer lines alters on contraction (Vibert, Haselmore, Lowy and Poulsen, 1972; Huxley, 1973). Parry and Squire (1973) and Huxley showed that these changes could have been produced by a movement of tropomyosin in the groove between the two strings of actin monomers in the thin filament. Confirmation for this idea has been provided by Wakabayashi, Huxley, Amos and Klug, (1975), using the three-dimensional reconstruction technique described on p. 234.

They compared the structure of thin filaments prepared from actin and tropomyosin only with those to which troponin T and troponin I has been added. The former will combine with myosin whereas the latter will not. The results showed that the position of the tropomyosin molecule was indeed different in the two types of filament. This leads to the suggestion that tropomyosin in the resting muscle is placed so as to prevent the myosin heads combining with their appropriate actin monomers, but that it can be moved out of the way by a conformational change in the troponin complex brought about by binding of calcium to troponin C. Fig. 11.35 shows the estimated positions of the tropomyosin in the active and resting states.

12 The dynamics of muscular contraction

In the previous chapter, we examined some of the evidence relating to the means whereby muscular contraction is brought about. We must now consider some of the mechanical and thermal properties of contracting muscle.

CONTRACTION AGAINST DIFFERENT TYPES OF LOAD

As we have seen, an *isometric* contraction is one in which the muscle length is held constant; the methods of measuring the tension produced by the muscle in this type of contraction have been described in the previous chapter. When a muscle in isometric tetanus is suddenly shortened, the tension falls abruptly, and then rises to a new maximum level which is determined by the isometric length–tension relation (fig.

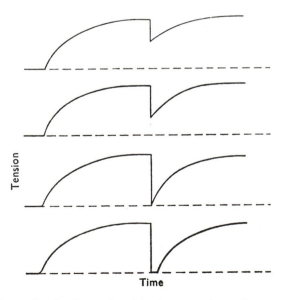

Fig. 12.1. The results of an isometric quick release experiment. See text for details.

12.1). This type of experiment is known as a *quick release* (Gasser and Hill, 1924). The amount by which the tension falls is roughly proportional to the length change during the quick release, up to the point at which the tension falls to zero. With quick releases of greater extents, there is an increasing delay between the time of release and the redevelopment of tension. The time course of the redevelopment of tension after the quick release is similar to, but not usually identical with, that of the development of tension at the beginning of the period of stimulation. The converse of this type of experiment occurs when the muscle is stretched during contraction.

In the measurement of *isotonic* contractions, the tension exerted by the muscle is maintained constant (usually by allowing it to lift a constant load) and its length change is measured. A *free-loaded* contraction is one in which the muscle is loaded in the resting state and then stimulated. This type of measurement is not much used, since the initial length of the muscle will vary with different loads. In an *after-loaded* contraction, the muscle is not loaded at rest, but must lift a load in order to shorten. The apparatus used for measuring an after-loaded contraction is shown schematically in fig. 12.2. The muscle is attached to a light, freely-pivoted isotonic lever so that it must lift a weight when it shortens. The movement of the lever can be recorded by means of a smoked drum or by the interruption of a beam of light focused on a photocell. When the muscle is relaxed, the lever rests against a stop, so

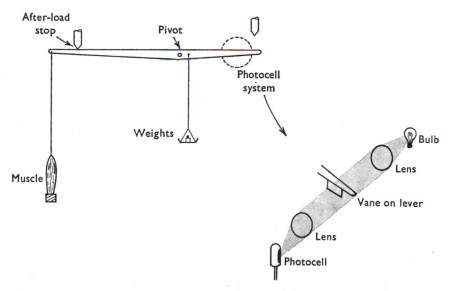

Fig. 12.2. Isotonic lever, arranged for use with after-loaded contractions. The inset shows the photocell system used to record the position of the lever.

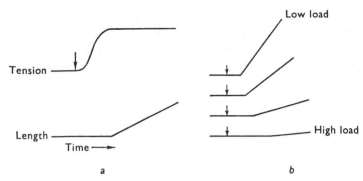

Fig. 12.3. After-loaded isotonic contractions. *a*: Tension and length changes in a single contraction. Repetitive stimulation starts at the time indicated by the arrow, and shortening is shown as an upward deflection of the length trace. *b*: Length changes in after-loaded isotonic contractions with different loads.

that the resting muscle does not have to support the load. Fig. 12.3 shows what happens when the muscle is stimulated tetanically. At the start of the stimulation period, the tension begins to increase so that, for a time the muscle is contracting isometrically. Then, when the tension produced by the muscle is equal to the load, the muscle begins to shorten. From now on, the load on the muscle (and therefore the tension developed by it) is constant, so that the contraction is isotonic. Notice that the velocity of shortening is fairly constant. With increasing loads, after-loaded contractions are affected in two ways (fig. 12.3*b*). First, the time between the beginning of the shortening period and the onset of shortening is longer; this is because the muscle takes longer to reach the higher tension required to lift the greater load. Secondly, the velocity of shortening is less; by plotting the velocity of shortening against the load, a *force–velocity curve* is produced, as is shown in fig. 12.11. If the load is further increased so that it is equal to the maximum isometric tension, shortening does not occur at all.

An alternative way of measuring isotonic contractions is by means of the *isotonic release* method. The apparatus used for this is shown in fig. 12.4. The muscle is attached to an isotonic lever as in fig. 12.2, but the lever can here be held up against the contraction of the muscle by means of a stop which can be immediately withdrawn at any desired time during the contraction. The results of a typical experiment are shown in fig. 12.5. At the beginning of the stimulation period the muscle contracts isometrically and the tension rises to its maximum value. When the release relay stop is withdrawn, the tension falls abruptly to a lower level which is equal to the value of the load. The change in lengths seems to consist of two components: first an abrupt shortening which is coincident with the change in tension, and secondly a steady isotonic

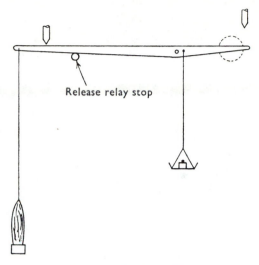

Fig. 12.4. Isotonic lever arranged for isotonic releases. The system is as in fig. 12.2 except that the lever is held up by a stop mounted on a relay. When the relay is activated, the stop is immediately withdrawn so that the muscle can now shorten.

shortening at more or less constant velocity. Since the mass of the lever and the load can never be negligible, the lever system possesses inertia, and so there is usually some oscillation at the change-over between these two phases. With increasing loads (fig. 12.5*b*), the initial length change is less, and the velocity of shortening during the isotonic phase is less.

Various other types of load can be used. An *elastic load* causes the shortening of a muscle to be proportional to its tension; in almost all 'isometric' contractions the load is, to some extent, elastic (see p. 212). An *inertial load* consists of a mass which is accelerated by contraction of the muscle. Loads of these types, although very commonly those which the muscle contracts against in nature, are not much used in experiments on the mechanics of muscle, since they do not allow either the length or the tension to be extrinsically determined during the course of the contraction.

THE HEAT PRODUCTION OF MUSCLE

Almost all chemical reactions, and a number of physical changes, are accompanied by the evolution or absorption of heat. Contracting muscles produce heat, and the measurement of this heat production is of considerable interest in that we might expect it to be related to the chemical reactions involved in the contraction process. Furthermore,

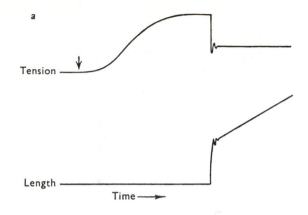

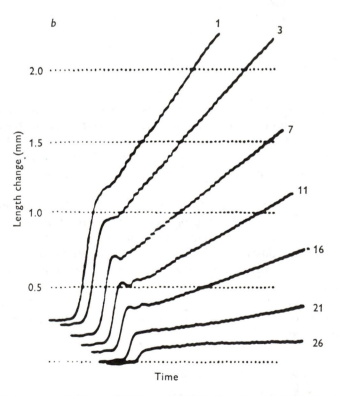

Fig. 12.5. Isotonic releases. *a*: Diagram of the length and tension changes during an isotonic release; shortening is shown upwards on the length trace. *b*: Length changes during a series of isotonic releases against different loads. Shortening is again shown upwards, the dots on the grid lines are at 1 msec intervals, and the figures opposite each trace indicate the load (g) on the muscle after release. From a frog sartorius muscle, with maximum isometric tension 32 g. (*b* from Jewell and Wilkie, 1958.)

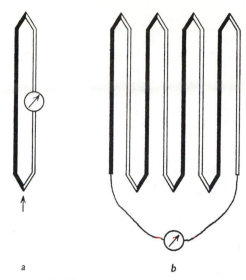

a   b

Fig. 12.6. *a*: A thermocouple. The junctions between the two different metal wires are indicated by arrows. If the temperature of the two junctions is different, current flows through the circuit and the galvanometer is deflected. *b*: A number of thermocouples in series, forming a thermopile.

measurement of the heat production is an essential prerequisite in any attempt to measure the energy expended by a contracting muscle. This energy (E) is the sum of the work done by the muscle on the load (W) and the heat produced during the contraction (H), or

$$E = H + W. \tag{12.1}$$

The technical problems involved in measuring the heat production of muscles are considerable, since the temperature changes involved may be extremely small. The methods used by Hill and his colleagues involve the use of a thermopile. If an electrical circuit is made up of two different metals (fig. 12.6*a*), an electromotive force is produced if the temperature of one of the junctions is different from that of the other; such a device is called a thermocouple. A thermopile consists of a large number of thermocouples in series (fig. 12.6*b*), and is therefore more sensitive than a single thermocouple. In Hill's experiments, the temperature changes produced by a moist muscle resting against the 'hot' junctions of the thermopile were measured by means of a galvanometer. The deflection of the galvanometer was determined by fixing a mirror to the galvanometer axis and projecting a thin beam of light via the mirror either onto photosensitive paper fixed to a revolving drum or onto a photocell whose output was displayed on an oscilloscope (fig. 12.7).

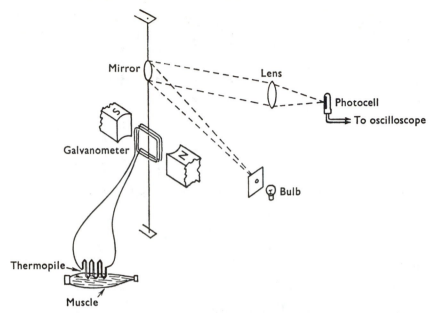

Fig. 12.7. Schematic diagram to show one method of measuring the heat production of a muscle.

The analysis of the records obtained by this procedure is rather complicated. The chief difficulty is that the galvanometer axis cannot change its position instantaneously when the current flowing through it changes. Other delays occur in the conductance of heat through the muscle and between the muscle and the thermopile, and in the time taken for the 'hot' junctions of the thermopile to reach the new temperature (which depends on the heat capacity of the thermopile). Prior to 1937, the apparatus was calibrated by heating a dead muscle with an electric current, followed by numerical analysis of the records to allow for the delays in the system. But with the development of more rapidly responding thermopiles and galvanometers it became possible to calibrate the system directly by measuring the sensitivity and heat capacity of the thermopile, the heat capacity of the muscle and of the Ringer's fluid between the muscle and the thermopile, and the heat loss from the system.

The main features of the heat production during isometric contractions were determined by Hill and Hartree in 1920. Later work has been much concerned with increasing the accuracy and time resolution of the measurements involved, and with the heat changes associated with shortening or lengthening of the muscle.

In resting muscle, various metabolic processes occur throughout the life of the muscle, and these processes are associated with the liberation of heat. This *resting heat* is about 0.2 mcal g^{-1} min^{-1} in frog sartorius muscles at 18 °C. The nomenclature of the heat changes produced by contracting muscles has varied somewhat since the original formulation by Hill and Hartree; we shall adopt the scheme given by Wilkie (1954), in which the heat produced during contraction and relaxation is called the *initial heat*, and the heat produced after the muscle has relaxed is called the *recovery heat*.

In an isometric tetanus, the initial heat consists of two phases (fig. 12.8): (i) the *activation heat* (or *maintenance heat*), which begins at a fairly high level soon after the onset of stimulation, falling to a steady level which lasts throughout the contraction, and (ii) the *relaxation heat*, which is produced while the muscle is relaxing. The activation heat must clearly be related to the chemical reactions involved in tension development. The relaxation heat seems to be rather more complex. During an isotonic contraction, the relaxation heat is negligible if the load is held up so that the muscle does not lengthen during relaxation, whereas it is large if the muscle is extended by the load during relaxation; this indicates that the major part of the relaxation heat is due to

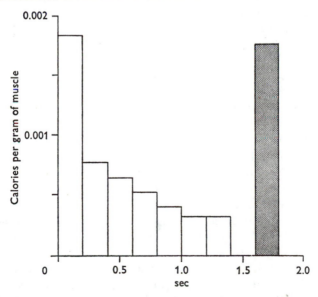

Fig. 12.8. The rate of heat production of a frog sartorius muscle during an isometric tetanus. The muscle was stimulated for a period of 1.2 sec; the heat production during this time is the activation heat. The burst of heat production after the end of stimulation (shaded) is the relaxation heat. (From Hill and Hartree, 1920.)

degradation of the work done by the load on the relaxing muscle. Minor components of the relaxation heat are provided by thermoelastic effects (see below) and possibly, as has been suggested by Davies (1964), heat production associated with the reaccumulation of calcium ions by the sarcoplasmic reticulum.

Elastic bodies usually undergo heat changes when they are stretched or allowed to shorten. Most substances, such as steel or wood, absorb heat when stretched and release heat when released; active muscle shows this type of behaviour (Hill, 1953; Woledge, 1961, 1963). Rubber, on the other hand, releases heat when stretched and cools on release; this type of behaviour is shown by resting muscle.

When an active muscle shortens, an extra amount of heat is released, in addition to the activation heat. This is called the *heat of shortening*, and we shall examine its properties in the following section. When a contracting muscle is stretched, its heat production is greatly reduced, so that it is actually absorbing energy during the period of stretch.

The high energy compounds which have been broken down during contraction have to be resynthesized. For a long period after the end of a contraction, the heat production of a muscle is higher than the normal resting level. This heat production is known as the *recovery heat*, and is thought to correspond to the respiratory processes involved in resynthesis of the high energy phosphate compounds; it is greatly reduced in the absence of oxygen.

ANALYSIS OF THE DYNAMICS OF MUSCULAR CONTRACTION

So far we have described some of the mechanical and thermal properties of contracting muscles without providing any theoretical scheme which will account for these properties and relate them to each other. The search for such a scheme has occupied muscle physiologists for the last half century (see Hill, 1965) and has still not yet been achieved. One of the best analyses was that produced by A. V. Hill in 1938, which has greatly influenced much of the work performed in this field since that time. Hence we must examine Hill's work in some detail, although, as we shall see, a number of later observations indicate that his analysis is incomplete.

Most early theories of the mechanism of muscular contraction suggested that, on stimulation, the muscle became effectively a stretched elastic body. But the force exerted by a purely elastic body when stretched is determined by its length, and is independent of the velocity of shortening. Thus the observation that, during isotonic shortening, force is determined by velocity indicates that the 'elastic' theory cannot be correct. It was then suggested (Gasser and Hill, 1924; Levin and Wyman, 1927) that some of the elasticity of the muscle is damped by an internal viscosity. The idea was that some of the tension in the elasticity was used up internally in working against the viscosity, more tension being so absorbed at high velocities, leaving less to appear at the ends of the muscle.

The visco-elastic theory was generally accepted until 1938, in spite of the fact that Fenn (1923, 1924) had provided evidence to show that it must be incorrect. Fenn measured the shortening of a muscle under various loads and also measured its heat production. He was then able to calculate the total energy release (from (12.1)), and found that more energy is released when the muscle is allowed to shorten during the contraction. This is not what one would expect from the elastic and visco-elastic theories, which postulate that the muscle contains a fixed amount of potential energy at the beginning of a contraction, which must be converted into heat and work during the contraction. The extra release of energy on shortening is frequently known as the 'Fenn effect'.

Let us now consider some of the details of A. V. Hill's analysis (Hill, 1938). The first stage in this work was the experimental measurement of heat production during isotonic contractions, when it immediately became obvious that extra heat production (known as the heat of shortening) occurs if the muscle is allowed to shorten. Fig. 12.9 shows the results of some of these experiments. In fig. 12.9c, isotonic releases were performed at different times after the beginning of a period of stimulation; the muscle was allowed to shorten by the same amount against the same load in each case. It is obvious from the diagram that the heat of shortening does not vary appreciably with time. Fig. 12.9a shows the heats of shortening when the muscle was allowed to shorten for different distances against a constant load after an isotonic release; the heat production increased with increasing shortening. Fig. 12.9b shows that the amount of the heat of shortening was apparently independent of the load when the muscle shortened over the same distance against various loads, although the rate of heat production (and, of course, the velocity of shortening) decreased with increasing loads. From these results, it was concluded that the extra heat released during shortening is proportional to the distance shortened, i.e.

$$\text{heat of shortening} = a \cdot x, \qquad (12.2)$$

where x is the distance shortened and a is a constant having the dimensions of a force. Now the work done by the muscle in an isotonic contraction is given by the product of the force exerted and the distance shortened, i.e.

$$\text{work} = P \cdot x,$$

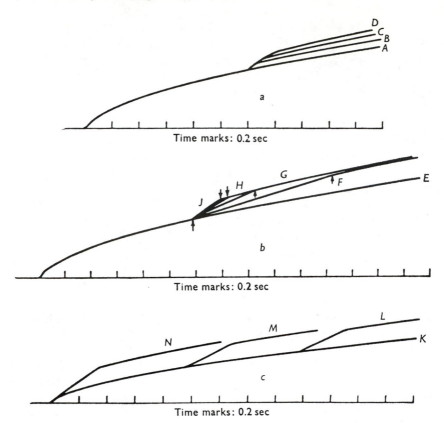

Fig. 12.9. Heat production in a frog sartorius muscle when it was allowed to shorten during a tetanus at 0 °C, to show the heat of shortening. The curves are galvanometer deflections proportional to the amount of heat produced by the muscle. For the upper traces (*a*), the muscle was released at 1.2 sec after the onset of stimulation and allowed to shorten various distances (*B*, 1.9; *C*, 3.6; *D*, 5.2 mm) against a constant load (3 g) before being stopped. For the middle traces (*b*), the muscle was released at 1.2 sec after the onset of stimulation and allowed to shorten a constant distance (5.2 mm) against various loads (*F*, 24.9; *G*, 13.9; *H*, 5.7; *J*, 3.0 g). For the lower traces (*c*), the muscle was released at various times (*L*, 1.8; *M*, 0.95 sec; *N*, immediately) after the onset of stimulation and allowed to shorten a certain distance (9.1 mm) against a constant load (2.9 g). Traces *A*, *E* and *K* show galvanometer deflections during isometric contractions. (From Hill, 1938, by permission of the Royal Society.)

where P is the load. The extra energy released by the muscle during an isotonic contraction (E_e) is given by

$$E_e = \text{heat of shortening} + \text{work}$$

$$= P \cdot x + a \cdot x$$

$$= (P+a)x.$$

Therefore the *rate* of extra energy liberation is given by

$$\frac{dE_e}{dt} = (P+a)V \tag{12.3}$$

where V is the velocity of shortening.

The next stage in the analysis was to determine how the rate of extra energy liberation varied with the load. Knowing a, this could be determined by measuring the velocity of shortening (V) at different loads (P), and then putting the appropriate values in (12.3). It was found that the rate of extra energy liberation decreased linearly with increasing tension (fig. 12.10), i.e.

$$(P+a)V = b(P_0 - P), \tag{12.4}$$

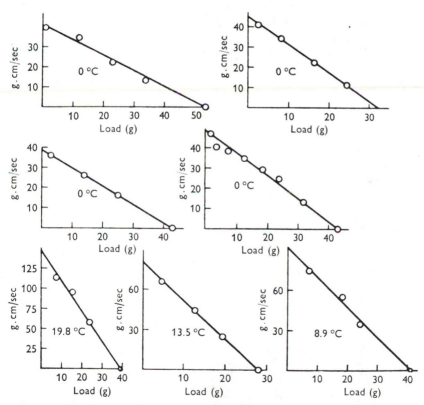

Fig. 12.10. The effect of the load on the rate of extra energy liberation in seven different muscles. The ordinate in each case is $(P+a)V$. The slope of the straight lines drawn through the experimental points is equal to $-b$. Further explanation in the text. (From Hill, 1938, by permission of the Royal Society.)

where P_0 is the isometric tension and b is a constant with the dimensions of a velocity.

Equation (12.4) can be expressed differently by the following procedure. Adding $b(P+a)$ to both sides, we get

$$(P+a)V+b(P+a)=b(P_0-P)+b(P+a),$$

i.e.
$$(P+a)(V+b)=b(P_0+a)=\text{constant}. \tag{12.5}$$

Equation (12.5) is a hyperbola in which the tension produced by the muscle is related to its velocity of shortening. Hence, if the analysis is correct, the equation should fit the force–velocity relationship of the muscle as determined by purely mechanical measurements, and, in particular, the values of a and b determined from such measurements should agree with those derived from measurements of the heat of shortening and the rate of extra energy release during shortening. Hill concluded that this was so: fig. 12.11 shows the force–velocity relation fitted by (12.5), and table 12.1 compares the values of a and b as determined by his thermal and mechanical measurements.

Hill's 1938 paper consists of three parts. The first part deals with the experimental methods used in determining the heat production and mechanics of contraction, and the second part contains the elegant

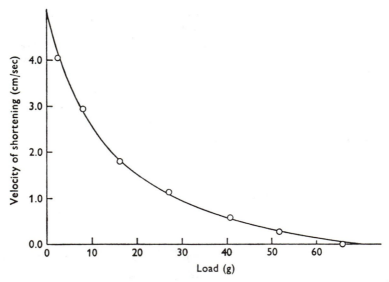

Fig. 12.11. The force–velocity curve of a frog sartorius muscle at 0 °C. The experimental points were determined from after-loaded contractions as in fig. 12.3. The curve is drawn according to (12.5) with $a = 14.35$ g, $b = 1.03$ cm/sec, and $P_0 = 65.2$ g. (From Hill, 1938, by permission of the Royal Society.)

TABLE 12.1. *Comparison of the values of a and b derived from thermal and mechanical measurements. Frog sartorius at 0 °C. (From Hill, 1938; see also Sandow, 1961.)*

		Thermal	Mechanical
a	g . wt/cm^2	255–550 in 14 expts	357
a/P_0		0.21–0.31 in 11 expts	0.22
b	muscle lengths/sec	0.28–0.41 in 11 expts	0.27

experiments and analysis outlined above. In the final section, Hill proposed a new mechanical model of contracting muscle to replace the visco-elastic model. He suggested that a contracting muscle consists of two components: (1) a *contractile component*, whose properties are determined by the force–velocity relationship, in series with (2) a *series elastic component*, whose properties can be described by a force–extension curve. This idea is illustrated diagrammatically in fig. 12.12.

Now let us see how this model is able to explain some of the mechanical properties of tetanically contracting muscles. Firstly, consider the time-course of the rise of tension during an isometric tetanus (fig. 11.2). In the resting condition, there is no tension on the muscle, therefore the series elastic component must be slack. With the onset of stimulation, the contractile component begins to shorten; since the tension is zero, it will initially shorten at its maximum velocity. But since

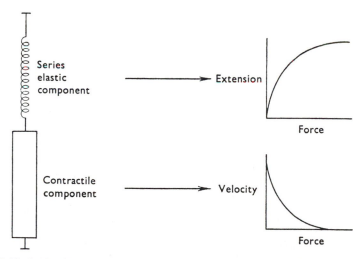

Fig. 12.12. A. V. Hill's two-component model of contracting muscle. The properties of the contractile component are determined by its force–velocity curve, and the properties of the series elastic component are determined by its force–extension curve.

the ends of the muscle are fixed, the shortening of the contractile component must produce a corresponding extension of the series elastic component, and so the tension will begin to rise according to the force–extension curve. This increase in tension results in a reduction in the velocity of shortening of the contractile component (according to the force–velocity curve), so the series elastic component will be extended less rapidly, and hence the tension will rise more slowly. These interactions will continue until the tension has reached its full isometric level (P_0), when the shortening velocity is zero and the series elastic component is fully extended.

The model can also provide an explanation for the quick release phenomena observed by Gasser and Hill (fig. 12.1). The quick release allows an abrupt shortening of the series elastic component, and therefore an abrupt fall in tension. Because of this fall in tension (which will be roughly proportional to the extent of the release), the contractile component now begins to shorten at a velocity determined by the new tension. In so doing, it will extend the series elastic component again, and so the tension will again rise to isometric level in the same way that it did at the beginning of the period of stimulation. The new isometric tension may be slightly different from the old one, since the muscle is now at a shorter length.

During after-loaded isotonic contractions (fig. 12.3), the series elastic component is at constant length and tension during the period of shortening, so that the velocity of shortening of the whole muscle is equal to that of the contractile component. In an isotonic release (fig. 12.5), the

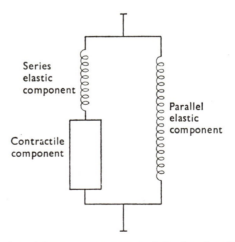

Fig. 12.13. Modification of the two-component model at long lengths to include a third component, the parallel elastic component, which is reponsible for the resting tension of a stretched muscle.

sudden reduction in tension from isometric to the value of the isotonic load causes an abrupt shortening of the series elastic component. Thereafter the series elastic component is at constant length and the muscle shortens at a velocity determined by the force–velocity relation.

The elastic elements

We must now consider the nature of the series elastic component. Part of this series elasticity is external to the muscle, being the elasticity of the connections between the muscle and the recording system. It is desirable to reduce this external elasticity as far as possible (by using steel wire or chain connections rather than cotton thread, for example) but it can never be entirely eliminated. Of the remaining series elasticity, about half is located in the muscle tendons and the rest in the muscle cells themselves (Jewell and Wilkie, 1958); we shall see later, in considering Huxley and Simmons' model of the contractile mechanism, just where in the muscle cells this is. The force–extension curve of the series elastic component can be measured in various ways; one of the most reliable methods involves measuring the extent of the elastic length change during isotonic releases.

At long lengths, the two-component model shown in fig. 12.12 is inadequate, since it does not account for the resting tension exerted by a stretched muscle. The model is then modified by including a third element, the *parallel elastic component*, as is shown in fig. 12.13. This is probably mainly composed of the connective tissue sheaths of the muscle, but it is possible that components of the muscle cells (such as the hypothetical '*S* filaments' described on p. 226) are also involved.

Alternative force–velocity equations

Hill's equation for the force–velocity relation was originally applied to muscles at rest length only. Abbott and Wilkie (1953) measured the force–velocity relation at various lengths below the rest length, and found that the results could be described by a simple modification of Hill's equation, so that, at any particular length,

$$(P+a)(V+b) = b(P'_0 + a), \tag{12.6}$$

where P'_0 is the full isometric tension at that length. The constants a and b did not appear to change with length. Equation (12.6) is merely an extension of (12.5).

The first force–velocity equation was that given by Fenn and Marsh (1935):

$$P = P_0 e^{-\alpha V} - kV,$$

where a and k are constants. An equation of somewhat similar form is that provided by Aubert (1956):

$$P = A\,e^{-V/B} \pm F,$$

where B and F are constants, and $A = P_0 + F$. Notice that Aubert's equation introduces a 'frictional element', F which decreases the tension on shortening and increases it on lengthening. Both these equations are largely empirical; they are not connected with any particular theory of muscular contraction.

Another force–velocity equation is that provided by Polissar (1952):

$$V = \text{constant}\,(A^{1-P/P_0} - B^{P/P_0-1}),$$

where A and B are constants. Polissar's equation was derived from a theoretical model of the muscle which has now been superseded by the sliding filament theory. It does not have as much predictive value as Hill's and Aubert's equations, since it contains four constants instead of three.

LIMITATIONS OF HILL'S ANALYSIS

It is clear that Hill's analysis provides an elegant theoretical scheme which fits many of the features of muscular contraction. In recent years, however, a number of discrepancies between the predictions of the theory and experimental observations have been discovered; we must now examine some of these discrepancies.

The effect of load on the heat of shortening

In Hill's original experiments, it appeared that the heat of shortening was independent of the load, being determined only by the distance shortened: (12.2). Hill (1964a) has since made a very careful re-investigation of this point, and concludes that the heat of shortening increases with increasing loads. The relation between the heat of shortening (now represented by α) and the load P is given by

$$\alpha/P_0 = 0.16 + 0.18\,P/P_0.$$

This conclusion invalidates the derivation of (12.5) from thermal measurements, but does not, of course, affect its applicability as an empirical description of the force–velocity relation.

The time course of the rise in tension during an isometric tetanus

According to Hill's model, the time course of the rise in tension during an isometric tetanus is precisely determined by the force–velocity curve of the contractile component and the force–extension curve of the series elastic component, as we have already seen (p. 277). In mathematical terms,

$$\frac{dP}{dt} = \frac{dP}{dx} \cdot \frac{dx}{dt}$$

or

$$\frac{dP}{dt} = V\frac{dP}{dx} \qquad (12.7)$$

where V is given by the force–velocity curve and dP/dx is the slope of the force–extension curve. Integration of (12.7) with respect to P gives

$$t = \int_0^{P_0} \frac{1}{V\,dP/dx}\,dP, \qquad (12.8)$$

where P_t is the tension at time t after the beginning of stimulation.

This relation has been very carefully examined by Jewell and Wilkie (1958). They determined experimentally the force–velocity and force–extension curves, and then calculated the expected form of the time course of tension development in an isometric contraction from (12.8) using numerical integration. This calculated curve could then be compared with the actual responses (both at the start of stimulation and after a quick release) of the same muscle. The results are shown in fig.

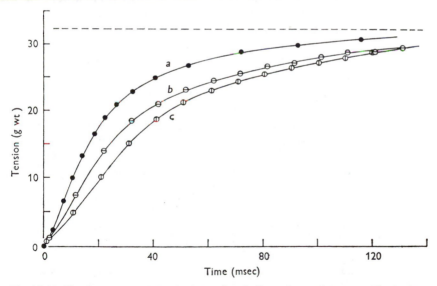

Fig. 12.14. The time-course of the rise in tension during an isometric tetanus. The broken line shows the final maximum tension. Curve *a* is calculated from the force–velocity relation of the contractile component and the force–extension curve of the series elastic component by means of (12.8). Curve *b* is the actual time-course of tension development after a quick release. Curve *c* is the actual time-course of tension development at the beginning of a period of stimulation. (From Jewell and Wilkie, 1958.)

12.14. It is clear that the tension does not rise as rapidly as the theory predicts; the time taken for the tension to reach a given level is about 80 per cent longer than expected for the initial development of tension, and about 50 per cent longer than expected after a quick release. Obviously, something must be wrong with the theory; Jewell and Wilkie suggested that the velocity of shortening is determined by the 'past history' of the muscle, as well as by the force acting on it at a particular time.

Transient changes in velocity of shortening after isotonic releases

During an isotonic release, there is a period of oscillation between the abrupt shortening ascribed to the series elastic component and the steady shortening that follows, as can be seen in fig. 12.5b. Some of this oscillation is due to the inertia of the lever system used for recording the length change. Any such inertial oscillation in length must also be accompanied by an oscillation in tension; thus any investigation of this phenomenon must include measurements of the tension of the muscle as well as its length. Podolsky (1960) carried out experiments of this type on sartorius muscles, and showed that the shortening velocity does not reach a steady value until some time after the oscillation in tension has stopped (fig. 12.15). Related results were obtained by Aidley (1965b) from frog rectus abdominis muscles (which contract very slowly and so enable the transient changes in velocity to be seen very easily) and by Armstrong, Huxley and Julian (1966) using single frog semitendinosus fibres in the spot follower apparatus described in the previous chapter. We shall examine the details of these transients, as measured by A. F. Huxley and his colleagues, in the following section.

All these results imply that the force–velocity relation is not instantaneously obeyed when the tension changes; shortening velocities seem to be higher than usual immediately after a fall in tension. In other

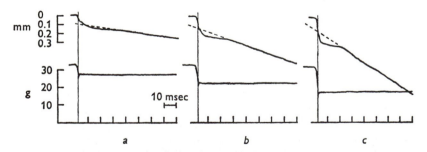

Fig. 12.15. Isotonic releases of frog sartorius muscle, showing transient velocity changes immediately after the reduction in tension. Shortening is shown as a downward deflection of the upper traces. The loads on the muscle after release, expressed as fractions of P_0, were: a, 0.84; b, 0.69; c, 0.55. (From Podolsky, 1960.)

words, the properties of the contractile component are not completely described by the steady-state force–velocity relation.

THE STRUCTURAL BASIS OF THE DYNAMICS OF CONTRACTION

In the previous chapter, we saw that the sliding filament theory provides an excellent explanation for the structural changes that occur when striated muscles contract, and for the form of the length–tension curve. To what extent can this theory account for the properties of contracting muscles which have been described so far in this chapter? That is to say, can the sliding filament theory account for the existence of the activation heat, the extra energy release on shortening (the Fenn effect), the force–velocity relation, and the transient velocity changes following change in tension? In order to answer these questions we shall consider three particular hypotheses of the mechanism of muscular contraction, all based on the sliding filament theory. They have been worked out quantitatively, giving in many cases good agreement with experimental measurements on muscle, but we shall here consider their qualitative aspects only.

A. F. Huxley's 1957 theory

The theory produced by A. F. Huxley in 1957 was a first attempt to devise a precise model based firmly on the sliding filament theory. It has proved to be most successful in the sense that it has suggested a number of experiments and provided the initial inspiration for several daughter theories. Although Huxley's suggestions regarding the particular mechanism whereby force is generated between the myosin and actin filaments may not be correct, some of the conclusions which he reached are of more general application.

He begins by assuming that shortening and the development of tension are produced by independent tension-generators (the cross-bridges) which can be effective only in the region of overlap of the thick and thin filaments. Within each contraction, each tension generator undergoes cycles of activity: attachment, pulling, and detachment, followed by reattachment (perhaps at a different site) and a new cycle. An essential point in what follows is the postulate that the probabilities of attachment and detachment of the cross-bridges are determined by their position.

The hypothetical tension-generating mechanism is shown in fig. 12.16. An active site M on the myosin filament oscillates by thermal agitation backwards and forwards along the length of the filament, but is restrained by elastic elements on each side (the 'springs' in the diagram),

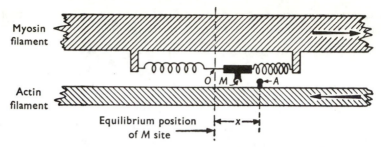

Fig. 12.16. The tension-generating mechanism in A. F. Huxley's model. The part of a fibril which is shown is in the right-hand half of a sarcomere. Details in the text. (From Huxley, 1957, by permission of Pergamon Press Ltd.)

its equilibrium position being denoted by O. On the actin filament is an active site A, at a longitudinal distance x from O. The A and M sites can become attached to each other (forming a cross-bridge), and this reaction has a rate constant f:

$$A + M \overset{f}{\to} AM. \tag{1}$$

This AM link can be broken by combination of the A site with a high energy phosphate compound XP,* the reaction proceeding with a rate constant g:

$$AM + XP \overset{g}{\to} AXP + M. \tag{2}$$

Finally the system is reset so that reaction (1) can occur again by dissociation of AXP and splitting of the high-energy phosphate bond (which supplies the energy needed to bring about this dissociation):

$$AXP \to A + X + \text{phosphate}. \tag{3}$$

The rate constants f and g are assumed to be dependent upon x, the position of the A site with respect to O (fig. 12.17). f is zero when x is negative (A to the left of O), and increases linearly with increasing x up to the point h, beyond which it is again zero. g is very high (and constant) when x is negative, but zero at O, and small when x is positive, increasing slowly with increasing x.

During shortening, a single cycle of reactions (1) to (3) occurs as follows. An A site approaches an oscillating M site from the right; reaction (1) cannot occur until it passes the point h. When x is less than h, there is a fairly high probability that reaction (1) will occur; assume that, in this case, it does. M is now drawn towards O by the elastic force

---

* Biochemical evidence now suggests that high energy phosphate (ATP) combines with myosin rather than with actin (fig. 11.27), but this detail does not affect the development and conclusions of Huxley's theory.

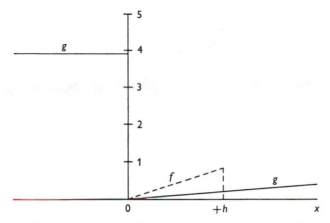

Fig. 12.17. The dependence of the rate constants f and g on x in A. F. Huxley's model. The unit of the ordinate scale is the value of $(f+g)$ when $x = h$. (From Huxley, 1957, by permission of Pergamon Press, Ltd.)

in the left hand 'spring'. During this time there is a low probability that reaction (2) will occur, but as soon as the link passes O (i.e. as soon as x becomes negative) g becomes very high, and thus the probability of reaction (2) occurring increases enormously, and so the link is broken. Then reaction (3) occurs, and the A site is therefore ready to interact with the next M site that it meets. It is assumed that similar reactions occur asynchronously along the length of the sarcomere (probably by means of a 'vernier' arrangement whereby the distance between successive M sites is different from that between successive A sites*). This means that there will always be some links formed at any one time, so that the filaments will slide smoothly past each other.

The tension generated at any one contraction site is the tension in the left hand 'spring' when the AM link is in existence. The tension produced by the whole muscle is the sum of the tensions at all contraction sites in a length of muscle equal to half a sarcomere (the forces generated in the two halves of a sarcomere, and in other sarcomeres along the length of the muscle, are in series with each other, and are therefore not additive). In an isometric contraction, some sliding occurs until the full isometric tension is reached and the series elastic component is fully stretched. All the AM links formed will then be to the right of O. There will be continual breakage of these links (reaction (2)) since g, though small, is not zero. Thus each pair of sites will be

* The distance between successive M sites in contact with the same actin filament is almost certainly 429 Å. The equivalent distance between successive A sites is probably 54 Å (the distance between successive units in one of the actin helices).

continually going through the cycle of reactions from (1) to (3), and hence the high energy phosphate compound XP will be continually broken down, with the release of energy. This accounts for the activation heat and the consumption of ATP in an isometric contraction.

When the muscle shortens, the average value of g will rise, since the A sites are continually moving to the left of O. Thus the rate at which the cycle of reactions occurs will also increase. This accounts for the extra energy release during shortening.

Another consequence of shortening of the muscle is that any one A site will only be in a position to react with the corresponding M site for a limited time; this time will decrease as the velocity of shortening increases. Hence, since the rate constant of link formation (f) is finite, the probability of a link being formed between any one pair of sites will decrease as the velocity of shortening increases. This means that, as the shortening velocity increases, the total number of links formed at any one time will decrease, and therefore the tension of the whole muscle will also decrease. Furthermore, as the velocity increases, more and more links will remain in existence when x is negative (since g is also finite); in these cases, the tension in the right hand 'spring' will pull against the shortening generated by other links where x is positive, so further reducing the tension developed by the whole muscle. These two effects account for the force–velocity relation.

The hypothesis also provides a partial explanation for the transient velocity changes which appear to follow the changes in tension during an isotonic release (see Aidley, 1965b, and Civan and Podolsky, 1966). We have seen that more links exist in the isometric condition than when the muscle is shortening at a constant velocity. Immediately after the reduction in tension, these 'extra' links will still be in existence, and so the muscle will be able to shorten more rapidly than usual. However detailed examination of the transients (see below) shows that they are more complex than Huxley's theory predicts; the two modifications of it which we shall consider shortly were produced in attempts to meet this situation.

Huxley's hypothesis is mainly concerned with the mechanism of contraction. However, it is desirable that it should be related to the properties of muscle under other conditions. In the resting muscle, we must assume that either reaction (1) or reaction (3) cannot occur; Huxley does not specify the activation mechanism which is necessary for contraction (we may assume that calcium ions are involved), but this is not an essential point in relation to the characteristics of contraction. At the end of a contraction, the reaction sequence ceases and the tension falls as reaction (2) proceeds to completion. When the muscle is in rigor, reaction (2) cannot take place, so that the AM links cannot be broken, which accounts for the inextensibility of the muscle in this condition.

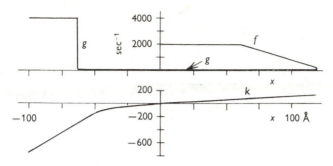

Fig. 12.18. Variation of the rate constants *f* and *g* with position in Podolsky and Nolan's model (upper graph). Compare with fig. 12.17. The lower graph shows how the force produced by a cross-bridge varies with position; in A. F. Huxley's model the corresponding curve would be a straight line through the origin. (From Podolsky and Nolan, 1971.)

Podolsky and Nolan's model

Podolsky and Nolan (1971, 1973) utilize the essential postulates of A. F. Huxley's model, but suggest a different relation between cross bridge position and the probabilities of formation and detachment, i.e. a different dependence of *f* and *g* on *x*. By choosing suitable relations (as in fig. 12.18) it is possible to predict satisfactorily the form of the transients and the properties of muscle. A consequence of the theory which is that the number of cross-links increases with increasing shortening velocity (the overall tension is still decreasing, however, since a high proportion of the links have *x* negative). This is just the opposite of what Huxley's theory predicts and, as the authors point out, the difference ought to be testable.

A. F. Huxley and Simmons's theory

This theory (A. F. Huxley and Simmons, 1971, 1973; Huxley, 1974) differs from the 1957 theory in its model of the tension-generating site. It takes into account our more recent knowledge of the structure of the myosin molecule and is especially concerned with explaining the mechanical transient responses. Before examining the details of the model, therefore, let us take a closer look at these transients.

Fig. 12.19 shows the two types of transient obtainable from single muscle fibres: the length changes which follow a step change in tension (the velocity transient) and the tension changes which follow a step change in length (the tension transient). The events in the tension transient fall into four fairly distinct phases. First there is a sudden drop in tension during the length change (phase 1). Next, immediately after the length change, there is a rapid rise in tension (phase 2). This is followed by a period of a few milliseconds during which the recovery of tension is greatly slowed or even reversed (phase 3). Finally the tension

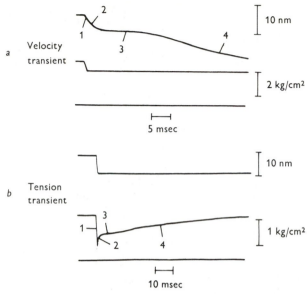

Fig. 12.19. Mechanical transient responses in frog muscle fibres during tetanic stimulation. *a*, the length change following a sudden change in tension; *b*, the tension change following a sudden change in length. In each case the upper trace shows length (shortening downwards), the middle trace tension and the bottom trace the tension zero baseline. The numbers 1 to 4 indicate corresponding phases in the two types of transient response, described in table 12.2. (From A. F. Huxley, 1974.)

TABLE 12.2. *Phases of the transient response to sudden reduction of length ('tension transient') or of load ('velocity transient').* (From A. F. Huxley, 1974.)

Phase	Time of occurrence	Events in 'tension transient'	Events in 'velocity transient'
1	During applied step	Simultaneous shortening	Simultaneous drop of tension
2	Next 1–2 ms	Rapid early tension recovery	Rapid early shortening
3	Next 5–20 ms	Extreme reduction or even reversal of rate of tension recovery	Extreme reduction or even reversal of shortening speed
4	Remainder of the response	Gradual recovery of tension, with asymptotic approach to isometric tension	Shortening at steady speed sometimes with superposed damped oscillation

The stated times are appropriate for frog muscle at 5 °C approx.

gradually climbs back to the isometric tension appropriate to its length (phase 4). A related set of four phases can be seen in velocity transients; table 12.2 summarizes these events.

Huxley and Simmons investigated the nature of the first two phases by making a series of length steps of different sizes and observing the height of the consequent tension changes. Their results are shown in fig. 12.20, where T_1 represents the tension at the end of the first phase and T_2 that at the end of the second. The T_1 curve is very nearly a straight line and this represents the behaviour of a passive elastic element. But the T_2 curve is quite different from this: tension falls very little for small length changes, after which the curve is roughly parallel to the T_1 curve with a displacement of about 6 nm per half sarcomere. This suggests that T_2 represents the properties of the tension-generators in the cross-bridges, each of which is capable of moving through about 6 mm while still exerting nearly maximum tension. To put it another way, each of the tension-generators seems capable of taking up about 6 mm of slack.

What structures contain the passive elasticity represented in the T_1 curve? The experiments were done with single fibres tied to the apparatus at points very close to their ends, and so the compliance of the tendons can be ruled out. This leaves either the thick and thin filaments or the cross-bridges between them. In order to distinguish between

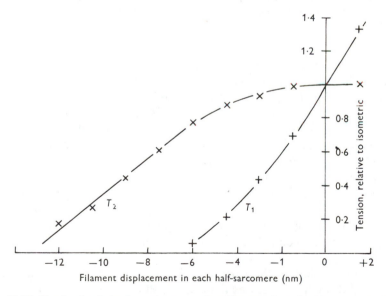

Fig. 12.20. Tension levels in the early stages of tension transients in a frog muscle fibre. The crosses show the tension at the end of phase 1 (the T_1 curve) and phase 2 (T_2), in a series of tension transients produced by different length changes. Further explanation in the text. (From A. F. Huxley, 1974.)

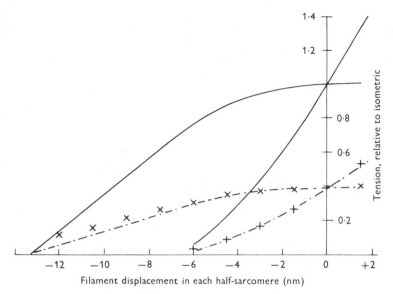

Fig. 12.21. T_1 and T_2 curves from the same frog muscle fibre at two different lengths. The continuous curves are those shown in fig. 12.20, obtained when the sarcomere length was 2.2 μm, at which all the cross-bridges would be overlapped by thin filaments. The crosses show T_1 and T_2 curves from the same fibre when stretched to give a sarcomere length of 3.1 μm, at which the overlap would be reduced to about 39 per cent. The interrupted curves are simply the continuous curves sealed down to 39 per cent. (From A. F. Huxley, 1974.)

these possibilities, Huxley and Simmons measured the T_1 and T_2 curves at different degrees of filament overlap. They found that both curves were scaled down in proportion to the decrease in overlap (fig. 12.21). Thus the stiffness of the elastic element whose properties are described by the T_1 curve is proportional to the number of cross-bridges that can be found and hence it is very reasonable to suggest that this elasticity actually resides in the cross-bridges.

It is now possible to link this analytical model of a cross-bridge with the structure of the myosin molecule. The tension generator is the S_1 subfragment, which can attach to a site on the actin filament and rotate about it. In doing so it pulls on the elastic element which is the S_2 subfragment that connects the S_1 head to the backbone of the myosin filament.

The time course of the early recovery of tension (phase 2) is much more rapid for large releases than for small ones, and even slower for stretches. Huxley and Simmons showed that this phenomenon could be accounted for if the movement of the tension generator took place as a small series of steps, so that the S_1 head could be in two, three or four

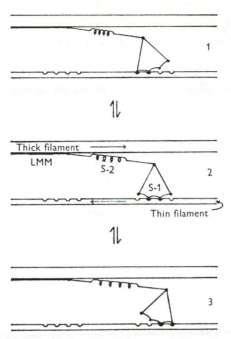

Fig. 12.22. Huxley and Simmons' model of the cross-bridge, incorporating an elastic element and a stepwise shortening element. Here the elastic element is equated with the S_2 portion of the myosin molecule and the stepwise shortening element with the S_1 portion and its combination with actin. (From A. F. Huxley, 1974.)

different positions while bound to the actin filament. The model shown in fig. 12.22 has three such positions.

The discovery that some of the series elasticity resides in the cross-bridges enables us to see why A. V. Hill's two-component model cannot accurately describe the mechanical properties of muscle. The more cross-links that are in existence, the stiffer the series elasticity will be, and so it cannot be regarded as an independent element with an invariant force–extension curve.

Many problems still remain with this model, as A. F. Huxley (1974) himself has pointed out. We have no hard evidence that the tension generator really is the S_1 fragment while the elasticity is the S_2 fragment; there are other possibilities. We do not know the molecular details of the tension generating mechanism, and we have very little idea as to how the splitting of ATP provides the energy for the process. But it seems likely that, as with its predecessors, Huxley and Simmons' theory will stimulate further experiments which will help our understanding of this most fascinating biological machine.

THE ANALYSIS OF TWITCHES

We have so far been mainly concerned with the dynamics of muscle during repetitive stimulation, when contraction is tetanic. We must now consider what happens during the response of the muscle to a single electrical stimulus.

Isometric and isotonic twitches

A possible explanation of the form of the isometric twitch in terms of A. V. Hill's two-component model is as follows (Hill, 1949*a*). In the resting condition, there is no tension in the muscle, and the series elastic component is slack. Soon after the stimulus, the contractile component begins to shorten. This extends the series elastic component (since the ends of the muscle are fixed) and so the tension begins to rise in accordance with the force–extension curve of the series elastic component. This rise in tension causes the shortening velocity of the contractile component to fall in accordance with the force–velocity curve and therefore the rate of rise in tension decreases. For a short time, the time course of tension development is indistinguishable from that at the start of a tetanus, but soon the twitch tension does not rise so rapidly as the tetanus tension. The reason for this is that the activity of the contractile component is less, in other words, its force–velocity relation is no longer maximal – the tension at any particular shortening velocity is less than that in a tetanus. Another way of stating this idea is to say that the tension which could be exerted by the contractile component if it were neither shortening nor lengthening (denoted by P_i and termed the 'active state') begins to fall. At the peak of the twitch, the series elastic component is at its fullest extent and the contractile component has just stopped shortening, so that the tension of the muscle is equal to P_i. But now P_i continues to fall, so that the contractile component is no longer capable of sustaining the tension in the series elastic component, so the series elastic component shortens, restretching the contractile component in the process; hence the tension begins to fall, and eventually reaches zero.

This explanation contains a number of assumptions. A major one is that the contractile component assumes the properties that it possesses during maintained stimulation very soon after the onset of contraction. Hill's evidence for this assumption depends upon experiments in which a muscle was rapidly stretched shortly after a stimulus. It was found (fig. 12.23) that tension rose rapidly during the stretch and fell a little immediately afterwards; after this the change in tension depended upon the extent of the stretch. For long stretches (trace *e* in fig. 12.23) the tension continued to fall. For shorter stretches (*b* and *c*), the tension

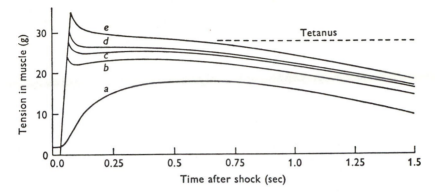

Fig. 12.23. Tension changes in a toad sartorius muscle following a rapid stretch applied just after a single stimulus. Trances *b* to *e* show the effects of stretches applied 34 msec after the stimulus. In *a*, the stretch was applied 70 msec before the stimulus. The final length of the muscle was the same in each case. (From Hill, 1949*a*, by permission of the Royal Society.)

rose for a short time, indicating that the contractile component was still able to shorten at these high tensions at this time. When the stretch was just sufficient to bring the tension up to the maximum tetanic tension, the tension remained steady for a short time before falling as the muscle relaxed (*d*). This indicates that the contractile component is capable of maintaining the full tetanic isometric tension at very short times after the onset of contraction.

In isotonic afterloaded twitches, shortening begins when the tension is sufficient to lift the load. The rate of shortening falls off until the peak of the twitch, when the load begins to extend the muscle again.

The onset of activity

A number of changes in the properties of a muscle occur just after the stimulus and before contraction begins. These changes are not fully understood, and in some cases it is not clear whether they are associated with the coupling process or the initiation of the contraction process.

The time between the stimulus and the beginning of the rise in tension during an isometric twitch is known as the *latent period*. It lasts about 15 msec in a frog sartorius at 0 °C, and decreases with increasing temperature. If the muscle is stretched so that there is some resting tension, the use of very sensitive recording methods reveals that there is a brief decrease in tension before the large increase that constitutes the contraction begins (fig. 12.24). This change is known as *latency relaxation* (Sandow, 1944). The amount of light diffracted by a muscle begins to change during the latent period, at about the time that latency relaxation begins (D. K. Hill, 1953). The resistance of the muscle to

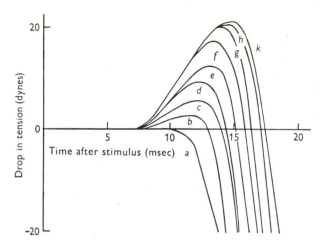

Fig. 12.24. Latency relaxation. The curves show the initial changes in tension during isometric twitches of a frog sartorius muscle at 0 °C. Muscle lengths (mm) as follows: *a* 27, *b* 30, *c* 31, *d* 32, *e* 33, *f* 34, *g* 35, *h* 36, and *k* 37. (From Abbott and Ritchie, 1951.)

stretching and the rate of heat production both begin to rise about half-way through the latent period (A. V. Hill, 1949*b*, 1950*b*). For a short period after the stimulus, an increase in the pressure on the muscle will increase the size of the subsequent twitch (Brown, 1936).

The 'active state'

A. V. Hill represented his ideas about what happens during an isometric twitch in the form shown in fig. 12.25. The curve P_n is the tension during

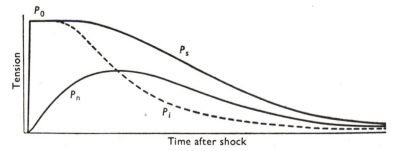

Fig. 12.25. The mechanical changes during a muscle twitch, according to the 'active stage' concept. P_n, time course of the tension change during an isometric twitch. P_s, time course of the tension changes during a twitch in which the muscle is rapidly stretched, by an amount just sufficient to extend the series elastic component to the length which it is during the plateau of an isometric tetanus, soon after the stimulus. This results in the tension attained immediately after the stretch being equal to the maximum tetanic isometric tension, P_0. P_i is the 'active state' curve deduced from these results. (From Hill, 1949*a*, by permission of the Royal Society.)

a normal twitch, P_s is that in a stretch which is just sufficient to cause the tension to remain constant for a short time after the end of a stretch (trace d in fig. 12.23); the tension at the beginning of this curve is equal to P_0, the maximum isometric tension in a tetanus. P_i, as we have seen, represents the tension which could be exerted by the contractile component if it were neither shortening nor lengthening. Originally, Hill used the term 'active state' merely to refer to the condition which a contracting muscle is in (in contrast to the 'resting state'), but the term later acquired quantitative connotations, being applied to the curve P_i in fig. 12.25.

The precise form of the P_i curve could not be determined from Hill's experiment; that given in fig. 12.25 is merely an intelligent guess. An attempt to perform a more accurate determination was made by Ritchie (1954). His method (fig. 12.26) involves the measurement of the maximum tension following a quick release at various times during an isometric twitch, thus giving a series of curves as is shown in fig. 12.27. At the peak of each of these curves, the tension is neither rising nor falling, therefore the contractile component is neither shortening nor lengthening, and therefore the tension is equal to P_i. Thus P_i can be plotted against time to give an 'active state' curve.

Is the time course of the 'active state' independent of length and tension changes? Consider fig. 12.28, which shows length changes following isotonic releases against the same load at different times during a twitch (Jewell and Wilkie, 1960). At late times after the

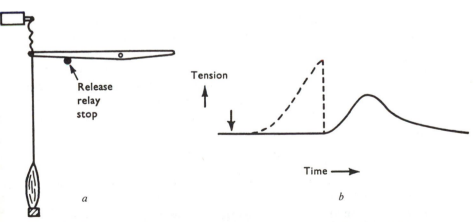

Fig. 12.26. Ritchie's method for determining the time course of the 'active state' during a muscle twitch. a: Schematic diagram of the apparatus; the quick release stop is withdrawn at different times after a single stimulus. b: Tension changes in the muscle; the full line indicates the tension recorded by the transducer, and the dotted line indicates the tension in the muscle up to the time of the quick release.

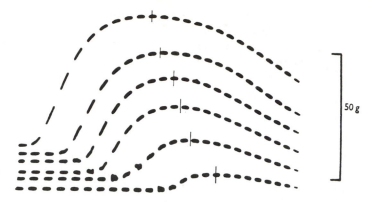

Fig. 12.27. Records obtained by the method shown in fig. 12.21, from a frog sartorius muscle at 0 °C. The top trace is an isometric twitch, the others show the redevelopment of tension after quick releases at various times after the stimulus. The final length of the muscle was the same in every case. The traces are modulated at 50 Hz. (From Ritchie, 1954.)

stimulus (at 0.6 sec, for example), the muscle is still shortening after a late release, but is lengthening after an early release. This indicates that, at this time, the 'active state' is greater than the load after the late releases but less than the load after the early releases. In other words, if the muscle is allowed to shorten, the duration of the 'active state' lessens. Further evidence that this is so has been provided by Penny-cuick (1964), from mechanical measurements, and by Hill (1964*b*), from heat measurements.

The results of these experiments lead to a rather curious situation. The time course of the 'active state' can only be measured by means of experiments in which the length and tension of the muscle are changed; indeed, as Pringle (1960) points out, the 'active state' is defined in such a way that it must be assumed that it is unaffected by changes in length and tension. Yet we have seen that such changes do alter the 'active state'! We must conclude that the concept of the 'active state' leads more to confusion than enlightenment.

A satisfactory substitute for the concept of the active state, that is to say a quantitative measure of the 'activity' of the muscle, is not yet available. A very promising start, however, has been made by Julian (1969, 1971; Julian and Sollins, 1973). His approach is via A. F. Huxley's 1957 theory, and he suggests that changes in 'activation' could involve changes in (*a*) the number of sites which are available for forming cross-bridges, and (*b*) the rate constant for breakage of cross-bridges. Both of these, he suggests, are dependent upon the myofibrillar calcium ion concentration. By postulating a plausible time course for

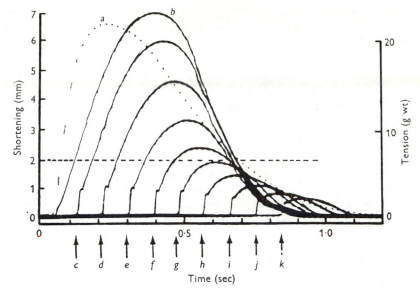

Fig. 12.28. An experiment which appears to indicate that the time course of the 'active state' is affected by length and tension changes in the muscle. Responses of a frog sartorius muscle at 0 °C to a single stimulus. Trace *a* shows the isometric twitch. Trace *b* shows an isotonic after-loaded twitch with a very light load. The other traces show the length changes occurring after isotonic releases against the same light load at different times after stimulation. (From Jewell and Wilkie, 1960.)

these changes following a single stimulus, it is possible to produce computer predictions of the behaviour of muscles during a twitch which agree closely with actuality.

CHEMICAL ENERGETICS OF CONTRACTION

We have seen that there is good evidence that ATP acts as the fuel for muscular contraction, and that it is possible to measure the total energy output (heat plus work) of a contracting muscle. Can this energy output be accounted for by the chemical energy released from ATP or creatine phosphate? It has not been very easy to answer this question, in spite of a considerable amount of experimental and theoretical work on the subject (some of the problems are described by Carlson and Wilkie, 1974, Curtin and Davies, 1973, Wilkie, 1976, Woledge, 1971 and Kushmerick, 1977).

Let us first examine a piece of work by Wilkie (1968) which apparently provides a clearcut correspondence between energy release and chemical breakdown. Wilkie used frog sartorius muscles which were poisoned with iodoacetate (to prevent glycolysis) and nitrogen (to prevent oxidative phosphorylation). Under these conditions we would expect any ADP formed from ATP breakdown to be immediately rephosphorylated by the breakdown of creatine phosphate, so that change in creatine phosphate content should represent the net chemical energy consumption of the muscle. Wilkie therefore measured the amount of creatine phosphate in individual muscles after they had performed a contraction in which he had also measured the work done by the muscle and the heat it produced. He also measured the creatine phosphate content of the unstimulated muscle from the other leg of the frog so as to estimate the change in creatine phosphate content due to activity. He repeated this experiment for a variety of different types of contraction: isometric and isotonic, twitches and tetani. The results are shown in fig. 12.29. Clearly the breakdown of creatine phosphate is directly proportional to the sum of the heat and work produced by the

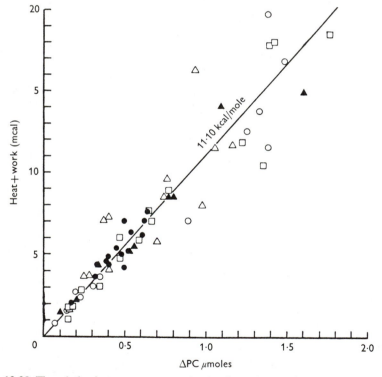

Fig. 12.29. The relation between energy production (heat plus work) and phosphocreatine breakdown in frog sartorius muscles poisoned with iodoacetate and nitrogen. Each point represents a determination on one muscle after the end of a series of contractions. The symbols indicate different types of contraction (isometric twitches and tetani, isotonic twitches, and isotonic tetani in which the muscle was stretched by the load). (From Wilkie, 1968.)

muscle, 46.4 kJ of energy being released per mole of creatine phosphate broken down.

Does all of this energy come from the breakdown of creatine phosphate? Although the figure of 46.4 kJ/mole (or 11.1 kcal/mole) did at first appear to be in reasonable agreement with the expected values for the heat of hydrolysis of creating phosphate, more recently doubts have arisen. Woledge (1973), as a result of careful calorimetric measurements, concludes that the expected value should be about 34 kJ/mole. So it seems as though there is some extra heat-releasing reaction which we do not know about.

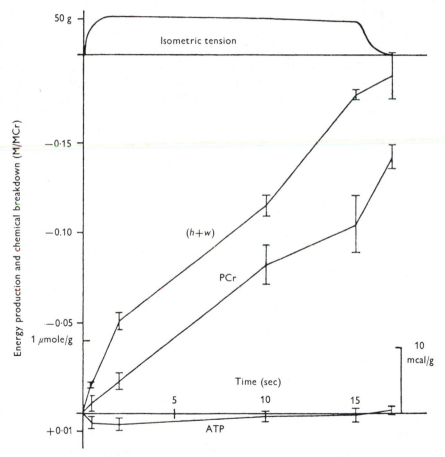

Fig. 12.30. The time course of heat production and phosphocreatine (PCr) breakdown in a frog sartorius muscle during an isometric tetanus. Chemical change is scaled as moles per mole of total creatine; heat production is expressed on the same scale by assuming a heat of reaction of 46.4 kJ/mole. (Gilbert *et al.*, 1971.)

Further complexities arise when the time course of chemical change is determined. In order to do this it was necessary to freeze muscles at different times during the contraction cycle, and hence to use a very rapid freezing technique. Kretschmar and Wilkie (1969) designed a rather dramatic apparatus to do this: the muscle was hit simultaneously from each side by two aluminium hammers which until that moment had been sitting in liquid nitrogen. Thus the creatine phosphate content of muscles could be determined at different times during a period of contraction, and compared with the heat production of other muscles. Using this method Gilbert, Kretschmar, Wilkie and Woledge (1971) found that the initial rate of heat production was notably higher than could be accounted for by chemical breakdown, even if a value of 46.4 kJ/mole was assumed for the energy release on creatine phosphate hydrolysis (fig. 12.30). In another investigation, Kushmerick and Davies (1969) measured the ATP breakdown during the contraction phase of muscles poisoned with DTNB. They found that here also the chemical changes could not account for all the heat produced by the muscle. Both of these sets of experiments indicate that here is another energy releasing process that we do not know about. Possibilities to be considered include heat changes associated with conformational changes in the contractile molecules and with the binding of calcium ions by them.

A. V. Hill (1960) tells how a newspaper report in 1928 alleged that he had solved The Mystery of Life. In so far as this relates to the energetics of the contractile mechanism, it is clear that his successors still have some way to travel before reaching that goal.

13 The comparative physiology of muscle

In the two previous chapters, we have been very largely concerned with the properties of vertebrate twitch skeletal muscle, exemplified by the sartorius of the frog. But it must be realized that the frog sartorius represents only one of a considerable variety of types of muscle, and a rather specialized one at that. In this chapter we shall attempt to broaden the picture a little by considering some of the variations in the processes of excitation and contraction in different muscles.

EXCITATION PROCESSES

Vertebrate skeletal muscles

The skeletal muscle fibres of frogs and toads fall into two distinct classes: *twitch* fibres and *tonic* fibres.* The sartorius is composed entirely of twitch fibres. Tonic fibres are found to varying extents in other muscles; the rectus abdominis is particularly rich in them.

Most twitch fibres have only one motor end-plate, although a few may have two. The cell membrane is electrically excitable and, as we have seen in chapter 6, will produce an all-or-none propagated action potential if it is sufficiently depolarized. The end-plate potential is sufficiently large to cause such a depolarization, so that, just as in the nerve axon, excitation consists of 'all-or-nothing' events.

The excitation process in tonic fibres is quite different (Kuffler and Vaughan Williams, 1953a, b). The fibres are electrically inexcitable, so that no propagated action potentials occur. There are a large number of neuro-muscular junctions on each fibre; this condition is known as *multiterminal innervation* (fig. 13.1b). Each motor impulse causes the release of only a relatively small amount of transmitter substance, so that the end-plate potentials are small in size. These end-plate potentials show pronounced summation on repetitive stimulation; thus the amount by which the fibre membrane is depolarized is dependent upon the frequency of the nervous input. There is negligible contractile

---

* Twitch and tonic fibres are frequently referred to as 'fast' and 'slow' fibres respectively. Since this nomenclature can lead to some confusion, we shall not adopt it here.

301

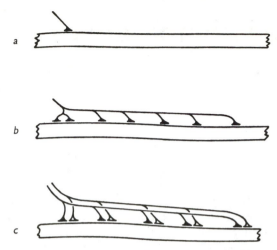

Fig. 13.1. Varieties of innervation in muscle fibres. *a*: Uniterminal innervation. *b*: Multi-terminal innervation. *c*: Multiterminal and polyneuronal innervation.

response to a single stimulus; with repetitive stimulation the tension rises with increasing stimulus frequency, reaching a maximum at about 50 impulses/sec. It seems probable that the tonic fibres are used for maintained low-tension contractions such as are involved in the maintenance of posture, whereas the twitch fibres are used for rapid movements.

The muscle fibres of fishes, with some exceptions, fall into two classes, twitch and tonic, as in amphibians (see Bone, 1964). The tonic fibres are usually reddish, since they contain the oxygen-storage pigment myoglobin, and located peripherally. Fish which spend much of their time 'cruising' at relatively low swimming speeds (such as salmonids and pelagic sharks) tend to possess a relatively greater development of these peripheral tonic fibres.

Birds also have two types of fibres, with uniterminal and multi-terminal innervation respectively, but those with multiterminal innervation are electrically excitable and may show propagated action potentials (Ginsborg, 1960).

Almost all the skeletal muscles of mammals are composed entirely of twitch fibres. Tonic muscle fibres are found in some of the muscles innervated by the cranial motor nerves, including the extraocular muscles which determine the direction of gaze, the tensor tympani of the middle ear, and some of the muscles of the larynx. The twitch fibre are of two types, named fast-twitch and slow-twitch according to their speeds and duration of contraction (Close, 1972; Buller, 1975). The fibres of any one motor unit are all of the same type, and sometimes whole muscles are composed of the same type of fibre. The soleus, for example, is a slow-twitch muscle while the gastrocnemius (which is larger and acts in parallel with it) contains a majority of fast-twitch fibres. Slow-twitch muscles are usually used in the control of posture, whereas fast-twitch muscles are used for locomotory and other movements.

Arthropod skeletal muscles

All arthropod muscles are multiterminally innervated. In many of them the cell membrane is electrically excitable, so that responses to nervous stimulation may look very similar to those recorded at the neuromuscular junction of a vertebrate twitch fibre (fig. 13.2a). However, on closer investigation, it is found that these electrically excited responses are not 'all-or-nothing' propagated action potentials, but graded responses whose size is roughly proportional to the initial depolarization (fig. 13.3). In this respect, these graded responses are similar to the subthreshold local responses of nerve axons. It seems reasonable to suggest that they are produced by a similar mechanism, i.e. that the increase in sodium conductance which follows depolarization is too small and proceeds too slowly to counteract the effects of the increase in potassium conductance. In some cases, it is clear that the electrically

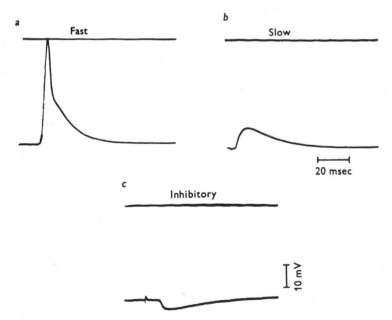

Fig. 13.2. Electrical responses of locust leg muscle fibres to stimulation of the 'fast' (a), 'slow' (b) and inhibitor (c) axons. The upper traces show the zero potential level. (From Usherwood, 1967.)

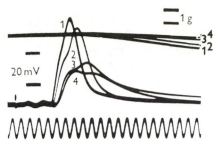

Fig. 13.3. The effect of a neuromuscular blocking agent (tryptamine) on the 'fast' response in a locust leg muscle. The upper traces show the tension developed during a twitch and, initially, zero membrane potential. The lower traces show the electrical response of a muscle fibre. Time signal 500 Hz. Trace 1 shows the normal response, traces 2 to 4 show responses at 5 sec intervals after application of tryptamine. Notice that the electrically excited component of the response is not an all-or-nothing phenomenon. (From Hill and Usherwood, 1961.)

excited responses are not dependent on an inflow of sodium ions; the inward current is carried by calcium ions in barnacle muscle fibres (Hagiwara and Naka, 1964) and, to some extent, by magnesium ions in muscle fibres of the stick insect (Wood, 1957).

The motor innervation of arthropod muscles differs strikingly from that of vertebrate muscles in three respects: (1) many muscles are supplied by an inhibitor axon, stimulation of which causes relaxation if the muscle is excited, (2) each muscle is innervated by only a small number of motor axons, and (3) *polyneuronal innervation* occurs, in which an individual muscle fibre may be innervated by two or more axons (fig. 13.1c). The second and third of these points are well illustrated by the jumping muscle (the extensor tibiae) of the hind leg of locusts (Hoyle, 1955a, b; Usherwood and Grundfest, 1965). This muscle is innervated by three axons: a 'fast' excitor, a 'slow' excitor and an inhibitor. The majority of the fibres are innervated by the 'fast' excitor. These fibres respond to stimulation of the 'fast' axon by means of a postsynaptic potential and a large electrically excited response, similar to that shown in fig. 13.2a; the mechanical response of the muscle is a rapid twitch. About a quarter of the fibres in the muscle, including some of those also innervated by the 'fast' excitor, are innervated by the 'slow' excitor. Stimulation of the 'slow' excitor produces postsynaptic potentials which are not large enough to elicit electrically excited responses (fig. 13.2b); the mechanical response of the muscle is scarcely detectable after a single stimulus, but repetitive stimulation produces a slow, fairly smooth contraction whose intensity increases with increasing stimulus frequency. Some of the fibres innervated by the 'slow' axon are also innervated by the inhibitor axon.

These fibres respond to inhibitory stimulation by hyperpolarizing potentials (IPSPs) similar to those of vertebrate motoneurons (fig. 13.2c); the tension produced by stimulation of the 'slow' excitor axon is reduced if the inhibitor is active at the same time.

Slight differences from this pattern are seen in other arthropod muscles. In the dorsal longitudinal flight muscles of locusts, the fibres are divided into five groups ('motor units'), each innervated by an axon which produces responses of the 'fast' type (Neville, 1963). The fibres of some muscles (such as locust spiracular muscles; Hoyle, 1959) are electrically inexcitable; the 'fast' response is then merely a larger version of the 'slow' response. The EPSPs of crustacean muscles frequently show marked summation and facilitation (fig. 13.4).

The electrical effects of postsynaptic inhibition in arthropod muscle fibres are essentially similar to those in vertebrate motoneurons. The sign and magnitude of the IPSPs can be altered by changing the membrane potential. The reversal potential is usually fairly near to the resting potential; it is practically unaffected by changes in external potassium ion concentration, but very sensitive to changes in external chloride ion concentration (Boistel and Fatt, 1958; Usherwood and Grundfest, 1965). Thus the action of the inhibitory transmitter (γ-aminobutyric acid) is to increase the chloride conductance of the membrane.

Presynaptic inhibition occurs in certain crustacean muscles, as has been shown very elegantly by Dudel and Kuffler (1961). Using an extracellular microelectrode, they were able to record the currents associated with EPSPs at single nerve terminals on the muscle fibres. It was found that these fluctuated in a discontinuous manner, suggesting corresponding fluctuations in the number of quanta of transmitter

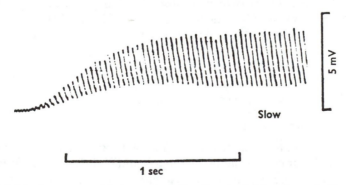

Fig. 13.4. Membrane potential changes in a *Panulirus* (rock lobster) muscle fibre during repetitive stimulation of the 'slow' excitor axon. Note the facilitation and summation of the responses. (From Hoyle and Wiersma, 1958.)

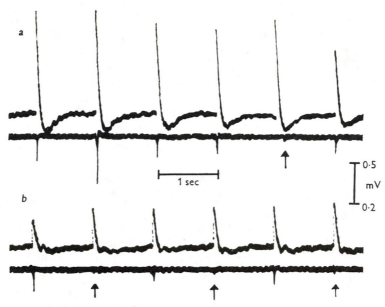

Fig. 13.5. Experiment showing presynaptic inhibition in a crayfish claw muscle. Upper traces show intracellular responses to stimulation of the excitor axon at a rate of 1/sec. Lower traces are simultaneous extracellular records showing activity at a single junctional region; the arrows indicate failure of transmission. *a*: Stimulation of the excitor axon alone. *b*: Stimulation of the inhibitor axon 2 msec before each excitatory stimulus. The decrease in the size of the intracellular responses in *b* shows that inhibition is occurring; the increase in the number of failures in transmission recorded by the extracellular electrode indicates that this inhibition must be at least partially presynaptic. (From Dudel and Kuffler, 1961.)

released (fig. 13.5*a*). When the inhibitory axon was also stimulated, the number of 'failures' at any one terminal increased, and the EPSP recorded intracellularly (which is caused by transmitter release from all the terminals on the fibre) decreased in size (fig. 13.5*b*). A statistical analysis of the type used by del Castillo and Katz on magnesium-treated frog muscle fibres (see chapter 6) was then carried out. It was found that the average number of quanta released per terminal per impulse was reduced from 2.4 to 0.6 during inhibition (fig. 13.6), which implies that the inhibitory process must be presynaptic.

Vertebrate heart muscles

The hearts of animals can be divided into two general types according to how their rhythmic contractile activity is initiated. In *myogenic* hearts, found in vertebrates and molluscs, excitation arises spontaneously in the heart muscle fibres themselves. In *neurogenic* hearts, found in many arthropods, the heart muscle fibres only contract in response to nervous

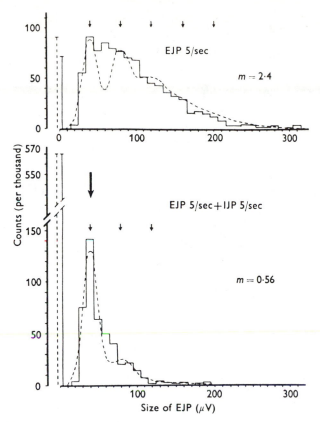

Fig. 13.6. Histograms giving the size distribution of excitatory postsynaptic potentials recorded extracellularly (as shown in the lower traces of fig. 13.5) from single junctions of a crayfish claw muscle fibre. Upper graph: in the absence of inhibition. Lower graph: each excitatory impulse is preceded by an inhibitory one 2 msec earlier. The broken curves are calculated Poisson distributions for a quantum size of 40 μV (standard deviation 10 μV) and average quantal content (m) equal to 2.4 (upper graph) and 0.56 (lower graph). Small arrows indicate units of quantum size; large arrow gives the average size of spontaneous miniature potentials. (From Dudel and Kuffler, 1961.)

stimuli. A full discussion of heart muscle physiology would be out of place here, but it is instructive to examine some of the properties of the excitation process in vertebrate heart muscle fibres.

Intracellular recordings from heart muscle fibres were first made by Draper and Weidmann (1951), using isolated bundles of Purkinje fibres from dogs. In mammalian hearts, the Purkinje fibres form a specialized conducting system which serves to carry excitation through the ventricle. After being isolated for a short time, they begin to produce rhythmic spontaneous action potentials, as is shown in fig. 13.7. The form of the

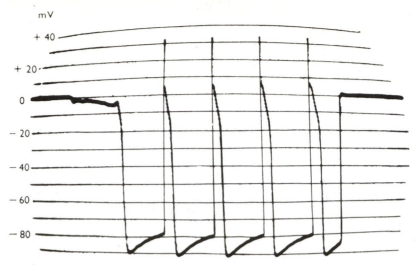

Fig. 13.7. Intracellular record of action potentials in a Purkinje fibre of a dog heart. The microelectrode entered the fibre soon after the start of the trace, and was withdrawn just before its end. The interval between successive action potentials was 1.4 sec. (From Draper and Weidmann, 1951.)

action potentials differs from those of nerve axons and vertebrate twitch muscle fibres in that there is a prolonged 'plateau' between the peak of the spike and the repolarization phase. The action potential is initiated when the slowly-rising *pacemaker potential* crosses a threshold level.

What is the ionic basis of these heart muscle action potentials? Draper and Weidmann showed that the peak membrane potential was reduced when the external sodium concentration was lowered, approximately in accordance with the Nernst equation for sodium ions, (5.2). This suggests that, as in the action potential of nerve axons, the initial rapid depolarization is brought about by a regenerative increase in the sodium conductance of the membrane. However, there is some evidence that this is not immediately followed by an increase in potassium conductance: the membrane resistance is quite high during the plateau (Weidmann, 1957) and, in the absence of sodium ions, the membrane resistance is higher for depolarizing currents than it is for hyperpolarizing currents (Hutter and Noble, 1960).

Direct measurements of membrane ionic currents need an effective voltage clamp system, but there are difficulties in producing one because of the complex geometry of the heart muscle. The 'sucrose gap' method has been much used in recent years. Fig. 13.8 illustrates the basics of the system. Part of the muscle bundle is enclosed in an isotonic sucrose solution; this penetrates the intercellular space between the cells and so

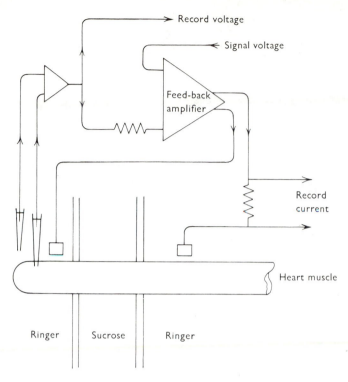

Record voltage

Signal voltage

Feed-back amplifier

Record current

Heart muscle

Ringer   Sucrose   Ringer

Fig. 13.8. Voltage clamp technique for heart muscle, using the sucrose-gap method.

renders them incapable of carrying electric current. It is therefore possible to pass current between the two Ringer-filled compartments and know that all of it is passing through the intracellular material; and hence it must all pass out across the cell membrane. Hence we have a measure of membrane current and a means of ensuring that this current flows across the cell membranes that we are interested in. These are those of a cell which is impaled with an intracellular microelectrode to record its membrane potential. The membrane potential is then used to control the current flow through the membrane by means of a feedback loop in the usual manner.

Another method has been used for Purkinje fibres, which are larger than other cardiac cells. Here a short length of fibre is cut out (the injured ends heal over) and two microelectrodes are inserted. Both methods are not very good at dealing with large currents.

One result of the use of the voltage clamp method is the discovery that depolarization results in an inward flow of calcium ions. Beeler and

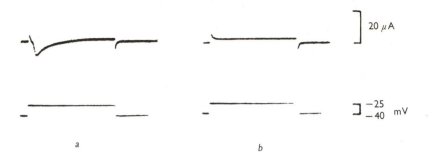

Fig. 13.9. Calcium current in voltage-clamped ventricular muscle from a dog heart. External calcium ion concentration was 1.8 mM in *a* and zero in *b*. The holding potential was -40 mV, which is sufficiently depolarized to inactivate the sodium current. Further depolarization to -25 mV for 570 msec resulted in the slow transient inward current shown in *a*. (From Beeler and Reuter, 1970.)

Reuter (1970) were able to demonstrate this by first depolarizing the fibres to -40 mV so as to inactivate the sodium current; further depolarization then resulted in a slow inward current (fig. 13.9). This slow current was unaffected by tetrodotoxin (which blocks the fast inward sodium current) but was abolished in calcium-free solutions.

Sorting out the full nature of the cardiac action potential has proved to be a complicated and as yet unfinished task (see Noble, 1975). A possible model, due to Noble, is shown in fig. 13.10. At the beginning of the trace the potassium conductance is relatively high and the sodium and calcium conductances very low; the membrane potential is thus at its most negative. But gradually the sodium conductance rises and the potassium conductance falls so that the membrane potential becomes slowly less negative. This is the pacemaker potential. After a time the pacemaker potential has depolarized the membrane sufficiently to start a large increase in sodium conductance and then we get the familiar runaway relation between membrane potential and sodium conductance so that, just as in the nerve axon, there is a massive inflow of sodium ions and a rapid overshooting depolarization. This increase in sodium conductance is rapidly inactivated so that the membrane potential becomes rapidly less positive. But now the model departs radically from the situation in nerve axons. The potassium conductance has fallen to a very low level and there is an elevated calcium conductance; hence the membrane potential remains near zero for some hundreds of milliseconds. This plateau declines gradually and is brought to an end as a result of an increase in potassium conductance, and the remaining calcium and sodium conductances are finally cut off during the repolarization phase.

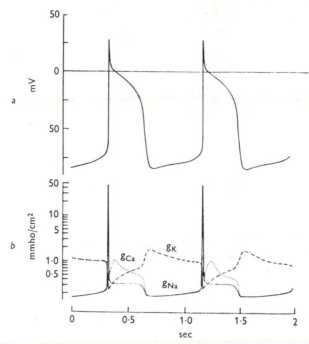

Fig. 13.10. A possible scheme whereby the form of the Purkinje fibre responses (*a*) can be explained in terms of changes in the sodium, potassium and calcium conductances of the cell membrane (*b*). (From Noble, 1974.)

At a much simpler level of technical expertise it is possible to measure the electrical activity of the human heart simply by attaching leads to the wrists and ankles of the subject. The resulting record is known as an electrocardiogram, or ECG for short. ECGs were first measured by Einthoven, using the string galvanometer which he invented for the purpose (see Einthoven, 1924), and their measurement (now usually by means of a hot wire pen recorder) has long been a standard procedure in medical practice. Fig. 13.11 shows a typical ECG, recorded between the right arm and left leg. The different peaks in the electrical cycle of events were labelled the *P, Q, R, S,* and *T* waves by Einthoven. The events in the heart cycle to which these electrical waves correspond can be worked out by recording with surface electrodes from exposed hearts in experimental animals. The heart beat is initiated by pacemaker activity in the cardiac cells of the sinuatrial node (see fig. 13.12). This excites the adjacent cells of the atria and a wave of depolarization sweeps over the whole of the atria. The currents associated with this are recorded in the ECG as the *P* wave. During the plateau of the atrial action potentials there is little current flow and so the level of the ECG

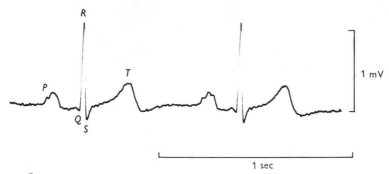

Fig. 13.11. A human electrocardiogram, recorded between electrodes applied to the right wrist and left ankle.

is not affected. On reaching the atrioventricular node the atrial depolarization excites the specialized conducting tissue of the ventricles, the Purkinje fibres of the bundles of His. The action potential passes down the left and right branches of this system, but the activity is not evident in the ECG because the number of cells involved is small and hence the current flow is small also. The bundle of His brings excitation to the mass of the ventricular muscle cells, beginning in the septum and then spreading from the apex of the ventricle up to the base. The net currents involved in the depolarization of all these ventricular cells are very large and are seen in the ECG as the *QRS* complex. The precise form of the *QRS* complex is explicable in terms of the detail of this

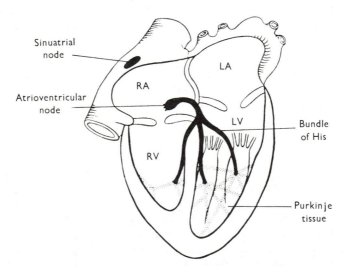

Fig. 13.12. Pacemaker and specialized conductile regions of the mammalian heart. (From Scher, 1965.)

three dimensional spread of activity in the ventricle, such as is described, for example, by Scher (1962). After this the whole of the ventricle is depolarized and there is very little electrical current flow (the ventricular muscle is contracting so as to pump blood out along the aorta and pulmonary artery). Then repolarization of the ventricular fibres occurs, at slightly different times in different places, and the current flow associated with this is seen as the *T* wave. After this the heart is electrically at rest except in the pacemaker regions, its muscles are relaxing and it is refilling with blood ready for the next cycle. The pacemaker potentials preceding the next *P* wave are not visible in the ECG.

A heart would be ineffective if the rhythmic action potentials and their associated contractions in any one fibre were independent of those of its fellows: it is obviously essential that the contractions of the fibres should be synchronized. Vertebrate heart muscle consists of a network of branching fibres which are connected to each other by *intercalated discs* (fig. 13.13). Under the electron microscope these are seen to be concentrations of dense material on each side of two cell membranes. Desmosomes, providing mechanical attachment, are present, and there are gap junctions (p. 171) between the cells (Fawcett and McNutt, 1969). It is probable that these gap junctions offer a low resistance to current flow, so that local circuits set up by an action potential in one cell can cross into the next cell and so excite it. The whole of the heart will then be 'driven' by those cells which have the most rapid spontaneous frequency. In the frog, these *pacemaker* fibres occur in the sinus venosus, and the excitation spreads from there into the auricle and

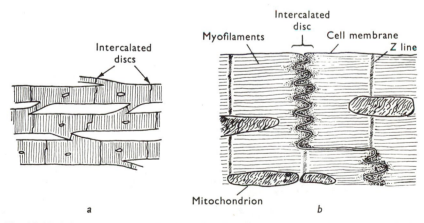

Fig. 13.13. Schematic diagrams to show the structure of vertebrate heart muscle. Light microscopy (*a*) shows uninucleate muscle cells separated by intercalated discs. Under the electron microscope (*b*) the intercalated discs are seen to be concentrations of dense material on each side of an intercellular boundary.

ventricle. In accordance with this view, it is found that certain fibres in the sinus venosus show pronounced pacemaker potentials, whereas these are absent in auricular fibres (fig. 13.14).

Although the heart-beat is myogenic in origin, its rate and amplitude can be modified by the action of extrinsic nerves. Stimulation of the parasympathetic nerve fibres (which release acetylcholine) causes hyperpolarization and a cessation of spontaneous activity. This inhibition is probably brought about by an increase in the potassium conduc-

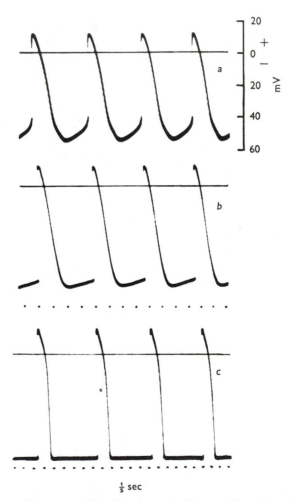

$\frac{1}{5}$ sec

Fig. 13.14. Intracellular records from frog heart muscle, *a* and *b* are from different fibres in the sinus venosus, *c* is from a fibre in one of the auricles. Notice the presence of pacemaker potentials (slow depolarizations preceding each action potential) in *a* and *b*, and their absence in *c*. (From Hutter and Trautwein, 1956.)

tance of the fibre membranes (Hutter, 1961). Stimulation of the sympathetic (adrenergic) nerve fibre causes depolarization and an increase in the firing frequency.

Vertebrate smooth muscle

Smooth muscles – muscles in which there are no oblique or cross striations – form the muscular walls of the viscera and blood vessels in vertebrates. They also occur in the iris, ciliary body and nictitating membrane in the eye, in the tracheae, bronchi and bronchioles of the lungs, and are the small muscles which erect the hairs in mammals. Bennett (1972) has provided an excellent account of the excitation of smooth muscles. Individual smooth muscle cells are uninucleate, and much smaller than the multinucleate fibres of skeletal muscles; they are usually about 4 μm in diameter and up to 400 μm long. The cells are held together in bundles, 20 μm to 200 μm in diameter, by thin sheets of connective tissue, and these bundles themselves are frequently interconnected. Adjacent cells are connected by a small flask-shaped protrusion from one into a pocket of the other, in which the opposing cell membranes are brought close together to form gap junctions; it seems very likely that current can flow from one cell to another through these gap junctions.

Many vertebrate smooth muscles show a great deal of spontaneous activity. This is particularly so in intestinal muscles, where the spontaneous contractions serve to mix and move the gut contents. The electrical activity consists of slow waves of variable amplitude and all-or-nothing action potentials (fig. 13.15). The fibres are depolarized and the frequency of the action potentials increases if the muscle is stretched (Bülbring, 1955). The spontaneous activity can be modified by the action of extrinsic nerves, by adrenalin, and, in uterine muscle, by hormones of the reproductive cycle. The smooth muscles of the iris, nictitating membrane and vas deferens are not spontaneously active.

Action potentials can be initiated in sheets or strips of smooth muscle by electrical stimulation; they will then propagate along the axes of the muscle cell and bundles and also, more slowly, across them.

The terminals of the sympathetic nerves supplying smooth muscle can be prepared for light microscopy by exposing freeze dried tissues to hot formaldehyde vapour; the catecholamines present are then converted to substances which fluoresce strongly under ultraviolet light. The relations of the terminals to the smooth muscle cells can be determined by electron microscopy. These methods show that the fine terminal branches have frequent swellings along their length, known as varicosities, which contain synaptic vesicles. In some muscles each varicosity is intimately apposed to an individual muscle cell (fig. 13.16*b*). In the spontaneously

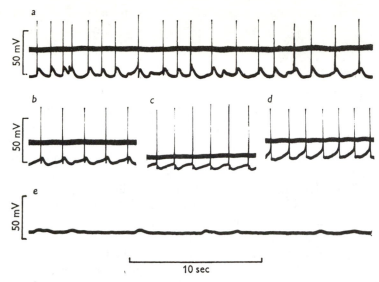

Fig. 13.15. Intracellular records of membrane potential in smooth muscle fibres of guinea pig intestine, showing various patterns of spontaneous activity. (From Bülbring, Burnstock and Holman, 1958.)

active muscles of the gut, the axon terminals remain in small bundles, associated with Schwann cells, and their varicosities do not form close contacts with individual muscle cells (fig. 13.16*a*). And in some muscles innervation is via both small axon bundles and close-contact varicosities.

Stimulation of the motor nerves innervating smooth muscle results in junction potentials (postsynaptic potentials) of various types (fig. 13.17). Excitatory nerves produce depolarizing potentials whereas inhibitory nerves produce hyperpolarizing ones. In muscles with innervation via close-contact varicosities, these have a relatively rapid time course,

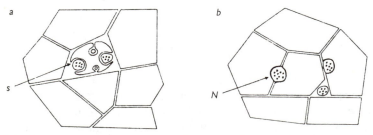

Fig. 13.16. Neuromuscular junctions in vertebrate smooth muscle. In *a* the axons occur in small bundles (*s*); they do not closely contact the muscle cells. In *b* the axons (*N*) occur singly and their varicosities make close synaptic contacts (with a 20 nm synaptic cleft) with the muscle cells. (From Bennett, 1972.)

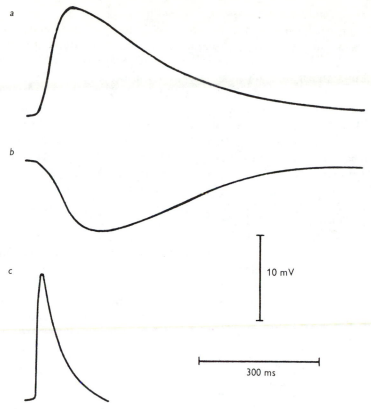

Fig. 13.17. Junction potentials in smooth muscle. *a*, An excitatory junction potential in guinea pig taenia coli muscle; *b*, an inhibitory junction potential in guinea pig taenia coli; *c*; an excitatory junction potential in mouse vas deferens. Innervation is by small axon bundles in the taenia coli (*a* and *b*), and by close-contact varicosities in the vas deferens (*c*). Notice that *c* has a much faster time course than *a* and *b*. (From Bennett, 1972.)

whereas in those with innervation via axon bundles only, the junction potentials may last for half a second or more.

THE GRADATION OF CONTRACTION IN SKELETAL MUSCLES

It is clearly necessary that the force exerted by a muscle should be capable of being regulated by its nervous input. There are two main ways in which this can be brought about; either the proportion of the fibres in the muscle that are excited is varied, or the degree to which the individual fibres are excited is controlled.

The first system is the main method in vertebrate skeletal muscles, indeed the only one in those muscles that consist entirely of twitch fibres. Each muscle is innervated by a large number of motor axons, and each axon innervates a group of muscle fibres known as a 'motor unit'. Since an action potential in a motor axon initiates propagated action potentials in each of the muscle fibres in the motor unit that it innervates, the excitation process in any one motor unit is all-or-nothing. The contractile response that follows excitation depends to some extent on the frequency of nervous stimuli (since the twitch tension is less than the tetanus tension), but the range over which this can occur is not great. For an isometric load, we might describe the contractile response as being 'half-or-more-or-nothing'. Thus fine control of tension development has to be brought about by variation in the number of active motor units. The excitation of different motor units in the muscle is usually asynchronous, so that contraction of the whole muscle is a fairly smooth process.

In arthropods, the mechanism of gradation is quite different. Here, as we have seen, the whole muscle is frequently innervated by only two or three motor axons, and the excitation of an individual fibre is definitely not an all-or-nothing process. The size of the electrical response in any fibre is dependent on (1) which axon, 'fast' or 'slow', is active, (2) the frequency of arrival of excitatory nerve impulses, which affects the degrees of membrane depolarization by summation and facilitation of the end-plate potentials, and (3) the activity of the inhibitor axon.

These two main types of gradation control may be combined in some cases, as in those vertebrate muscles which contain a large number of tonic fibres.

THE EXCITATION–CONTRACTION COUPLING PROCESS

It seems to be generally true that, in all muscles, contraction is initiated by depolarization of the cell membrane, whether this is brought about by nervous activity, spontaneous activity, drugs, or an increase in the external potassium ion concentration. Similarly, calcium ions are generally implicated in the excitation–contraction coupling process. Within this pattern, however, there are a number of differences in the details of the coupling process in muscles of different types.

The sarcoplasmic reticulum and the T system are well developed in muscle fibres that contract and relax rapidly, such as vertebrate twitch fibres and many arthropod muscles. These systems are much less evident, though still present, in vertebrate tonic fibres and in cardiac muscle. Thus, in fibres of large diameter, the degree of development of the T system and the sarcoplasmic reticulum seems to be related to the speeds of contraction and relaxation of the muscles. These internal membrane systems are practically absent from the thin fibres of smooth muscles, and also from the thin striated fibres of *Amphioxus* (Peachey,

1961). Peachey suggests that, in these cases, the fibres are activated by calcium ions either released from the membrane or entering the cell from the outside, the diffusion distances involved being small (0.5 μm in *Amphioxus* fibres, 2 to 4 μm in visceral muscle fibres).

As we have seen in chapter 11, potassium contractures may be of two types: those which are maintained for a considerable time, and those in which the muscle relaxes after a few seconds even though its cell membrane is still depolarized. Vertebrate twitch fibres show phasic contractures, while those of vertebrate tonic fibres are maintained. Arthropod muscles may show either type; the potassium contracture of the jumping muscle in locusts is phasic, but that of the homologous muscle in the second leg is maintained (Aidley, 1967). Phasic contractures seem to be associated with a high degree of development of the sarcoplasmic reticulum, but we do not know just what the difference in the coupling processes of the two types of muscle is.

The regulatory proteins

In vertebrate skeletal muscles, as we have seen in chapter 11, contraction is regulated by the calcium–sensitive troponin–tropomyosin system on the thin filaments. Similar thin–filament linked regulatory systems are found in amphioxus and mysid crustaceans and in the fast muscles of decapod crustaceans (Lehman, 1976).

In molluscan muscles, however, Kendrick-Jones, Lehman and Szent-Györgyi (1970) found that there is no troponin present. Molluscan thin filaments will react with rabbit myosin so as to split ATP, in the absence of calcium ions. But there is a calcium-sensitive mechanism on the thick filaments: molluscan myosin will only combine with pure actin and split ATP in the presence of calcium ions. Similar myosin linked regulatory systems are found in echinoderms and in a number of minor groups.

In insects and most other arthropods, and in annelids and nematodes, both types of regulation are present simultaneously.

THE ORGANIZATION OF THE CONTRACTILE APPARATUS

Striated muscles

The structure of striated muscles has been dealt with in chapter 11. The skeletal and cardiac muscles of vertebrates and arthropods, and the visceral muscles of arthropods, are almost all striated. Striated muscles also occur locally in many other animals; some examples are the swimming muscles of jellyfish, the pedicellariae muscles of echinoderms, and certain muscles in the heads of some annelids.

The sarcomere length, or, to be more precise, the length of the *A* band, is not always the same in different muscles. What effect will the length of the *A* band have on the contraction of the muscle? Let us assume that an individual cross-bridge always produces the same amount of tension in an isometric contraction. Cross-bridges acting between any one pair of actin and myosin filaments act in parallel, more of such cross-bridges can be formed with longer *A* bands, and therefore the maximum tension per unit cross-sectional area is proportional to the length of the *A* band. The forces generated in different half-sarcomeres are in series with each other, and so the total force is unaffected by the number of sarcomeres in a fibre. Hence we would expect muscles with long *A* bands to have a higher maximum tension per unit cross-sectional area than those with short *A* bands. A corollary of this idea is that, other things being equal, the velocity of shortening should decrease with increasing length of the *A* band. However, it seems that other factors are involved in the determination of shortening velocities; the sarcomere lengths of frog twitch and tonic fibres are the same, but their speeds of contraction are very different.

We have seen that, in vertebrate skeletal muscles, the myofilament array is such that each *A* filament is surrounded by six *I* filaments and each *I* filament by three *A* filaments, so that there are two *I* filaments per *A* filament in each half-sarcomere. Different arrangements occur in arthropod muscles. In most insect flight muscles each *A* filament is surrounded by six *I* filaments, but each *I* filament is placed between two *A* filaments so that the *I* to *A* ratio is three to one (fig. 13.18). In thoracic muscles of the cockroach there are eight or nine *I* filaments round each *A* filament and the *I* to *A* ratio is four to one (Hagopian and Spiro, 1968). Finally, in some insect leg and visceral muscles there are twelve *I* filaments round each *A* filament and the *I* to *A* ratio is six to one (Hagopian, 1966; Smith, Gupta and Smith, 1966).

A possible explanation of the functional significance of these 'extra' *I* filaments is as follows. We have seen that, other things being equal, the

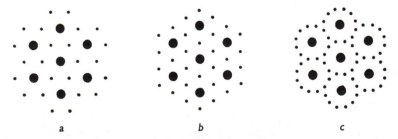

a   b   c

Fig. 13.18. Myofilament arrays in vertebrate and insect striated muscles. The diagrams show the arrangement of myosin (large spots) and actin (small spots) filaments as seen in transverse sections through the overlap region. *a*: Vertebrate muscles. *b*: Insect flight muscles. *c*: Insect leg and visceral muscles. (From Smith, Gupta and Smith, 1966.)

TABLE 13.1. *The relation between the I-to-A filament ratio and the A band length in various striated muscles of insects and vertebrates*

Muscle	A band length (μm)	I/A filament ratio
Stick insect visceral muscle[1]	6 to 7 ⎫	
Cockroach leg muscle[2]	4.5 ⎭	6:1
Locust flight muscle[3]	3.1 ⎫	
Giant water bug flight muscle[4]	2.3 ⎭	3:1
Vertebrate muscles	1.6	2:1

[1] Smith, Gupta and Smith, 1966; [2] Hagopian, 1966;
[3] Weis-Fogh, 1956; [4] Ashhurst, 1967.

maximum isometric tension produced by a muscle is proportional to the length of the *A* band. Hence the longer the *A* band, the greater the strain on the myofilaments during contraction. The presence of 'extra' *I* filaments is associated with long *A* bands, as is shown in table 13.1, so it is possible that the greater total number of *I* filaments is needed to reduce the strain on each one of them.

Insect asynchronous flight muscles contain another contractile protein, paramyosin (Bullard, Luke and Winkelman, 1973). This is a rod-like protein, molecular weight about 100 000, which aggregates readily to form structures with a repeat periodicity in the region of 720 Å. It was first found in molluscan muscles, where it is a major component of the thick filaments. In insect flight muscles, it may form a core in the thick filaments, and could conceivably connect them to the *Z* lines.

Oblique striated muscles

Muscles of this type possess striations which follow a helical or oblique course within the fibres. They are found in annelids, molluscs, nematodes and some other invertebrates. The fibres contain two types of filament, corresponding to the *A* and *I* filaments of striated muscles. The oblique striations arise from a staggered arrangement of the thick filaments, as can be seen in electron micrographs of tangential longitudinal sections (fig. 13.19). During contraction the thin filaments slide between the thick ones, so shortening the distance between successive striations. Since the muscle cell is a constant volume system, shortening is accompanied by widening, hence the angle of the striations alters: they become less oblique (Lanzavecchia, 1968; Mill and Knapp, 1970).

Paramyosin occurs in the thick filaments of the oblique striated muscles of annelids and molluscs.

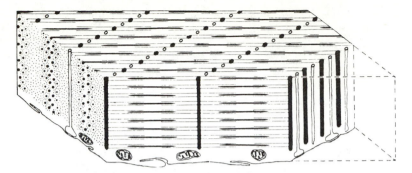

Fig. 13.19. Diagram showing the arrangement of myofilaments in an oblique striated earthworm muscle fibre. The thin filaments are attached to Z rods, shown in black here. Alternating with the Z rods are tubules of the sarcoplasmic reticulum. The angle of the oblique striations is exaggerated. (From Mill and Knapp, 1970.)

Vertebrate smooth muscles

Vertebrate smooth muscles contain the major contractile proteins actin and myosin and the actin occurs as thin filaments which are readily seen by electron microscopy. The organization of the myosin however has been a matter of some controversy (*see* Shoenberg and Needham, 1976). Although myosin is present in the muscle, no thick filaments were seen in the first electron micrograph sections. More recently, it has been possible, under the right conditions, to observe aggregation of myosin either as thick filaments 120 to 300 Å in diameter or as ribbons measuring 80 Å by 200 to 1100 Å in section. Ribbons are more likely to be seen in sections from cold muscles which do not show contractile activity. X-ray diffraction shows a strong 143 Å reflection in cold muscles at 12 to 16 °C, but a much weaker one in muscles maintained at 37 °C (Lowy, Poulson and Vibert, 1970; Shoenberg and Haselgrove, 1974). Shoenberg and Haselgrove conclude that the strong 143 Å reflection is produced by myosin in ribbons whereas the weaker one, produced by muscles apparently in conditions similar to those in the body, is produced by myosin in filaments that are probably about 200 Å in diameter.

The functional significance of this lability of organization of the myosin is quite obscure. Indeed it is not clear whether or not the myosin must be arranged in filaments in order for contraction to take place.

Many of the filaments apparently lie free in the cell. Others are attached at one end to *dense patches* which adhere to the cell membranes. In many preparations there are also a number of *dense bodies* in the body of the cell. Anchored to these is a set of *intermediate filaments* 100 Å in diameter; Shoenberg and Needham suggest that these form a supporting framework for the thick and thin filaments.

Invertebrate smooth muscles

Many muscles in invertebrates show no signs of striation when examined by light microscopy. Electron micrograph sections show two types of filaments, thick and thin, but these are not transversely or obliquely aligned (Hanson and Lowy, 1960).

The 'catch' muscles of molluscs (p. 333) have been particularly investigated (Lowy and Hanson, 1962). The thin filaments are similar to the *I* filaments of striated muscles and show the same F-actin double helix structure when isolated. But the thick filaments vary in thickness from 150 to 1500 Å, and appear to consist of a number of closely-packed ribbon-like elements. The muscles contain a high proportion of paramyosin, and it is this that forms the major part of the thick filaments. Myosin can be extracted from the thick filaments, when an array of roughly triangular pits are seen on their surface (Szent-Györgyi, Cohen and Kendrick-Jones, 1971). Neighbouring triangles all point in the same direction, but this reverses in the middle of the filament, so that the two halves have opposite polarities, rather as in the myosin filaments of vertebrate striated muscles. Many of the thin (actin) filaments are attached to dense bodies. They form 'arrowheads' when decorated with heavy meromyosin, and these are of opposite polarity on each side of a dense body, just as are those on each side of a *Z* band in striated muscles.

Does the sliding filament theory apply to smooth muscles?

For those smooth muscles which contain two types of filaments, thick and thin, we may speculate that shortening occurs by sliding of the filaments past each other, as is shown in fig. 13.20. Since the filaments are not transversely aligned with each other, it is rather difficult to test this idea. We may see what is apparently the 'end' of a filament in an electron micrograph, but it is just as likely that the filament has passed out of the plane of the section, hence it is not possible to measure

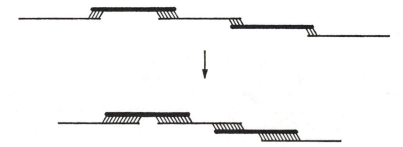

Fig. 13.20. A possible sliding-filament mechanism for invertebrate smooth muscles.

filament lengths with any accuracy. Points in favour of the applicability of the sliding filament theory are (1) that cross-bridges between thick and thin filaments can sometimes be seen, and (2) that the axial periodicities determined by X-ray diffraction appear to be constant at different muscle lengths. The situation in vertebrate visceral muscles, where the myosin molecules may not always be aggregated into filaments, is much less clear.

MECHANICAL PROPERTIES

The mechanical properties of muscles are related to the functions that they subserve in the body. For example, muscles that are involved in rapid movements have greater speeds of contraction than do those involved in the maintenance of posture. The muscles of small animals tend to contract more rapidly than those of larger animals, since their locomotory movements are relatively more rapid (see Hill, 1950c). In some cases, specialization is carried further, and apparently involves the development of new mechanical properties. We shall consider two examples of these specializations: the fibrillar muscles of insects, and the 'catch' muscles of molluscs.

Insect fibrillar muscles

The wing-beat frequency of many insects is very high; the record is held by the midge *Forcipomyia*, in which it is about 1000 beats per second. This means that the tension in the flight muscles in *Forcipomyia* must rise and fall within a millisecond, which seems almost impossibly fast. Pringle (1949) measured the electrical responses of the flight muscles of a flying blow-fly (*Calliphora*) and showed that their frequency was very much less than that of the wing-beat. This indicates that the individual contractions of the muscles, which produce the wing-beat, are not 'twitches' analogous to the twitches of other muscles. Pringle suggested that the function of excitation is to bring the contractile apparatus into a state of activity (analogous to the tetanus condition in other muscles) in which rhythmic contractions are possible. Further work by Roeder (1951) and others (see Pringle, 1957) showed that this asynchronous relation between the neurally excited muscle potentials and the contraction frequency occurs in the flight muscles of flies, beetles, bees and wasps, and certain bugs; it is also found in the tymbal (sound-producing) muscles of some cicadas (Pringle, 1954). These muscles are described as 'fibrillar' (since their fibres contain closely packed myofibrils) or 'asynchronous'. The flight muscles of more primitive insects, such as locusts, dragonflies and moths, are not of the fibrillar type, and show the familiar one to one ratio between the electrical and mechanical responses to nervous stimulation.

A closer investigation of the mechanical properties of fibrillar muscle was carried out by Machin and Pringle (1959), using a flight muscle from the rhinoceros beetle *Oryctes*. The muscle was partly dissected from the insect and connected to a piezo-electric force transducer at one end and to a moving-coil vibrator at the other. The position of the vibrator arm was measured by means of a light beam and a phototransistor, so as to give the length of the muscle. The load on the muscle (produced by the action of the vibrator) was electrically controlled from the output of the phototransistor, allowing regulation of its stiffness, damping and mass (inertia), as is shown in fig. 13.21. Examination of the load control circuit shows that it contains two differentiating units; if the output of the phototransistor is V, then stiffness is proportional to V, damping is proportional to $\mathrm{d}V/\mathrm{d}t$, and mass is proportional to $\mathrm{d}^2V/\mathrm{d}t^2$. The outputs of the length and tension transducers were connected to the X and Y amplifiers of an oscilloscope, giving a direct display of the length–tension diagram.

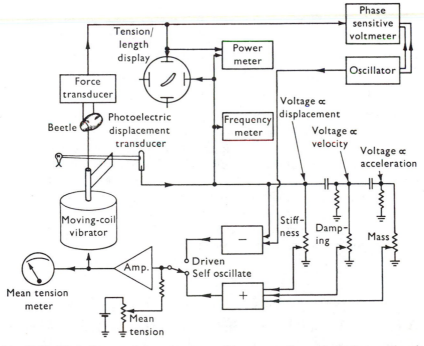

Fig. 13.21. Block diagram of the apparatus used to measure the mechanical properties of rhinoceros beetle flight muscles. (From Machin and Pringle, 1959, by permission of the Royal Society.)

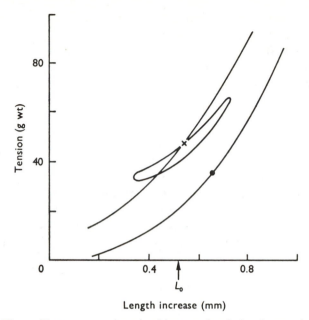

Fig. 13.22. The oscillatory contraction of a rhinoceros beetle basalar muscle. The lower sloping line shows the length–tension relation of the unstimulated muscle, the upper sloping line shows the length–tension relation when the muscle is stimulated under isometric conditions. The loop shows the oscillatory contraction produced by stimulation when the load possesses suitable inertial and damping components; it is traced out in an anticlockwise direction and at a frequency in the region of 25 Hz. (From Machin and Pringle, 1959, by permission of the Royal Society.)

In the resting condition, the muscle was very stiff, giving a steep passive length–tension curve. With an isometric load, stimulation of the motor nerve resulted in the development of an extra steady maintained tension, just as in other muscles. But with an inertial load, obtained by a suitable setting of the load parameters, the muscle would undergo an oscillatory contraction, giving a length–tension loop such as is shown in fig. 13.22. Now, what determines the frequency of this oscillatory contraction? If the oscillation frequency (f) is the same as the mechanical resonant frequency of the muscle and its load, it should be given by the equation.

$$4\pi^2 f^2 = \frac{\text{total stiffness of muscle plus load}}{\text{total mass of muscle plus load}}.$$

Hence the relation between f^2 and the stiffness of the load (mass being constant) should be a straight line, and the relation between $1/f^2$ and the mass of the load (stiffness being constant) should also be a straight line. Fig. 13.23 shows that this is so, from which we must conclude that

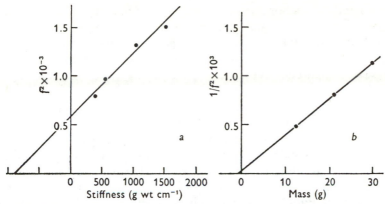

Fig. 13.23. Effect of load parameters on the oscillation frequency of an *Oryctes* flight muscle. *a*: Effect of varying the stiffness at a constant mass (30 g). *b*: Effect of varying the mass at a constant stiffness (400 g wt/cm). *f* is the oscillation frequency in c/sec. Further explanation in the text. (From Machin and Pringle, 1959, by permission of the Royal Society.)

the frequency of oscillation is equal to the resonant frequency of the system. In a flying insect, the load is determined by the inertia of the wings, the aerodynamic forces on the wings, and the stiffness of the thoracic skeleton by whose movements the wings are moved. Thus the wing-beat frequency of insects with fibrillar flight muscles can be altered by altering the resonant frequency of the wing–thorax system, such as by cutting off the tips of the wings (when the frequency increases) or by loading them with small amounts of wax (the frequency decreases).

Much information about the properties of fibrillar muscle has been obtained by using the method of sinusoidal analysis. The muscle, or a bundle of glycerol-extracted fibres, is subjected to a sinusoidal length change of small amplitude, and the resulting sinusoidal change in tension is measured. The process is then repeated with length changes of the same amplitude but of different frequencies. In experiments on intact muscles, the apparatus shown in fig. 13.21 can be used if the switch is placed in the 'driven' position.

The results obtained from this type of experiment are usefully expressed in a diagram known as the *Nyquist plot*, which is constructed as follows. Referring to fig. 13.24*a*, we see that the tension output is a sinusoidal waveform of the same frequency ($1/b$) as the length input, but that it suffers a phase shift *a*. The magnitude of this phase shift can be expressed either as a lead or lag of so many units of time, or as the *phase angle*, which is given by a/b revolutions, $360a/b$ degrees or $2\pi a/b$ radians. We then construct a graph whose axes are both in units of tension and to the same scale, the abscissa being labelled 'in-phase component' and the ordinate 'quadrature component'. A point is plotted on this graph at a distance from the origin equal to the amplitude of the tension trace and at an angle to the abscissa equal to the phase angle (fig. 13.24b). Similar measurements are then made with length inputs of the same amplitude but different frequencies, to give a complete Nyquist plot as is shown in fig. 13.24c.

It is essential to use length changes of small amplitude in these experiments, since the analysis depends on the tension change being almost perfectly sinusoidal; this will not be so for large length changes. The process of analysis can be made easier by using a resolved components indicator, an electronic device which gives a direct measure of the in-phase and quadrature components of the tension change. Results from different muscles can be compared by converting the in-phase and quadrature components of tension into elastic and viscous moduli respectively, by means of the formulae

$$E_e = \frac{\Delta F_e \cdot L}{A \cdot \Delta L}$$

and

$$E_v = \frac{L}{A \cdot \Delta L}$$

where E_e and E_v are the elastic and viscous moduli, ΔF_e and ΔF_v are the in-phase and quadrature components of the tension change, L is the length of the muscle, ΔL is the amplitude of the length change, and A is the cross-sectional area of the muscle (Machin and Pringle, 1960). E_e and E_v are then conveniently expressed in kg/cm^2, and the Nyquist plot has now been converted into a *vector modulus plot*.

Vector modulus plots for a beetle flight muscle in the resting condition and when stimulated are shown in fig. 13.25a. If we compare these with that obtained from the mechanical visco-elastic system shown in fig. 13.25b, it is clear that the resting muscle behaves very similarly to the visco-elastic model. The active muscle, however, differs from the model in that its viscous modulus is negative at low frequencies; this indicates that tension changes are lagging behind length changes (so giving the negative phase angle) and that the muscle is therefore doing work on the apparatus. Nyquist and vector modulus plots of this type

---

Fig. 13.24. How a Nyquist plot is constructed. *a* shows the imposed length change at one particular oscillation frequency, and the consequent tension change. The amplitude of this tension change and its phase angle with respect to the length change are then plotted on the Nyquist graph as shown in *b*. Similar measurements are made for other frequencies, to give the complete Nyquist plot shown in *c* (the curve shown is such as is obtained from an active fibrillar muscle). Further explanation in text.

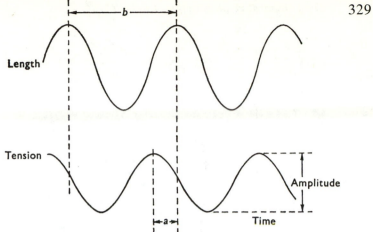

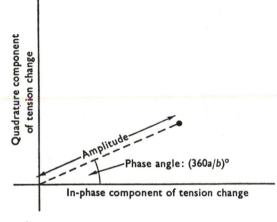

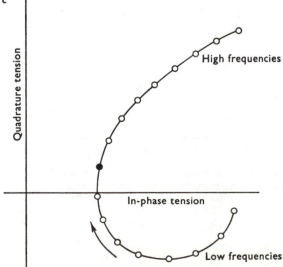

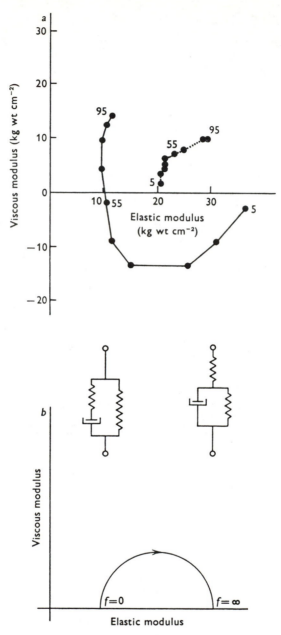

Fig. 13.25. Vector modulus plots of a beetle flight muscle (*a*) and a visco-elastic model (*b*). In *a*, the continuous curve shows the results from a stimulated muscle and the dotted curve shows those from an unstimulated one; the points are shown at frequency increments of 10 Hz, starting with 5 Hz. (From Machin and Pringle, 1960, by permission of the Royal Society.)

are characteristic of insect fibrillar muscle; driven oscillations at negative phase angles are equivalent to free oscillations against a suitable load.

The work output per cycle of an oscillating fibrillar muscle is greatest when the frequency is such that the viscous modulus is at its maximum negative value. Maximal power output occurs at a slightly higher frequency, when the product of the frequency and the viscous modulus is at its maximum negative value. Since the wing-beat frequency of a flying insect is determined by the mechanical resonant frequency of the wing–thorax system, we would expect the Nyquist plot of the flight muscles to be such that maximum power output occurs at or near this frequency. If measurements are made at the temperature of the flight muscles in the flying insect, this is found to be so (Machin, Pringle and Tamasige, 1962).

Jewell and Rüegg (1966) showed that glycerinated fibres from the wing muscles of giant water bugs (Belostomatidae) showed the typical

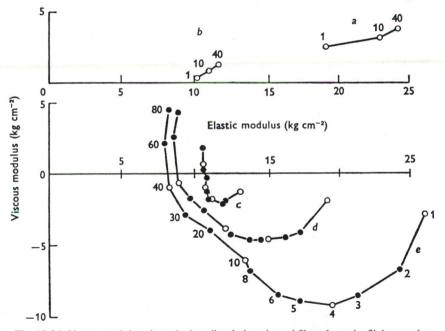

Fig. 13.26. Vector modulus plots of a bundle of glycerinated fibres from the flight muscles of a giant water bug, *Lethocerus*, in response to oscillatory length changes of constant amplitude. Curve *a* shows the response with the fibre in a buffered salt solution containing neither ATP nor calcium ions (the rigor state). Curve *b* shows the response in the presence of ATP (5 mM) but the virtual absence of calcium ions. Curves *c* to *e* show responses in the presence of both ATP and calcium ions, the calcium ion concentrations being: *c*, 5.2 × 10^{-8}M; *d*, 1.4 × 10^{-7}M; and *e*, 2.1 × 10^{-7}M. The frequencies for curves *c* and *d* were the same as for curve *e*. (From Jewell and Rüegg, 1966, by permission of the Royal Society.)

oscillatory phenomena in the presence of suitable concentrations of ATP and calcium ions. This indicates that the oscillation really is a property of the contractile apparatus, not of the excitation or coupling processes, which merely serve to keep the muscle in the active condition. Fig. 13.26 shows, in the form of vector modulus plots, the effects of ATP and calcium ion concentration on the response of these glycerinated fibres to imposed oscillations. A feature of these results is that the frequency at which the viscous modulus is maximally negative (2 to 5 Hz) is much lower than the wing-beat frequency of the intact insects (20 to 30 Hz); the discrepancy can be explained partly in terms of temperature effects.

The relations between mechanical factors and the ATPase activity of glycerinated water bug flight muscle fibres has formed the subject of an interesting investigation by Rüegg and Tregear (1966). They found that, in the presence of calcium ions, more ATP is split when the fibres are subjected to oscillatory length changes, and that this 'extra' ATPase activity reaches a maximum at a frequency which is apparently identical with that at which the power output is maximal (fig. 13.27). Another interesting feature is that the calcium-activated ATPase activity is greatly increased if the muscle is stretched. A possible explanation for this effect is that stretching the muscle also stretches the A filaments, so exposing more ATPase sites; such a stretch could be due to 'connecting filaments' linking the A filaments to the Z line, but the evidence for the existence of such filaments is equivocal.

From an evolutionary point of view, it is interesting to speculate on whether the oscillatory characteristics of fibrillar muscle are dependent

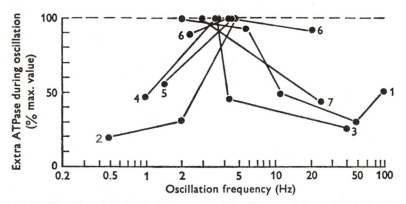

Fig. 13.27. The effect of oscillation frequency on the ATPase activity of glycerinated giant water bug flight muscle fibres. The curves were obtained from seven different experiments, the maximal value being plotted as 100 per cent. Notice that the ordinate shows the *extra* ATPase activity resulting from oscillation. (From Rüegg and Tregear, 1966, by permission of the Royal Society.)

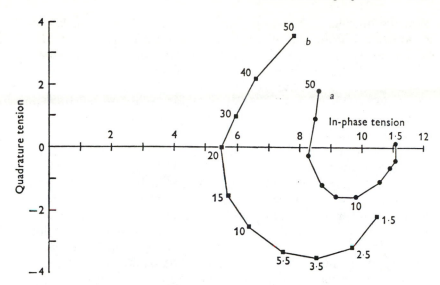

Fig. 13.28. A comparison of the oscillatory activity, as shown by Nyquist plots, of glycerinated fibres from *Fidicina* tymbal muscles (curve *a*) and *Lethocerus* flight muscles (curve *b*). Measurements were made at the same frequencies in each case: these are given (in Hz) in full for curve *b* and in part for curve *a*. The units of the tension scales are mg peak-to-peak per fibre per percentage length change. The solutions in which the fibres were bathed contained adequate amounts of ATP, magnesium and calcium. Notice that the cicada tymbal fibres (which are neurogenically controlled in life) exhibit oscillatory power output at frequencies below 30 Hz. (From Aidley and White, 1969.)

on some basically new properties of the muscle, or whether they arose by enhancement of properties already present to some extent in other muscles. Some evidence in favour of the second possibility is provided by the properties of the sound-producing muscles in the cicada *Fidicina*: although these muscles are neurogenically activated, with the conventional one-to-one relation between nerve impulse and muscle contraction (Aidley, 1969), their glycerinated fibres show all the mechanical properties of glycerinated water bug flight muscles (Aidley and White, 1969). Fig. 13.28 shows Nyquist plots for *Fidicina* and *Lethocerus* glycerinated fibres, showing oscillatory power output from the fibres at frequencies below 30 Hz. These results suggest that the work on insect fibrillar muscles (reviewed by Pringle, 1967, and Tregear, 1975) may help in understanding how more conventional muscles work.

Molluscan 'catch' muscles

Certain muscles of molluscs, of which the most closely investigated is the anterior byssus retractor of the mussel *Mytilus*, are remarkable in that their rate of relaxation is dependent upon the nature of the excitatory stimulus. In 'phasic' responses, the relaxation following an isometric contraction is complete within a few seconds, whereas in 'tonic' or 'catch' responses, the relaxation phase lasts for several minutes or sometimes hours. Phasic responses are usually produced by electrical stimulation using repetitive brief pulses or alternating current. Tonic responses are usually produced by continuous direct current stimulation or by treatment with acetylcholine. Tonic responses are abolished by 5-hydroxytryptamine.

The mechanical responses of the *Mytilus* anterior byssus retractor muscle have been investigated by Jewell (1959) and Lowy and Millman (1963), among others. Jewell found that the behaviour of the muscle during stimulation was not greatly different from that of frog leg muscles, except that the time course of contraction was much slower. During the tonic response, however, the ability to shorten and to redevelop tension after a quick release was very much reduced (fig. 13.29c). If the muscle was restretched after a quick release during this time, the full 'catch' tension was re-established.

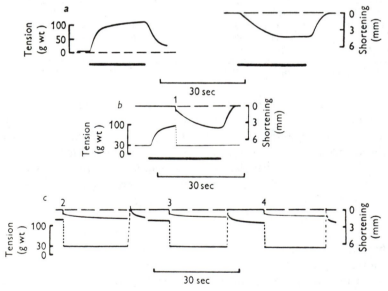

Fig. 13.29. Experiments showing some of the mechanical properties of a *Mytilus* anterior byssus retractor muscle. *a* and *b*, Phasic responses, induced by repetitive simulation; *c*, tonic response, produced after washing the muscle for a short time in an acetylcholine solution. The left record in *a* shows the isometric tension developed during the phasic response, the right record shows the shortening produced during an after-loaded isotonic contraction against a 30 g load. Record *b* shows an isotonic release during phasic stimulation. In record *c*, the muscle first undergoes an isotonic release and is then restretched, and the process is then repeated twice; notice the maintenance of tension but absence of the ability to shorten during the tonic response. (From Jewell, 1959.)

The structural basis of the 'catch' mechanism has been a matter of some controversy. It is generally agreed that contractions during stimulation are probably brought about by the action of actin–myosin cross-bridges. One school of thought (see Lowy and Millman, 1963; Lowy, Millman and Hanson, 1964) suggests that the tonic contraction is also due to such linkages, whose rate of breakage after tonic stimulation is extremely slow. Developing this idea, Szent-Györgyi *et al.* (1971) suggest that a structural transformation takes place in the paramyosin core of the thick filament, which slows the rate of detachment of myosin from actin. The other school (see Johnson, 1962; Rüegg, 1968) suggests that the 'catch' mechanism is due to a separate system, possibly involving linkages between paramyosin molecules, in parallel with the actin–myosin system.

14 The electric organs of fishes

A number of different types of fish are capable of producing appreciable electric currents in the water surrounding them. These currents are produced by special organs known as *electric organs*. The position of the electric organs in various electric fish is shown in fig. 14.1.

Electric organs are composed of columns of cells called *electroplaques* (sometimes called electroplates, electroplaxes or electrocytes), each of which is innervated by an excitor nerve. These electroplaques (with one exception – see p. 345) appear to be modified muscle cells which have lost their contractile function. In this chapter we shall examine the means whereby the electroplaques produce the electric currents constituting the discharge of the electric organs.

THE ELECTROPLAQUES OF THE ELECTRIC EEL

The electric eel, *Electrophorus electricus*, is capable of producing an electric discharge of over 600 volts (the record seems to be 866 volts), consisting of about half a dozen pulses each lasting two to three msec. The gross structure of the electric organ is shown in fig. 14.2. Each electroplaque is about $100\,\mu$m thick (longitudinally), 1 mm wide (vertically) and 10 to 30 mm long (radially). The nerve endings are restricted to the posterior face. The anterior (non-innervated) faces are much folded, giving rise to numerous papillae. The high-voltage discharges are produced by the main organ; Sachs' organ gives much smaller discharges. In the main organ, up to 6000 rows of electroplaques are arranged in series with each other. This series arrangement of the electroplaques leads to addition of the voltages produced by each one of them; for example, if a discharge of 600 V were produced by an organ containing 4000 rows of electroplaques, each electroplaque would have to produce a potential of 150 mV across it.

The electrical properties of the electroplaque were investigated by Keynes and Martins-Ferreira (1953), using intracellular electrodes. They used the electroplaques from the organ of Sachs for their experiments, since these electroplaques can be more easily isolated than can those of the main organ. In the resting condition, there is a resting

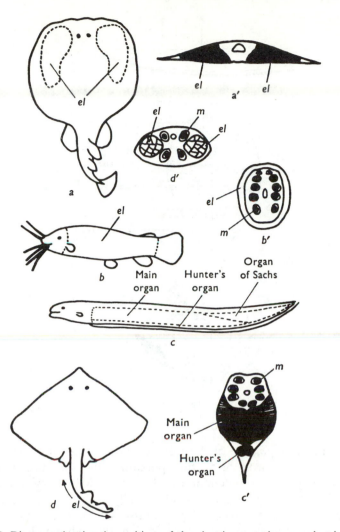

Fig. 14.1. Diagrams showing the positions of the electric organs in some electric fish. *a*, *Torpedo*; *a′*, transverse section through its electric organs. *b*, *Malapterurus*; *b′*, transverse section. *c*, *Electrophorus*; *c′*, transverse section through the main organ. *d*. *Raia*; *d′*, transverse section through its tail. *m*, muscles; *el*, electric organs. (From Keynes, 1957.)

potential of about −80 to −90 mV. Electrical stimulation of the electroplaque results in an all-or-nothing action potential appearing across the innervated (posterior) face, but there is no potential change across the non-innervated (anterior) face. The experimental evidence for this statement is as follows. In fig. 14.3, trace *a* is recorded from two microelectrodes placed just outside the innervated face; there is, of course, no potential difference between them either at rest or during

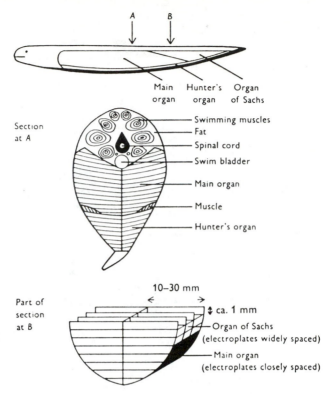

Fig. 14.2. The gross structure of the electric organs in the electric eel. (From Keynes and Martins-Ferreira, 1953.)

activity. For trace *b*, one of the microelectrodes was inserted into the cell; this then records the resting potential, and, on stimulation, an 'overshooting' action potential of about 150 mV is observed. This means that the innervated face is electrically excitable, just as the cell membranes of nerve axons and vertebrate twitch muscle fibres are. In trace *c*, one of the electrodes is pushed right through the electroplaque so that the potential across the whole electroplaque is recorded. There is no steady potential at rest, indicating that the resting potential across the non-innervated face is equal to that across the innervated face. On stimulation, an action potential of the same size and shape as that across the innervated face appears across the whole electroplaque. This suggests that the non-innervated face is electrically inexcitable, and that the whole of the discharge is accounted for by the activity of the innervated face.

Confirmation of this view is provided by the experiment shown in fig. 14.4. Here, the potential across the non-innervated face (trace *b*) is not

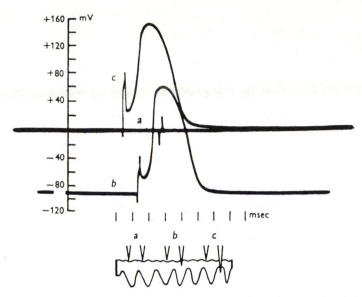

Fig. 14.3. Responses of a Sachs' organ electroplaque of *Electrophorus* to electrical stimulation. The position of the recording electrodes for each trace is shown in the lower diagram, in which the innervated face of the electroplaque is uppermost. (From Keynes and Martins-Ferreira, 1953.)

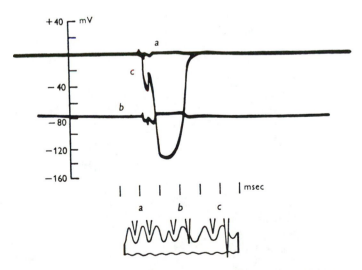

Fig. 14.4. Results of an experiment similar to that of fig. 14.3, but this time the non-innervated face of the electroplaque is uppermost. (From Keynes and Martins-Ferreira, 1953.)

affected by stimulation (except for some small electrotonic changes), whereas the potential across the whole electroplaque (trace *c*) shows the familiar action potential.

Thus the voltage produced by a stimulated electroplaque is due to the asymmetry of the responses of its two faces, as is shown diagrammatically in fig. 14.5. In the complete electric organ, the electroplaques are arranged in series so that, as we have seen, their voltages are additive. Notice that the innervated face becomes negative to the non-innervated face during the discharge; this situation occurs in many other electric fish, and is sometimes known as 'Pacini's law' – but by no means all electric fish obey this rule. In *Electrophorus*, since it is the posterior face of the electroplaques that is innervated, the head end becomes positive to the tail during the electric discharge.

The ionic basis of the electroplaque action potential is much the same as that of the action potentials of nerve axons and twitch muscle fibres. Keynes and Martins-Ferreira showed that its size is dependent on the external sodium ion concentration, and Schoffeniels (1959), from radioactive trace measurements on isolated electroplaques, has shown that the sodium inflow across the innervated membrane increases greatly during activity.

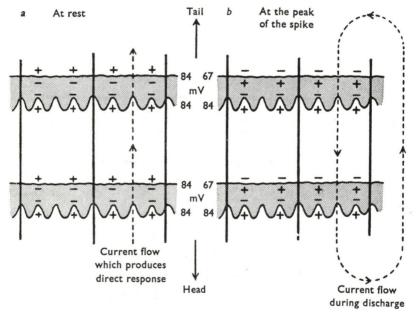

Fig. 14.5. Diagram to show membrane potentials across the innervated (uppermost in the diagram) and non-innervated faces of electroplaques in the Sachs organ of the electric eel, at rest (*a*) and at the peak of the discharge (*b*). (From Keynes and Martins-Ferreira, 1953.)

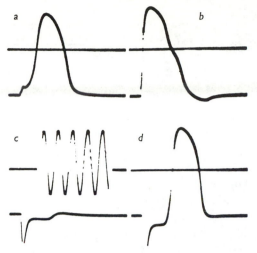

Fig. 14.6. Comparison of electroplaque responses to direct and nervous stimulation in the electric eel. *a*: Direct stimulation at moderate intensity. *b*: Direct stimulation at high intensity; note the extremely low latency of the response. *c*: Nervous stimulation at low intensity, producing a postsynaptic potential which is not large enough to elicit an action potential. *d*: Nervous stimulation at high intensity; note the appreciable latency of the response. Upper traces show zero potential levels and, in *c*, a 100 mV peak-to-peak 1000 Hz calibration waveform. (From Altamarino, Coates and Grundfest, 1955.)

In the living animal, discharge of an electroplaque is not, of course, initiated by direct electrical stimulation, but by excitation via the efferent nerves innervating it. Altamarino, Coates and Grundfest (1955) found that stimulation of the motor nerves elicits an excitatory postsynaptic potential across the innervated surface of the electroplaque, which then induces the action potential (fig. 14.6). This system is biophysically similar to that whereby vertebrate twitch muscle fibres are excited, except that each electroplaque is innervated by a number of nerve fibres; if only a small proportion of these fibres is stimulated (as in fig. 14.6*c*), the resulting postsynaptic potential is not large enough to elicit an action potential.

THE ELECTROPLAQUES OF SOME OTHER ELECTRIC FISH

Marine electric fish

This group includes the electric rays *Torpedo*, *Narcine* and *Raia*, and the 'stargazer', *Astroscopus*. In all these fish, the innervated faces of the electroplaques are not electrically excitable, and the response to nervous stimulation consists solely of a depolarizing postsynaptic potential (Bennett and Grundfest, 1961*a*, *b*; Bennett, Wurzel and

Grundfest, 1961; Brock, Eccles and Keynes, 1953), as is shown in fig. 14.7. Hence the maximum potential across the active electroplaque cannot be as large as it is in *Electrophorus*; in fact, it is usually less than 70 mV.

Now why should the electroplaques of marine electric fishes be electrically inexcitable? If it were not for the existence of *Astroscopus*, we could assume that the situation was of no functional significance, since the marine rays and skates are phylogenetically distinct from the fresh-water teleosts; but *Astroscopus* is much more closely related to the fresh-water forms than to the rays, yet it too has electrically inexcitable electroplaques. There is no generally agreed answer to this question, but a possible explanation is as follows. We can regard the excited electroplaque membrane of a fresh-water fish as consisting of a battery V_{Na} in series with a resistance R_{Na}. The current produced by the discharge flows through the resistance R_{ex}, which consists mainly of the resistance of the external medium in which the fish swims; the voltage produced by the membrane is V. This rather simplified scheme is shown in fig. 14.8a. In this circuit, it is evident that

$$V = V_{Na}\left(\frac{R_{ex}}{R_{Na} + R_{ex}}\right).$$

Thus if R_{ex} is reduced, V also falls. But we know that, in an electrically excitable system such as this one is, R_{Na} rises as V falls, which will itself produce a further fall in V, and so on. Hence it follows that an electrically excitable system, based essentially on a positive feedback relation

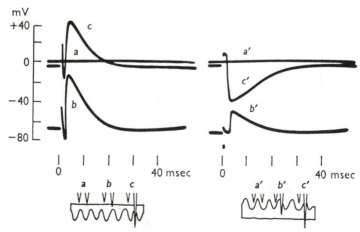

Fig. 14.7. Responses of *Raia* electroplaques, as recorded by electrodes in various positions. (From Brock, Eccles and Keynes, 1953.)

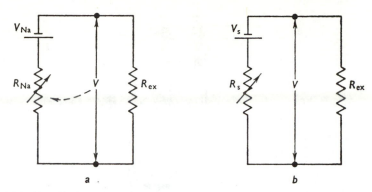

Fig. 14.8. Simplified equivalent circuits of the electric organs in fishes whose electroplaques are (*a*) and are not (*b*) electrically excitable.

between membrane potential and sodium conductance, is ineffective if the external resistance is sufficiently low. An analogous circuit for an electrically inexcitable electroplaque is shown in fig. 14.8*b*. In this system the resistance of the excited synaptic membrane (R_s) is independent of the membrane potential, hence the membrane potential during excitation is not disproportionately lowered if the external resistance is low. Since the external resistance is very much lower for fishes living in sea water, we might expect such fish to rely on purely synaptic excitatory mechanisms for the discharge of the electroplaques, as indeed they do.

Gymnotus carapo

This is a South American fresh-water fish which produces weak electric discharges. The electroplaques lie in eight tubes, four on each side of the animal. Those in the dorsal pair of tubes are innervated on the anterior surface, whereas those in the other tubes are innervated on the posterior surface. The physiology of the discharge has been investigated by Bennett and Grundfest (1959). They found that both faces of the electroplaques are electrically excitable, but, when the motor nerves are stimulated, the innervated surfaces are excited before the non-innervated surfaces, so that the potential change across the whole electroplaque is diphasic. During the natural discharge, the upper tube on each side (tube 1) fires first, and then the lower tubes (2 to 4) fire together. The result of this is that the potential change over the whole fish is triphasic (fig. 14.9), the sequence of events being: (*a*) excitation of the innervated faces in tube 1; (*b*) excitation of the non-innervated faces in tube 1 and of the innervated faces in tubes 2 to 4; and (*c*) excitation of the non-innervated faces in tubes 2 to 4.

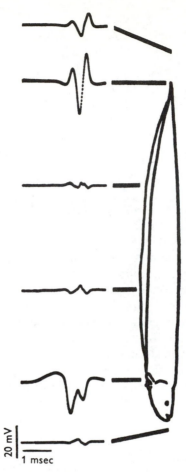

Fig. 14.9. Waveforms of a single pulse from the repetitive discharge of *Gymnotus carapo*, recorded at different positions near the body surface. Notice (*a*) that the potential is triphasic near the tail, rather more complex elsewhere, and (*b*) that the potentials are maximal near the tail and the head, suggesting that the electric organ acts as a dipole generator. (From Bennett and Grundfest, 1959.)

Malapterurus

The strong electric discharge of *Malapterurus electricus* (the electric catfish) is of considerable interest in that it does not obey 'Pacini's law': the electroplaques are innervated posteriorly, yet the head becomes negative to the tail during activity. At one time it was thought that the electroplaques in this fish were derived from glandular tissue rather than muscle, and, in the absence of physiological observations, it was suggested that the innervated face became hyperpolarized during activity. More recent observations have shown both these ideas to be wrong.

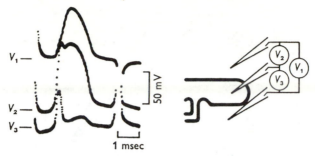

Fig. 14.10. Responses of a *Malapterurus* electroplaque to nervous stimulation, recorded across the whole electroplaque (V_1), across the anterior face (V_2) and across the posterior face (V_3). (From Keynes, Bennett and Grundfest, 1961.)

The electroplaques of *Malapterurus* are rather unusual in that innervation occurs at the end of a stalk which is produced from the centre of the posterior face, as is indicated in fig. 14.11 (a somewhat similar arrangement occurs in mormyrids; see Bennett and Grundfest, 1961c). The sequence of events during discharge of an electroplaque has been determined by Keynes, Bennett and Grundfest (1961), using intracellular microelectrodes. Excitation arises in the stalk and then passes into the electroplaque disc, where both faces are excited simultaneously. However, the action potential across the anterior ('non-innervated') face lasts only for about 0.3 msec (figs. 14.10 and 14.11). Hence, during the later stages of the response, only the anterior face is active, so that there is a net potential difference across the whole electroplaque.

Sternarchus

The electric organ of *Sternarchus* is apparently unique in that it is composed of tissue derived from nerve axons rather than from muscle cells (Couceiro and de Almeida, 1961; Bennett, 1971). The electromotor axons leaving the spinal cord are swollen distally to form

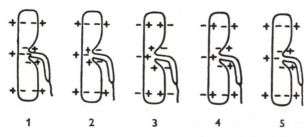

Fig. 14.11. Schematic diagram showing the sequence of potential changes during the discharge of a *Malapterurus* electroplaque. (From Keynes, Bennett and Grundfest, 1961.)

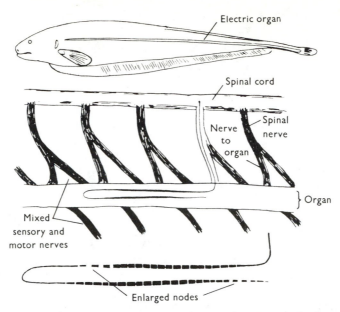

Fig. 14.12. The anatomy of the electric organ of *Sternarchus*. The middle diagram shows the course of one of the modified nerve fibres which constitute the organ and the lower one shows its structure. (From Bennett, 1971.)

'electrotubes' which are analogous in function to the electroplaques of other electric fishes. Fig. 14.12 shows the structure of these swollen axons. After entering the electric organ each axon passes forwards for several millimetres, and then turns back on itself and runs backwards for an approximately equal distance. Each arm of this loop has very narrow nodes in the swollen region, large nodes in the region just proximal to it, and very large nodes in the region just distal. Clearly such an arrangement would greatly affect the lines of current flow in the axon (the space constant of the enlarged regions must be very great, for example), but the full details have not yet been worked out. Since there are no synapses between the units of the electric organ and the nerves which supply them, the discharge of the organ is unaffected by injection of curare.

THE FUNCTIONS AND EVOLUTION OF ELECTRIC ORGANS

The powerful electric organs of *Electrophorus* and *Torpedo* are used both offensively, in temporarily paralysing smaller fishes which can then be eaten, and defensively, in deterring predators. The discharges of *Astroscopus* and *Malapterurus* are also able to stun smaller fishes, and are probably used for offensive purposes.

The function of the weak discharges of the electric organs of *Gymnarchus, Gymnotus,* sternarchids and mormyrids was, until quite recently, obscure. Then Lissmann (1951) showed that *Gymnarchus* could detect the presence of electrical conductors and non-conductors in its environment, and was sensitive to weak electric currents. He suggested that the electric organs functioned as the 'transmitter' of a direction-finding system, the 'receiver' being electro-sensitive sense organs capable of detecting changes in the pattern of current flow around the fish. We shall examine these electro-receptors in a little more detail in chapter 16; suffice it to say, here, that Lissmann's hypothesis is now well substantiated (Lissmann and Machin, 1958).

In discussing various objections to his theory of evolution by natural selection, Darwin (1859) mentioned the problem of the electric organs of fishes. At that time only the powerful electric organs (such as those of *Electrophorus* and *Torpedo*) were known, and it was impossible to see how these could have gradually developed through ancestors with weaker discharges, since such weak discharges would not be effective as weapons of offence or defence and would therefore give no selective advantage to a fish possessing them. Lissmann (1958) suggests that the offensive–defensive functions of strong electric organs are a development of the direction-finding function of weak electric organs. The first stage in the evolution of the direction-finding system must have been the development of electroreceptors; these may have been capable of detecting muscular action potentials, and the sensitivity and accuracy of the system could then have been increased by modification of the muscle cells to form electric organs.

15 The organization of sensory receptors

All animals are sensitive to some extent to changes in their environment. In all except the most lowly animals, special parts of the body are developed, which are responsive to some of these changes and which feed information concerning them into the central nervous system; in the higher animals, information about the workings of the animal's own body may be acquired in a similar fashion. These specially sensitive structures are known as sense organs or sensory receptors. In this and the next two chapters we shall examine some of their properties. Since there is an enormous variety of different types of receptor, and since our understanding of how receptors work is still very incomplete, the following pages must necessarily be no more than a rather brief introduction to sensory physiology. This chapter attempts to give a general outline of the properties of receptors, chapter 16 surveys a variety of different types, and chapter 17 deals with a particular sense organ, the vertebrate eye, in a little more detail.

The methods used for investigating receptors fall into two main categories. Firstly, there is the behavioural, or psychophysical, approach, where the receptor is investigated indirectly by observation of the response of the animal to a sensory stimulus. For example, suppose we are interested in the ability of an animal to discriminate colours. In the case of man, the experiments are not too difficult, since we can ask the subject if two colours look different to him. With other animals, however, we have to devise experiments in which the behaviour of the animal depends upon the ability to discriminate colour; for example, we can train a fish to feed from a red container and then see if it always chooses the red container from among a series of greys. Secondly, there is the electrophysiological method, in which the receptor is investigated by means of electrodes placed in its vicinity or on the sensory nerves leading from it. This type of study gives precise, quantitative information about the properties of the receptor, but it does not always enable us to decide which of these properties are relevant to its normal function. For example, we may be able to show by electrophysiological means that a receptor can be excited both by mechanical stimuli and by changes in temperature, but we cannot from these experiments alone determine how the animal interprets this information. Hence a thorough investigation of a sensory system must involve both behavioural and electrophysiological studies.

Now let us consider how sensory systems can be classified. A moment's thought will show that the popular list of 'the five senses' – sight, hearing, touch, taste and smell – is incomplete. We are also sensitive to changes in temperature, to our position with respect to gravity, to the movements of our bodies, and so on. These various categories of sensation are called sensory *modalities*. Some of these modalities can further be divided into *qualities*: for example, we can determine the pitch as well as the loudness of a sound, the colour as well as the brightness of a light source.

The sense organs themselves can be classified according to the type of stimulus which normally excites them. Thus *mechanoreceptors* are excited by mechanical stimuli, *photoreceptors* are sensitive to light, *thermoreceptors* are temperature-sensitive, *chemoreceptors* are sensitive to the chemical composition of the surrounding medium, and *electroreceptors* are sensitive to very weak electric currents. It is important to realize that the specificity of receptors is not absolute; most receptors, for instance, can be excited by strong electric currents.

An alternative classification is based on the position of the receptors and the stimulus. Thus *exteroceptors* are sensitive to stimuli originating outside the body, and *interoceptors* are excited by stimuli inside the body. Exteroceptors can be divided into *distance receptors* (such as those involved in sight and hearing), which are able to detect phenomena at a distance, and *contact receptors* (such as those involved in touch and taste) which can only be stimulated by contact with the stimulus. Interoceptors are divided into *equilibrium receptors*, which give information about the movement and position of the whole body, *proprioceptors*, which give information about the relative positions and movements of the muscles and skeleton, and *visceroceptors*, which monitor the conditions in the rest of the body.

Examples of these various categories are shown in table 15.1. In addition to these, there is a rather ill-defined sensation known as pain. Since pain is a subjective phenomenon, it is very difficult to study it in animals other than man. Stimuli which produce reactions which appear to involve the sensation of pain in animals are sometimes called *nociceptive*.

The anatomy of sense organs shows a great deal of variety. All sense organs are innervated by sensory neurons. In the simplest type, there is but a single sensory neuron whose peripheral end is excited by the sensory stimulus. In the Pacinian corpuscle (fig. 15.1) for example, there is a single axon whose terminal is unmyelinated and surrounded by a

TABLE 15.1. *Types of receptors.*

| Type of receptor | Exteroceptors | | Interoceptors |
	Contact	Distance	
Mechanoreceptors	Touch	Hearing (phonoreceptors)	All proprioceptors All equilibrium receptors
		Lateral line organs of fishes	Some visceroceptors, e.g. baroreceptors of carotid body (sensitive to blood pressure)
Photoreceptors	—	Sight. May or may not be image-forming and/or colour sensitive	—
Chemoreceptors	Gustatory (taste)	Olfactory (smell)	E.g. chemoreceptors of carotid body
Thermoreceptors	Most	Sensitivity to radiant heat. E.g. facial pit of crotaline snakes	E.g. hypothalamic thermoreceptors
Electroreceptors	—	Found in electric and some other fishes	—

series of lamellae. The mechanoreceptive hairs of arthropods (fig. 15.2) are also innervated by a single sensory hair, although in this case the neuron soma is placed distally instead of in the central nervous system. Notice that, in each case, the neuron ending is associated with *accessory structures* (the lamellae in the Pacinian corpuscle and the cuticular hair in the arthropod touch receptor). In some cases, no specialized accessory structures can be seen, and the nerve endings are 'free' in the tissues.

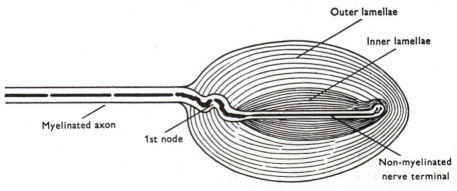

Fig. 15.1. Diagrammatic section through a Pacinian corpuscle. (From Quilliam and Sato, 1955.)

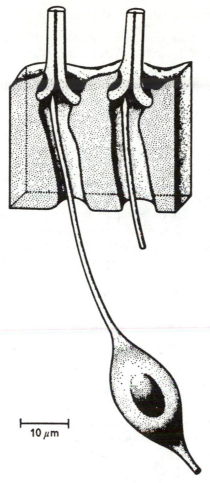

10 μm

Fig. 15.2. Diagram showing how the sense cell is attached to the base of a mechano-receptive hair in the hair-plate organ in a honey-bee's neck. (From Thurm, 1965.)

A more complex type of organization occurs when the sensory neuron is connected to a specialized *receptor cell* (or 'secondary sense cell'), so that excitation of the sensory neuron is preceded by excitation of the receptor cell. In the taste buds of the mammalian tongue, for example, there are a number of receptor cells, each connected, either alone or with others, to a sensory neuron. In the very complex sense organs, such as the vertebrate eye and ear, there are large numbers of receptor cells and sensory neurons, and the accessory structures (such as the lens and the iris in the eye) may reach a high degree of sophistication.

THE CODING OF SENSORY INFORMATION

We must now consider how sensory information is passed from the receptor to the central nervous system. It is very easy to show that the information pathway involves the sensory nerves, since, after cutting these nerves, an animal no longer responds to stimulation of the receptor which they innervated. The first records of electrical activity in sensory nerves in response to sensory stimulation were made by Adrian in 1926. Using frog skin and muscle nerves, he showed that sensory stimulation was followed by action potentials in the sensory nerves. A more delicate analysis followed the recording of the electrical activity in a single sensory nerve fibre by Adrian and Zotterman (1926), whose experiments we shall now consider.

The skeletal muscles of tetrapod vertebrates contain sense organs known as muscle spindles, which are responsive to stretch. Each spindle (fig. 15.3) consists of a bundle of modified muscle fibres whose central region is innervated by a sensory nerve fibre. The sternocutaneous muscle of the frog, a small muscle which is attached at one end to the sternum and at the other to the skin, contains three or four of these muscle spindles. Adrian and Zotterman cut the nerve supplying the muscle so that it was connected to the muscle only, and placed it across their recording electrodes. The skin was cut so as to leave a small piece attached to the muscle, which could then be stretched by means of small weights hung from a thread tied to this piece of skin. They found that stretching the muscle produced a series of action potentials in the nerve. By progressively removing strips of muscle, they were able to reach a situation in which only one muscle spindle remained connected to the sensory nerve. Under these conditions the action potentials produced by stretch appeared at fairly regular intervals and were always the same size, thus showing that the all-or-nothing law applied to sensory nerve impulses.

The next step was to see how this rhythmical discharge was affected by the intensity of the stimulus. Adrian and Zotterman discovered that

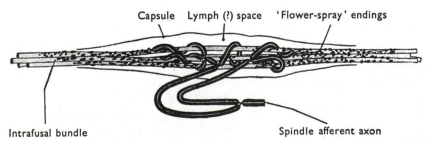

Fig. 15.3. Capsular portion and its sensory nerve ending in a frog muscle spindle. (From Gray, 1957, by permission of the Royal Society.)

the frequency of impulse discharge increased with increasing stretch of the muscle (fig. 15.4). This was a most important result, which has since been verified for a wide variety of different receptors. We can therefore say that, in a single sensory nerve fibre, information as to the intensity of the sensory stimulus is carried as a *frequency code* – the frequency of the all-or-nothing action potentials in the fibre.

With many sensory stimuli, a stimulus of constant intensity produces a sensation which declines with time. This phenomenon, which is known as *sensory adaptation*, can readily be seen by looking at a bright light or by resting light weights on the back of the hand. It is obviously of some interest to see if adaptation is accompanied by any change in the nervous output of the sensory organ. Using their frog muscle spindle preparation, Adrian and Zotterman found that the impulse frequency in the sensory nerve fibre declined during a constant stretch, as is shown in fig. 15.5. This observation provides a partial explanation for the existence of adaptation, and is further evidence for the frequency code hypothesis.

Adrian and Zotterman summarized their observations on the frog muscle spindle in the following words: 'The impulses set up by a single end-organ occur with a regular rhythm at a frequency which increases

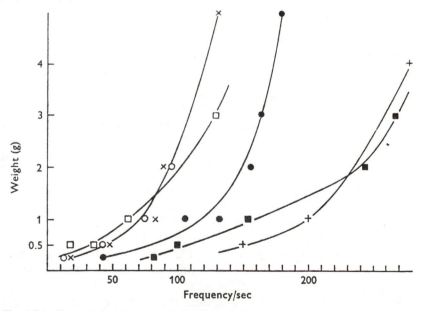

Fig. 15.4. The relation between the load applied to a frog muscle and the impulse frequency in the sensory nerve fibre from one of its muscle spindles. Results from six different experiments, denoted by different symbols; in each case the weight had been applied to the muscle for 10 sec before the measurement of nerve impulse frequency was made. (From Adrian and Zotterman, 1926.)

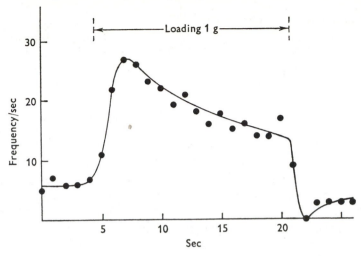

Fig. 15.5. Adaptation in a frog muscle spindle. (From Adrian and Zotterman, 1926.)

with the load on the muscle and decreases with the length of time for which the load has been applied.' It is a remarkable tribute to the quality of their investigation that much of the work done in sensory physiology since that time has been concerned with quantifying this statement for a wide variety of different sense organs.

THE INITIATION OF SENSORY NERVE IMPULSES

Having seen that sensory information is conveyed to the central nervous system by means of impulses in the sensory nerve fibres, the next problem for us to consider is how these impulses arise. Using the frog muscle spindle preparation with an external electrode placed as near the sensory nerve ending as possible Katz (1950) found that a local depolarization of the terminals appeared when the muscle was stretched, as well as the impulses in the sensory nerve axon (fig. 15.6). Treatment of the preparation with procaine prevented the production of impulses but did not affect the maintained depolarization. Finally, the degree of depolarization increased with increasing stretch, and the impulse frequency in the sensory nerve was linearly proportional to the extent of the local depolarization (fig. 15.7). Katz concluded from these results that the local depolarization is an intermediate link between the stimulus and the action potentials in the sensory nerve fibre.

We may call this local depolarization the *generator potential*, since it generates nerve impulses. In the receptor neurons of crustacean stretch receptors (fig. 5.8), Eyzaguirre and Kuffler (1955a) were able to record generator potentials with intracellular electrodes; their results, shown in

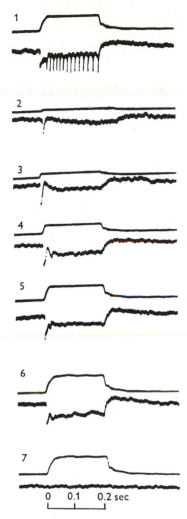

Fig. 15.6. The receptor/generator potential recorded extracellularly from a frog muscle spindle in response to various degrees of stretch. The upper trace in each record indicates the change in length of the muscle and the lower trace records the activity of the sensory axon. Record 1 is from an intact preparation, records 2 to 6 from the same preparation after treatment with procaine, and record 7, obtained after crushing the sensory axon, shows that the potential changes in 1 to 6 are not artefacts. (From Katz, 1950.)

fig. 15.9, are in full agreement with those of Katz. Generator potentials have since been recorded from a wide variety of sensory nerve endings. The size of the generator potential is always roughly proportional to the intensity of the stimulus; in other words, the generator potential is a graded potential change, not an all-or-nothing one.

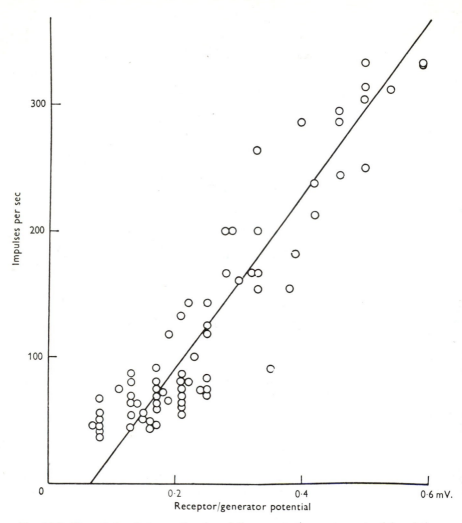

Fig. 15.7. The relation between the size of the receptor/generator potential and the frequency of impulse discharge in a frog muscle spindle. (From Katz, 1950.)

We shall define a *receptor potential* as a potential change, in the receptor cell or (in those cases where the sensory stimulus excites the sensory nerve terminal directly) in the sensory nerve terminal, induced by the action of the stimulus.* In the case of the stretch receptors of frogs and crayfish, it is clear that the receptor potential is the same as the

---

* There have been some differences in the use of the terms 'generator potential' and 'receptor potential' by different authors. The terminology adopted here is that suggested by Davis (1961).

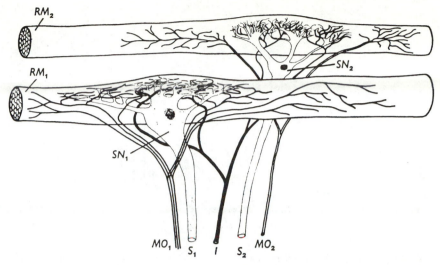

Fig. 15.8. Diagram to show the structure of the abdominal muscle receptor organs in the crayfish. RM_1, muscle bundle of the tonic receptor neuron, SN_1. RM_2, muscle bundle of the phasic receptor neuron, SN_2. MO_1 and MO_2, motor fibres of, respectively, the tonic and phasic muscle bundles. I, an inhibitory axon. (From Burkhardt, 1958.)

generator potential, since the stimulus produces a potential change which then generates impulses. However, where there is a separate receptor cell, it is evident that the generator potential is confined to the nerve terminal, and that any potential change that occurs in the receptor cell must be a receptor potential. It seems clear that the microphonic potentials of the ear (see p. 398) are receptor potentials in this sense.

The mechanism whereby the energy of the stimulus produces excitation of the receptor cell or sensory neuron terminal is known as the *transducer mechanism*. In photoreceptors, we know that the primary stage in this process is the capture of light quanta by, and the consequent photochemical changes in, the visual pigment molecules (see chapter 17). In other cases, however, we have very little idea what the nature of the transducer mechanism is. Where receptor potentials have been detected, such as in many mechanoreceptors and some chemoreceptors, it seems reasonable to suggest that the stimulus causes an increase in the ionic permeability of the receptor membrane, possibly by mechanical distortion of the membrane so as to open new ionic channels. In many cases, the transduction process acts as an amplification stage, in that the energy change comprising the receptor potential is much greater than the incident energy which produces it. A system whereby the incident energy alters the ionic permeability of the receptor membrane is, of course, ideally suited to provide such amplification.

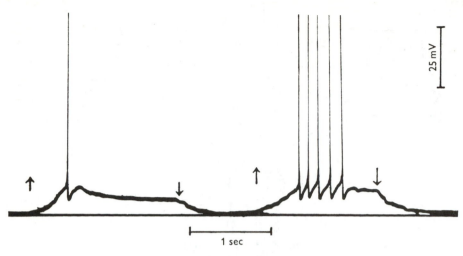

Fig. 15.9. Intracellular records from a crayfish stretch receptor cell. The receptor muscle was stretched between the times marked by the arrows. (From Eyzaguirre and Kuffler, 1955*a*.)

In those systems where there are separate receptor cells, there must be some mechanism whereby excitation of the receptor cell can produce the generator potential in the sensory neuron terminal. In other words, there is a synapse between the receptor cell and the neuron terminal, and our problem is to decide what the mechanism of transmission at this synapse is. There is some evidence that a chemical transmission process is involved. Vesicles, which appear to be very similar to synaptic vesicles, can be seen in the presynaptic regions of the rods and cones of the vertebrate eye (de Robertis and Franchi, 1956) and in the mechanoreceptive hair cells of the vertebrate labyrinth; vesicles of a slightly different type are found in the receptor cells of mammalian taste buds (de Lorenzo, 1963). Fuortes (1959) has shown that the generator potential in the eccentric cell of a *Limulus* ommatidium is accompanied by an increase in membrane conductance; this suggests that transmission between the retinula cells (the receptor cells) and the eccentric cell (the sensory neuron) is chemical rather than electrical.

The general theme of this section – that there is a definite control sequence of events in the excitation of receptors – is illustrated diagrammatically in fig. 15.10. We must now take a closer look at the final stages in this sequence, the generator potential and the initiation of action potentials.

There is some evidence that generator potentials are produced, in part, by an increase in the sodium conductance of the membrane. Diamond, Gray and Inman (1958) investigated the effects of sodium ion

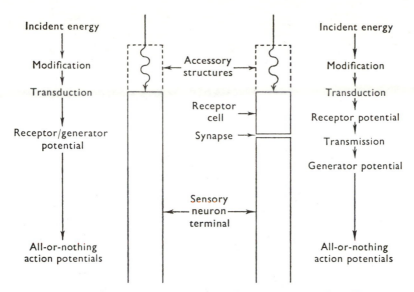

Fig. 15.10. Sequences of events in the excitation of sensory nerve endings. The sequence on the left applies where there is no specialized receptor cell, that on the right applies where there is such a cell.

concentration on the response of Pacinian corpuscles to mechanical stimulation. Since the receptor region – the sensory neuron terminal – is closely surrounded by the lamellae of the corpuscle, the experimental solutions had to be perfused through the arterial supply of the corpuscle. They found that the production of action potentials was quite rapidly abolished on perfusion with sodium-free solutions, and that the generator potential fell to about 10 per cent of its initial value. With intermediate sodium ion concentrations, the size of the generator potential increased with increasing sodium ion concentration (fig. 15.11). These results suggest that the generator potential normally results from an inflow of sodium ions, but that other ions are probably involved as well, since the response is not completely abolished in sodium-free solutions. Similar results have been obtained by Ottoson (1964), on the generator potential of the frog muscle spindle.

We have seen that the generator potential carries sensory information in the form of an 'amplitude code' in which the degree of depolarization is related to the intensity of the stimulus. We must now consider how this amplitude code is converted into the frequency code of action potentials in the sensory nerve fibres. The precise details of this process are not fully agreed upon (see Hodgkin, 1948; Fuortes and Mantegazzini, 1962), but the following explantion is probably somewhere near the truth.

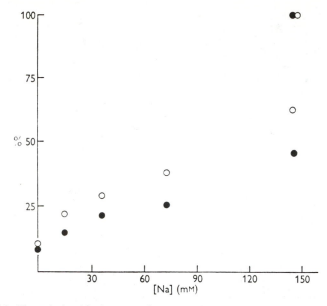

Fig. 15.11. The relationship between the external sodium ion concentration and the receptor/generator potential of a Pacinian corpuscle. The ordinate shows the amplitude (white circles) and rate of rise (black circles) of the receptor potential, each expressed as a percentage of the initial value in 150 mM Na. (From Diamond, Gray and Inman, 1958.)

Consider a sensory nerve terminal which can be functionally divided into two regions (fig. 15.12): a receptor region, which is particularly sensitive to the stimulus and responds to it by means of a graded receptor/generator potential, and a conductile region whose activity consists of all-or-nothing action potentials; at the boundary between these two regions is the impulse initiation site. We assume that the stimulus causes an increase in the ionic permeability of the membrane so that there is an inward flow of current across the cell membrane in the receptor region; this will produce an outward flow of current at the

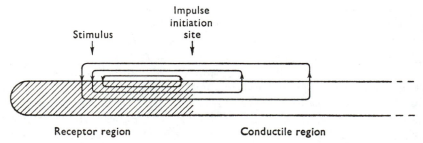

Fig. 15.12. Local circuit currents producing impulse initiation in a sensory nerve terminal.

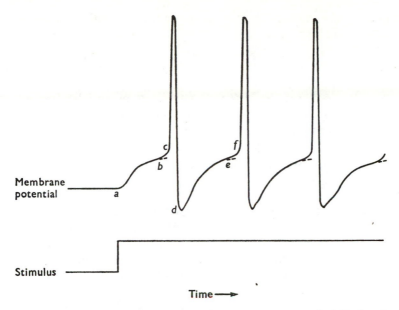

Fig. 15.13. Generation of the frequency code in a sensory nerve terminal. Explanation in text.

impulse initiation site. We further assume that the stimulus is maintained and that the current it produces is constant. The electrical response at the impulse initiation site to this constant generator current is shown in fig. 15.13. The initial depolarization, from *a* to *b*, is a purely electrotonic potential. Then, from *b* to *c*, the depolarization is sufficient to cause a local response of the membrane, i.e. a further depolarization caused by an increase in sodium conductance. At *c*, the membrane potential crosses the threshold, the sodium conductance increases regeneratively, and an action potential results. After the action potential, there is a positive phase (*d*) produced by the high potassium conductance of the membrane. From *d* to *e*, the combined effects of the decrease in potassium conductance and the generator current produce depolarization again; at *e* the sodium conductance begins to increase, and the process becomes regenerative, producing another action potential, at *f*. After this action potential, the sequence repeats itself for as long as the generator current persists.

If the intensity of the generator current is increased, the depolarization between *d* and *f* will occur more rapidly. This means that the time interval between successive impulses will be less, so that the impulse frequency is higher. At relatively high impulse frequencies, the threshold for production of an action potential, which is crossed at *f*, will be raised, since the second action potential will fall within the refractory period of the first. This effect sets an upper limit to the discharge frequency.

Ripley, Bush and Roberts (1968) have described an unusual phenomenon in the coxal muscle receptor organs of crabs. Here the sensory nerve fibres do not conduct action potentials. Instead the receptor potential spreads electronically along the nerve fibre (which is a few millimetres long) into the thoracic ganglion.

FURTHER ASPECTS OF SENSORY CODING

Adaptation

Different receptors show different adaptation rates. The rate and extent of adaptation is related to the function of the receptor, depending upon whether the animal needs information about the steady level of the stimulus or about changes that occur in the stimulus intensity. We find that the response of many receptors consists of two parts, a tonic response, which is proportional to the intensity of the stimulus, and a phasic response, which is produced by changes in stimulus intensity. An example of a case in which the response is predominantly tonic is in receptors indicating the position of the animal with respect to gravity. In touch receptors, on the other hand, the tonic response is very small and the phasic response is rapidly adapting; a circumstance which probably accounts in part for our not feeling the presence of the clothes we are wearing. Muscle spindles provide an example of a receptor which has both tonic and phasic responses.

The presence of adaptation probably has a number of different advantages. In many cases, it is much more important for an animal to be able to detect changes in its environment than to measure accurately the static properties of the environment. For example, objects moving in the visual field are likely to be associated with danger or (when the animal is a carnivore) food. Adaptation thus serves to reduce the amount of unimportant information which reaches the central nervous system. Adaptation also enables a receptor to supply information about the rate of change of a stimulus, and can thus make a receptor much more sensitive to small changes in stimulus intensity without sacrificing the ability to respond over a wide range of stimulus intensities.

An example of a receptor system in which adaptation is extremely rapid occurs in the pad of the cat's foot (Armett *et al.*, 1962). Here, the receptors sensitive to pressure on the skin only respond if the displacement of the skin reaches the threshold amplitude in less than 0.7 msec, and the response of a receptor consists of a single nerve impulse. It is obvious that such a system cannot use the normal frequency code for measuring the intensity of the stimulus. Instead, stimulus intensity seems to be signalled by means of the number of receptors excited.

Adaptation can occur at a number of stages in the control sequence of sensory excitation. In the Pacinian corpuscle, the accessory structures (the lamellae) are involved in the process; when pressure is applied to the outside of the corpuscle, the lamellae are temporarily deformed, so deforming the nerve terminal, but the inner lamellae rapidly move back to their original positions, so that excitation of the nerve terminal is not maintained (Hubbard, 1958). In other cases, such as in many photoreceptors, the generator potential shows adaptation. Finally, we might expect to see some adaptation at the impulse initiation site, by means of a mechanism similar to that of accommodation in nerve fibres; a process of this kind probably occurs in Pacinian corpuscles (Gray and Matthews, 1951).

The sensory threshold

Suppose we perform an experiment in which a human subject listens to a series of sound pulses whose intensity is varied from quite loud to very soft, and is asked to state whether or not he can hear each pulse. At very low sound intensities, he will not be able to detect the pulses, whereas at higher intensities he will. In an intermediate range, he will be able to detect the pulses in a fraction of the times each pulse is presented, and this fraction will increase with increasing intensity. We may call the mid-point of this range, i.e. the sound intensity at which 50 per cent of the pulses are detected, the sensory threshold. Obviously, this is a rather arbitrary concept: we could equally well call the sensory threshold the

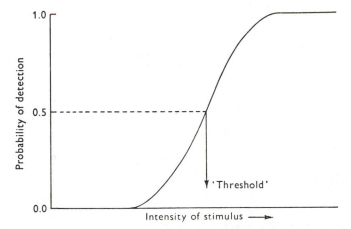

Fig. 15.14. Diagram to show how the probability of detection of a stimulus rises with increasing stimulus intensity. The threshold is here arbitrarily defined as that stimulus intensity which is detected 50 per cent of the times it is presented.

intensity at which there is a 10 per cent or 75 per cent probability of detection. But the important point here is that, with a stimulus whose intensity is variable over a continuous range (this proviso excludes vision and olfaction, where the stimulus is made up of discrete quanta or molecules), the probability of detection is a continuous function of the stimulus intensity; there are no discontinuities in the relationship (fig. 15.14).

The Weber–Fechner relation

Let us now consider the question, 'what is the relation between the intensity of the stimulus and the intensity of the sensation produced by it?' Weber, in 1846, found that human subjects could distinguish two weights, as long as the difference between them was about $2\frac{1}{2}$ per cent, irrespective of the absolute magnitude of the weights (this value – the minimum difference that can be detected – is known as the *increment threshold* or difference threshold). This conclusion was apparently confirmed by Fechner (1862), who described the results in the form of an equation:

$$\Delta I/I = k\,\Delta S \qquad (15.1)$$

where I is the stimulus intensity, ΔI is the increment threshold, k is a constant, and ΔS is a 'unit of sensation'. Fechner then suggested that this equation could be converted into a differential equation by making $\Delta I = dI$ and $\Delta S = dS$, so that

$$\frac{dS}{dI} = \frac{1}{kI}$$

which, when integrated, gives

$$S = a \log I + b \qquad (15.2)$$

where a and b are constants. Equation (15.2) is known as the 'Weber–Fechner law', and states that the intensity of the sensation is proportional to the logarithm of the intensity of the stimulus.

   In fact, it is found that the Weber–Fechner law is not universally applicable. In the human eye, for example, while it is approximately true for the medium range of light intensities, it does not apply at very low or very high light intensities. Nevertheless, it does provide 'a convenient rule of thumb' (Granit, 1955) for describing the relation between stimulus and response in a number of sensory systems.

Spontaneous activity

A number of receptors are spontaneously active: they produce nerve impulses in the sensory nerves in the absence of stimulation. An example of this is in the lateral line organs of fishes (Sand, 1937). In this case,

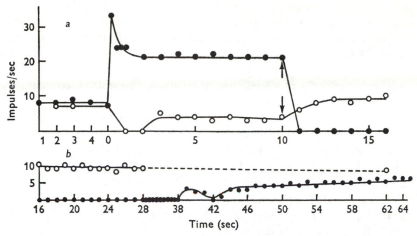

Fig. 15.15. Impulse frequency in a single lateral line fibre of a ray in response to perfusion of the hyomandibular canal. Black circles, headward perfusion (between 0 and 10 sec); white circles, tailward perfusion. Graph *b* is a continuation of *a*. (From Sand, 1937, by permission of the Royal Society.)

it is clear that the spontaneous activity enables the mechanoreceptors in the lateral line canals to be sensitive to movements of the canal fluid in both directions. In fig. 15.15, headward perfusion of the canal causes an increase in the discharge frequency, and tailward perfusion causes a decrease; thus the receptor is analogous to a centre-reading voltmeter, which can register both positive and negative voltages, as opposed to a voltmeter in which the zero position is at one end of the scale.

An alternative function of spontaneous activity may be to increase the sensitivity of a receptor. If the receptor is spontaneously active, then there is no such thing as a subthreshold stimulus, since any increase in stimulus intensity will change the frequency of impulse discharge. The problem for the animal is then the detection of this change in sensory nerve impulse frequency. The facial pit receptors of crotaline snakes show spontaneous activity, which is modified by very small changes in the radiant heat impinging upon them; it seems probable that the existence of this spontaneous activity contributes to the very great sensitivity of these organs (Bullock and Dieke, 1956).

Finally, there is considerable evidence that the activity of sense organs, spontaneous or induced, plays an important part in determining levels of activity in the central nervous system.

The use of multiple channels

A large number of sense organs send sensory information into the central nervous system via a number of nerve fibres. Such an increase in the number of transmission lines will obviously increase the amount of information that can be carried by the system. It is useful at this point to introduce the concept of a *receptor unit* (Gray, 1962); this consists of a single sensory nerve fibre together with its branches and any receptor cells which may be connected with it. A multiple channel system is thus a sense organ, or a group of closely associated receptors, consisting of more than one receptor unit.

A multiple channel system can increase the sensitivity of a sense organ, as the following theoretical argument shows. Consider a sense organ consisting of n identical receptor units excited independently of one another, and assume that each unit has a probability x of firing in response to a small constant stimulus. Then the probability of any unit not firing is $(1-x)$, and the probability of no unit firing is $(1-x)^n$. Hence the probability (p) of one or more units firing (which, we shall assume, constitutes detection of the stimulus) is given by

$$p = 1-(1-x)^n. \tag{15.3}$$

Fig. 15.16 shows how p varies with x for various values of n. It is evident

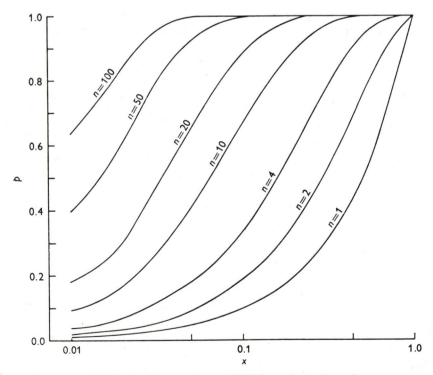

Fig. 15.16. Graphical solutions of (15.3) for various values of n.

that an increase in the number of channels can produce (*a*) a consider-able increase in sensitivity (just how much will depend on how *x* varies with the size of the stimulus), and (*b*) an increase in the sharpness of the threshold, since the slopes of the curves increase with increasing values of *n*. For the particular case where $n = 2$, (15.3) becomes

$$p = 2x - x^2. \tag{15.4}$$

This equation is fairly easy to test by psychophysical methods, since a number of sense organs are paired. Thus Pirenne (1943) found that the threshold for binocular vision in man is lower than that for monocular vision, and Dethier (1953) showed that the threshold of the tarsal taste receptors in flies followed a similar relationship (fig. 15.17).

Most information channels are 'noisy' to some extent, i.e. there is some random activity (noise) in addition to the signal. In a sensory nerve fibre, this noise will be seen as random fluctuations in the interval

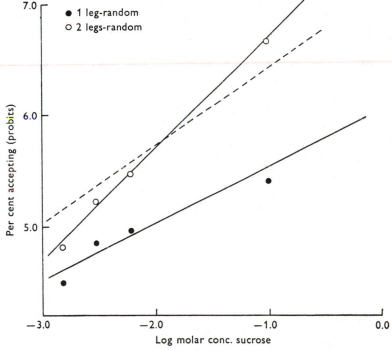

Fig. 15.17. The greater sensitivity of flies with two legs in contact with a sucrose solution as compared with those having only one leg so applied. The broken line shows the expected relation between sucrose concentration and the probability of response for two-legged flies, as calculated from application of (15.4) to the results for one-legged flies. The units (probits) on the ordinate are equal to one standard deviation; 5 probits correspond to 50 per cent acceptance. (From Dethier, 1963.)

between successive nerve impulses. How, then, is the animal to detect a small change in nerve impulse frequency against this noisy background? One way of doing this would be to measure the average frequency over a sufficiently long time, both before and during the stimulus. Use of this system must necessarily mean that the latency of detection will be appreciable, and that there will be no detection of very brief stimuli. However, by using a number of channels the same information can be obtained in a fraction of the time. Suppose that, in a single noisy sensory channel, the time needed to determine whether or not a new average frequency is significantly higher than the original one is t; then the equivalent time for n identical channels is t/n. Hence an increase in the number of sensory channels can reduce the response time. Another way of looking at this problem of detection is to consider the signal-to-noise ratio of the system. If the signal-to-noise ratio of a signal channel is q, then the signal-to-noise ratio of n identical channels is $q\sqrt{n}$. Hence increase in the number of channels increases the sensitivity of a receptor to changes in stimulus intensity. It seems very probable that considerations of this type apply to the large number of spontaneously active sensory channels found in the facial pit sense organs of crotaline snakes.

A further use of multiple channel systems is to ensure that the sense organ is sensitive to the direction or location of the stimulus. The most obvious example of this occurs in image-forming eyes, in which different photoreceptors respond when the stimulus (a light source, for example) is in different parts of the visual field.

We have so far considered the advantages of multiple channel systems only for cases in which the channels have identical receptor properties. A considerable increase in the information derived from a sense organ can be obtained by using a multiple channel system whose receptor units differ, quantitatively or qualitatively, in their sensitivity. A common type of organization is what may be called *range fractionation* (Cohen, 1964), in which different receptor units respond to different parts of the range of stimulus intensities or qualities covered by the whole sense organ. This leads to an increase in the sensitivity of the sense organ, since a small change in the stimulus can produce a large change in the response of a particular receptor unit, whereas if that receptor unit had to be responsive over the whole of the stimulus range, its sensitivity over a fraction of the range would be correspondingly reduced. An example of this type of organization is shown in fig. 15.18, where the different receptor units of the Golgi organs in the ligaments of a cat's knee have different restricted sensitivity ranges, but between them can provide information about the position of the knee joint over the full range from full flexion to full extension. Another example is in the vertebrate eye,

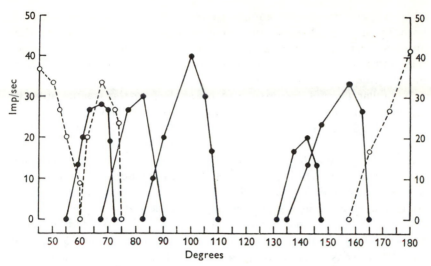

Fig. 15.18. The relations between the impulse discharge frequency in a number of cat knee joint receptors and the angle of the knee joint. (From Skoglund, 1956.)

where the rods and cones deal with low and high light intensities respectively.

Receptor units in a system which utilizes range fractionation frequently show bell-shaped response curves (figs. 15.18 and 15.19). It follows that there must be some ambiguity in the information supplied by any one receptor unit; in the unit whose response is shown in curve *a*

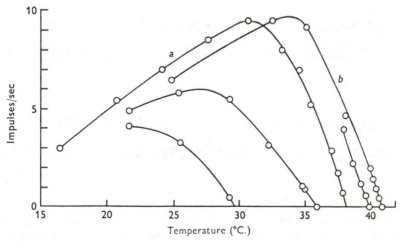

Fig. 15.19. Steady impulse discharge frequencies of five 'cold' fibres in a cat's tongue at different temperatures. Curves *a* and *b* are referred to in the text. (From Hensel and Zotterman, 1951.)

of fig. 15.19, for example, an impulse frequency of 7/sec is produced at both 24° and 35°. This ambiguity can be resolved by comparison with another receptor unit with a different, but overlapping, response range; thus, in fig. 15.19, the unit whose response is shown by curve *b* has an impulse frequency of 6/sec at 24° and 9/sec at 35°.

A similar type of argument can be applied to the fractionation of receptor units with respect to the qualitative aspects of the stimulus. Cohen, Hagiwara and Zotterman (1955) measured the responsiveness of different gustatory receptor units in the afferent nerves of the cat's tongue. They found that many fibres were responsive to more than one of the modalities which psychophysical measurements suggest can be distinguished, and concluded that the different sensations arise from comparisons of the activity in a number of different receptor units. For example, they suggested that the 'sour' sensation is produced by simultaneous activity in the 'water', 'salt' and 'acid' fibres; if the 'water' and 'acid' fibres cease their activity, the sensation is then one of 'salt'. Thus the specificity of the response is sharpened by utilizing the multiple channel system. The trichromatic colour vision system in man is another example of such sharpening of the specificity of sensation. An extreme

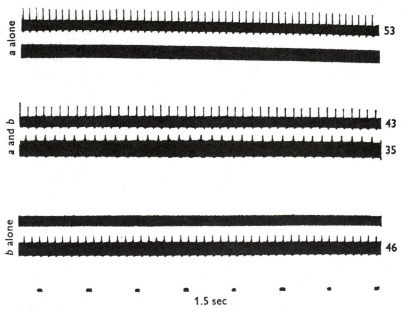

Fig. 15.20. Simultaneous records of the sensory discharges in two *Limulus* optic nerve fibres serving adjacent ommatidia (*a* and *b*), in response to illumination of either *a* or *b* alone or both together. The records last 1.5 sec in each case, and the figures to the right of them show the number of action potentials occurring in that time. (From Hartline and Ratliff, 1957.)

example of this type of system is found in the olfactory receptors in the antennae of the silkmoth (Schneider, Lacher and Kaissling, 1964). In over 50 cells whose response to various organic compounds was tested, no two cells showed the same response spectrum (fig. 16.32).

Interaction between receptors: *lateral inhibition*

A very interesting phenomenon occurs in the compound eye of *Limulus*, known as lateral inhibition (Hartline, Wagner and Ratliff, 1956; see also Hartline, Ratliff and Miller, 1961). Hartline and his colleagues dis-covered that the discharge rate of a single eccentric cell decreases if the adjacent ommatidium is illuminated, indicating that adjacent ommatidia are mutually inhibitory (fig. 15.20). This phenomenon leads to two interesting features in the pattern of information produced by the eye.

Firstly, the eye must become much more sensitive to edges. Consider the model system shown in fig. 15.21, where we have a row of omma-tidia *A* to *J* of which *A* to *E* are exposed to a relatively high light intensity, and *F* to *J* are partially shaded. The extent of the inhibition produced by each cell on its immediate neighbours is then indicated by the diagonal arrows in the diagram, full arrows indicating strong inhibition and broken arrows light inhibition. Cells *A* to *D* are mutually inhibited to some extent, and hence their discharge frequencies are considerably less than they would be if there were no lateral inhibition. But cell *E* is only slightly inhibited by cell *F*, hence its discharge rate is higher than those of cells *A* to *D*. Similarly, cell *F* is more strongly

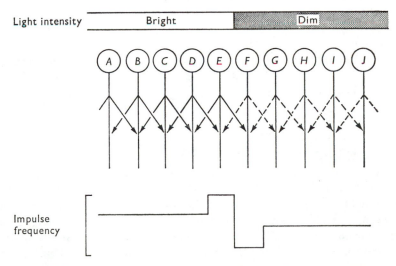

Fig. 15.21. A model system to show lateral inhibition. Explanation in text.

Fig. 15.22. An optical illusion which may be caused by lateral inhibition: there is an appearance of grey spots at the intersections of the white lines.

inhibited than are cells *G* to *J*, hence its discharge rate is lower than theirs are. The net effect of these interactions is to emphasize the presence of the edge. A similar process appears to form the basis of edge perception in our own eyes; it may also account for the well-known optical illusion shown in fig. 15.22 – we may suggest that the grey spots seen at the junctions of the white lines are produced by lateral inhibition from the surrounding receptors produced by the image of the white lines.

The second feature of lateral inhibition is that it can increase the resolving power of the eye. The 'fields of view' of adjacent ommatidia show some considerable overlap. Lateral inhibition has the effect of cutting down this overlap, so that each optic nerve fibre only responds to the illumination of a small part of the visual field (Reichardt, 1962).

TRANSMITTER–RECEIVER SYSTEMS

The energy for the activation of most distance exteroceptors is derived from the environment; a crotaline viper, for example, perceives the radiant heat emitted by its prey, and the prey (if it is lucky enough) hears the sounds produced by the approaching viper. However, there exist some specialized sensory systems in which the energy that stimulates the receptors is produced by the animal itself, being reflected back from or modified by objects in the environment.

One of the most well-known of such 'transmitter–receiver systems' is the echolocation system used by bats (see Griffin, 1958). It has long been known that blinded bats can avoid obstacles and catch insects in flight just as well as normal bats. Pierce and Griffin (1938) discovered that flying bats emit a continuous series of ultrasonic cries, and Griffin later showed that the frequency with which these cries are made is much increased when a bat approaches an obstacle. Griffin and Galambos (1941) measured the ability of bats to fly through a barrier of metal wires spaced just wide enough apart to let them through. They found that deaf or dumb bats frequently hit the wires whereas normal or blind bats rarely did so. The conclusion from these experiments was that bats detect obstacles by emitting ultrasonic sound pulses and hearing the sound waves reflected back to them; the position of an object is determined by localizing the source of the reflected sound. Echolocation systems which are similar in principle to those used by bats are also found in porpoises, certain cave-dwelling birds and possibly some other animals.

A different type of transmitter–receiver system is seen in the object location system used by some electric fish. Here the fish produces an electric field in the water by means of its electric organs, and can detect changes in this field with its electroreceptors. We shall examine these electroreceptors in the following chapter.

Finally, there are a number of deep-sea fishes which are bioluminescent. Since they have well-developed eyes, it seems probable that one of the functions of their luminescence is to produce light by which the fish can see.

CENTRAL CONTROL OF RECEPTORS

The efferent nerve fibres innervating mammalian limb muscles are of two types: the large, rapidly-conducting α fibres, and the small, more slowly-conducting γ fibres. The γ fibres innervate the muscle fibres (intrafusal fibres) in the muscle spindles, whereas the α fibres innervate the extrafusal fibres and so are responsible for the contraction of the muscle (Leksell, 1945; Kuffler, Hunt and Quilliam, 1951). Stimulation of the γ fibres causes contraction of the intrafusal fibres so that the middle regions (which are not contractile) are stretched; this excites the sensory ending (fig. 15.23). Thus the output of the sensory fibres can be modified by the action of the γ motor fibres; in other words, it is to some extent under the control of the central nervous system.

Such central nervous control of receptor output via efferent nerve fibres is not uncommon. The efferent fibres may act at various points in the sensory excitation control sequence. In the muscle spindles they act

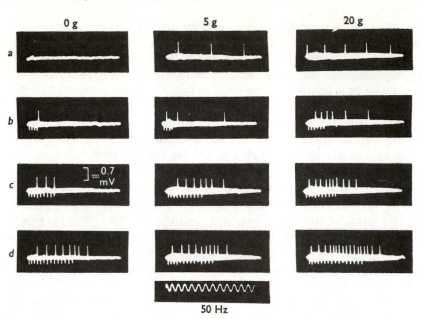

Fig. 15.23. The effect of γ motor stimulation on the afferent discharge from a cat muscle spindle. Upward deflections on the records are sensory nerve impulses; downward deflections are stimulus artefacts produced by stimulating a single fibre at 200/sec. Columns show different loads on the muscle. Rows show effects of stimulation: *a*, no stimulation; *b*, 4 to 6 stimuli; *c*, 9 to 11 stimuli; and *d*, 14 to 16 stimuli. (From Kuffler, Hunt and Quilliam, 1951.)

on the accessory structures; other examples of this type of action are the reflex control of pupil diameter in the vertebrate eye, and the action of the stapedius and tensor tympani muscles in the mammalian middle ear, which decrease the sensitivity of the ear to loud sounds. Direct efferent control of the size of the generator potential occurs in crustacean muscle receptor organs, where the sensory neuron is inhibited by an inhibitor axon (fig. 15.8) which, when stimulated, depresses the generator potential and prevents the initiation of sensory impulses (Eyzaguirre and Kuffler, 1955*b*).

Loewenstein (1956) showed that the sensitivity of mechanoreceptors in frog skin is increased if the sympathetic nerve supply to the skin is stimulated. Similar effects are produced by application of the sympathetic neurotransmitters, adrenaline and noradrenaline, to the skin. Livingston (1959) suggests that such sympathetic actions on sense organs are likely to be widespread.

Efferent nerve fibres occur in the auditory nerves of mammals (Rasmussen, 1953), and stimulation of them suppresses the afferent response in the auditory nerve to sound clicks. Electron microscopy shows that the receptor cells (hair cells) of the cochlea are innervated by two types of nerve ending, one of which contains synaptic vesicles and is therefore probably efferent (Engström and Wersäll, 1958). Similar apparently efferent endings are found on the homologous receptor cells in other parts of the ear and in the lateral line organs of fishes. Mammalian taste buds are also innervated by nerve endings which contain synaptic vesicles (de Lorenzo, 1963).

16 Some particular sense organs

Having considered some general properties of sensory receptor processes, we shall now examine the physiology of some particular sense organs in a little more detail. It is not possible in a book of this nature to deal with more than a small fraction of the different types of sense organ that have been described, hence this chapter merely provides a brief look at a vast and complex subject.

THE ACOUSTICO–LATERALIS SYSTEM OF VERTEBRATES

In vertebrates, the receptors involved in hearing, equilibrium reception, and the detection of water currents are all of one basic type and are probably of common evolutionary origin. In each case the receptor cells (fig. 16.1) possess cilium-like processes which are embedded in a matrix and whose movement leads to modification of the sensory output. The receptor cells are all connected to sensory neurons which enter the central nervous system in the brain stem. The whole complex is known as the acoustico–lateralis system. The fascination of this system is that we can see how a relatively simple type of receptor cell can be used to respond to a whole variety of different mechanical stimuli, solely by elaboration of the accessory structures in the various sense organs.

The lateral line organs of fishes

The sensory cells of the 'ordinary' lateral line system* are grouped together in organs called *neuromasts*. Each neuromast (fig. 16.2a) consists of a number of receptor cells and supporting cells situated in the epidermis; and a gelatinous projection, the *cupula*, into which the sensory hairs of the receptor cells project. Neuromasts may be situated either freely on the outer surface of the animal or, typically, in canals which open to the surface at intervals through small pores. The distribution of these canals in a typical fish is shown in fig. 16.3.

* The 'ordinary' lateral line system (Dijkgraaf, 1963) is that part of the system which contains neuromasts and is primarily responsive to water movements. Some fishes possess, in addition, numbers of 'ampullary' organs associated with the lateral line system, which appear to be electroreceptors (p. 418).

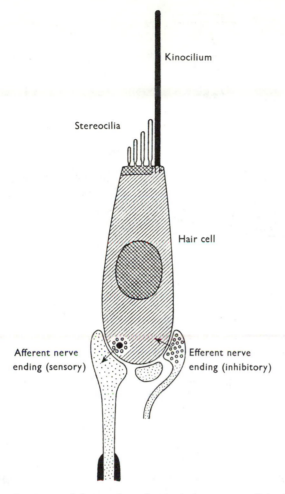

Fig. 16.1. The structure and innervation of a typical receptor cell in the vertebrate acoustico–lateralis system. (From Flock, 1971.)

If a water current impinges on the cupula of a neuromast it is moved to one side, and it is this movement ('detected' by the sensory hairs of the receptor cells) which acts as the effective stimulus. A fish whose lateral line system has been partially denervated no longer responds to water currents applied to the denervated region. This sensitivity to water movements enables the fish to detect moving objects in its immediate vicinity by means of the water currents which they produce (Dijkgraaf, 1934); fig. 16.4 shows how an approaching object produces movements of neuromast cupulae.

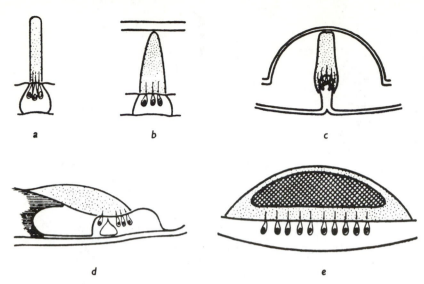

Fig. 16.2. Different accessory structures associated with receptor cells of the vertebrate acoustico–lateralis system. *a*: Free neuromast of the lateral line system. *b*: Neuromast in a lateral line canal. *c*: Ampullary sense organ of a semicircular canal. *d*: Organ of Corti in the mammalian cochlea. *e*: Otolith organ. (From Dijkgraaf, 1963.)

The electrical responses of the lateral line nerves to stimulation of the neuromasts by water currents of controlled velocity were investigated by Sand (1937), using a perfusion technique on the hyomandibular canal of the ray. He found that there was a considerable discharge in the nerve in the resting state, which was much increased by water movements in either direction. After dissecting the nerve so that only a single active unit remained, the spontaneous discharge in this unit was increased by perfusion in one direction but decreased by perfusion in the other direction (fig. 15.15). Some units were excited by headward perfusion of water in the canal, others by tailward perfusion.

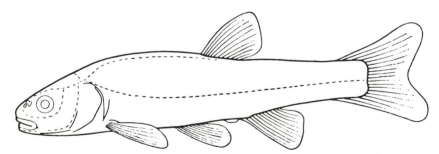

Fig. 16.3. Distribution of the lateral line canals on the body surface of a fish. (From Dijkgraaf, 1952.)

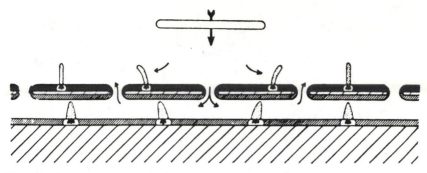

Fig. 16.4. The movements of neuromast cupulae produced by the water currents (small arrows) set up by an approaching object. (From Dijkgraaf, 1952.)

The semicircular canals

The inner ear, or labyrinth, of vertebrates is a complex organ concerned with equilibrium reception and hearing. It consists of (i) the membranous labyrinth, which is a series of interconnected sacs and tubes containing the receptor cells and filled with an aqueous fluid called the *endolymph*, and (ii) the osseous labyrinth, which is a cavity in the bone of the skull in which the membranous labyrinth is situated. The inner wall of the osseus labyrinth and the outer surface of the membranous labyrinth are separated by a thin layer of another fluid, the *perilymph*. The membranous labyrinth arises embryologically by an invagination of the ectoderm, and its receptor cells are very similar to those of the lateral line neuromasts.

The structure of the membranous labyrinth is shown in fig. 16.5. There are three semicircular canals, set in planes approximately at right angles to one another. The horizontal canals of the two labyrinths in the head lie in the same plane; the anterior vertical canal on one side is in approximately the same plane as the posterior vertical canal on the other, and *vice versa*. The canals are connected at each end to a sac called the utriculus, so that endolymph can move between canal and utriculus. At one end of each canal is a swelling, the ampulla, in which a group of receptor cells and supporting cells, the crista, is found (fig. 16.2c). Sensory nerve fibres leave the crista to join the auditory nerve.

The sensory hairs of the crista are inserted into a gelatinous cupula. In sections of preserved material the cupula appears as a rather small structure, but in the living animal it extends right across the ampulla to form a fairly close-fitting seal with the opposite wall. This fact, which has important connotations for the mode of action of the semicircular canals, was discovered by Steinhausen (1931) by observing the

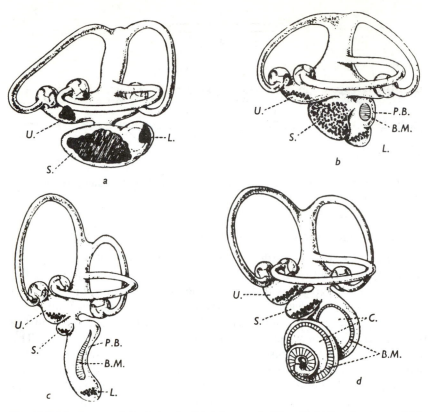

Fig. 16.5. Membranous labyrinths of a fish (*a*), a reptile (*b*), a bird (*c*) and a mammal (*d*). *U.*, Utriculus; *S.*, sacculus; *L.*, lagena; *P.B.*, papilla basilaris; *B.M.*, basilar membrane; *C.*, cochlea. (From von Frisch, 1936.)

appearance of the horizontal semicircular canal in a pike after indian ink had been injected into the endolymph.

It has been known for many years that the semicircular canals are concerned with the perception of rotational movements of the head. Such movements produce compensatory reflexes, especially in the extraocular muscles of the eyes, which are abolished by bilateral extirpation or denervation of the canals.

When a vertebrate is rotated at a constant angular velocity in the horizontal plane, its eyes move so as to compensate for this movement. If the head is passively turned to the left, for example, the eyes move to the right, so keeping the visual image fixed in position on the retina. As the rotation continues, the eyes flick rapidly to the left, and then revert to moving to the right again. This pattern of movements, which is known as *nystagmus*, may be repeated for a time. With prolonged rotation however, the nystagmus dies away and the eyes remain in their normal position with respect to the head. On cessation of a prologed rotation, the eyes show *after-nystagmus* for several seconds, i.e. they behave as though the head were being rotated in the opposite direction.

Can we relate these eye movements to the mechanics of the semicircular canals? At one time it was thought that the action of the semicircular canals was as follows. When the head was rotated, endolymph would flow freely through the canals by inertial forces, and the cupula extending only part-way across the ampulla, would respond to the velocity of this stream in the same way that the free neuromasts on the surface of a fish respond to water currents. This scheme (which we may call the 'flow' theory) could not account for the duration of nystagmus and after-nystagmus since the friction between the endolymph and the walls of the canal would bring the endolymph to rest with respect to the canal within a fraction of a second of the change in rotational velocity of the canal.

Steinhausen's observation that the cupula extends across the whole width of the ampulla in the living animal is incompatible with the 'flow' theory. In a further series of experiments, Steinhausen (1933) was able to show that when the cupula is displaced by a sudden change in the velocity of rotation, it takes several seconds to return to its resting position. As a result of these observations, the 'flow' theory was replaced by the 'displacement' theory, according to which the displacement of the cupula is strictly linked to the displacement of the endolymph, not to its velocity of flow. An elegant piece of evidence for this theory was provided by Dohlman (1935) who, using Steinhausen's technique for observation of the movements of the cupula, was able to inject a small oil droplet into a semi-circular canal of a cod. He found that the droplet only moved a short way round the canal during full displacements of the cupula (fig. 16.6); hence there can be very little flow of endolymph past the cupula.

Let us now see how, according to the displacement theory, the semicircular canal works. The essential stimulus for the hair cells of the crista is the angular deviation of the cupula from its normal position in the canal. This deviation is produced by the combination of three forces: (i) an inertial force, proportional to the moment of inertia of the endolymph and the angular acceleration of the canal, (ii) a viscous (damping) force, proportional to the viscosity of the endolymph, the dimensions of the canal and the velocity of movement of the endolymph through it, and (iii) an elastic restoring force in the cupula, proportional to its angular deviation from the resting position.

Now consider what happens if the canal is subjected to a period of rotation at a constant angular acceleration. In this case the inertial force

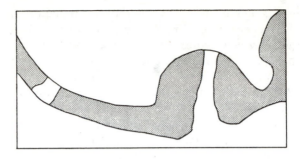

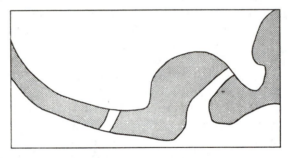

Fig. 16.6. The appearance of part of a fish semicircular canal after injection of indian ink into the endolymph followed by a further injection of an oil droplet into the canal. The upper diagram shows the system at rest, the lower diagram shows the displacement of the cupula and oil droplet after rotation. (Drawn from photographs by Dohlman, 1935.)

is constant and the viscous and elastic forces decrease and increase respectively as the cupula moves smoothly to a position determined by the acceleration. The behaviour of the system is more complex when it is subject to a long period of rotation at a constant angular velocity (fig. 16.7). At the start of this rotation the acceleration is very high and so the inertial force produces a rapid deflection of the cupula. Then, since there is no more acceleration, the inertial force falls and the cupula begins to return to its resting position under the influence of the elastic force. In order to do this it has to act against the viscous force by pushing the endolymph back through the canal, so that it is some time (about 30 sec) before the cupula reaches its resting position. If the rotation is now stopped, the endolymph tends to continue on its way, so that there is for a short time a large inertial force which pushes the cupula in the opposite direction to that in which it was previously deflected; it is this 'overswing' which produces after-nystagmus. Finally, over the next 20 to 30 sec, the cupula returns to its resting position, again by pushing the endolymph through the canal against the viscous force.

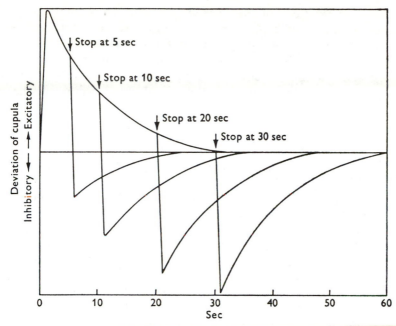

Fig. 16.7. Diagram to show the deflection of the cupula of a semicircular canal in response to rotations at constant velocity for various periods of time. The resting position is indicated by the horizontal line. The curves apply equally well to the changes in sensory nerve discharge produced by constant velocity rotations. (From Lowenstein and Sand, 1940a, by permission of the Royal Society.)

Excellent further evidence for Steinhausen's 'displacement' theory has been provided from two sources: by psychophysical measurements on the sense of rotation in man, and by electrophysiological measurements on the nervous output from the hair cells of the crista in cartilaginous fish. The analysis of the cupula movements in terms of three forces, given above, can be expressed in terms of a differential equation as follows:

$$A\frac{d^2x}{dt^2}+B\frac{dx}{dt}+Cx=0$$

where x is angular displacement, and A, B and C are constants. Van Egmond, Groen and Jongkees (1949) made a series of measurements on the sensation of rotation in human subjects for different values of x and its derivatives, and showed that the results were consistent with this equation.

Lowenstein and Sand (1936, 1940a) measured the nervous discharge from the horizontal semicircular canals of the dogfish and the ray in response to rotations in the horizontal plane. They found that the changes in the nerve impulse frequency in the sensory fibres followed the same time course as the displacement of the cupula as measured by Steinhausen. These sensory fibres showed a steady spontaneous discharge when the cupula was in its resting position, which increased when the cupula was deflected towards the utriculus and decreased when it

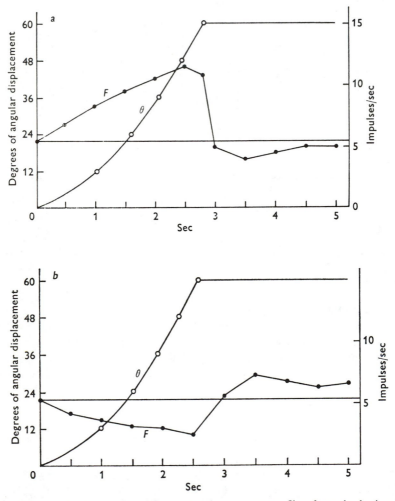

Fig. 16.8. Impulse frequencies (*F*) recorded from a sensory fibre from the horizontal semicircular canal of a ray in response to a constant angular acceleration (shown by the angular displacement curves (*θ*)) resulting in deflections of the cupula towards (*a*) and away from (*b*) the utriculus. (From Lowenstein and Sand, 1940*a*, by permission of the Royal Society.)

was deflected away from the utriculus (fig. 16.8). Similar features were seen in the responses of the vertical canals (Lowenstein and Sand, 1940b), except that here excitation occurred when the head was rotated so as to deflect the cupula away from the utriculus.

We must now consider a little further the nature of the effective stimulus for the canals. Linear accelerations will have no effect: the inertial forces produced in all parts of the endolymph will be of equal magnitude and in the same direction, hence there will be no tendency for the endolymph to rotate. (A linear acceleration could produce a response if the cupula and endolymph were of different densities. The cupula is probably very slightly denser than the endolymph in fact, but this can safely be neglected for our purposes.)

The mechanical response of a particular canal to rotation about an axis which is perpendicular to the plane in which the canal lies depends only on the magnitude and direction of the angular acceleration; it is independent of the position of the axis. For example, the response of a horizontal canal to a given angular acceleration of the head about a vertical axis passing through the midline is the same as if the axis passed through the centre of the canal. The reason for this is that the accelerations produced by rotation about an eccentric axis can be resolved into two components: an angular acceleration about the centre of the canal and a linear acceleration at right angles to the line between the centre of the canal and the axis of rotation (fig. 16.9a). If the angular acceleration is α rad/sec, the radius of the semi-circular canal is r and the distance between the centre of the canal and the axis of rotation is s, then the rotational component of the acceleration is $r\alpha$ and the translational component is $s\alpha$. A geometrical proof of these assertions is as follows. Referring to fig. 16.9b, the canal is rotated in the plane of the page about a point A at an angular acceleration α. The 'centre' of the canal (or, for that matter, any other point in the plane) is the point O. Then a point P on the canal will be subjected to an acceleration shown by the vector PR, equal to $\alpha . AP$, the angle $\angle APR$ being a right angle. We resolve PR into two components, one of which (PQ) is at right angles to OP and of length $\alpha . OP$ (i.e. PQ represents the acceleration at P that would be produced by an angular acceleration of α about O). The other component of PR, by the triangle of forces theorem, is then RQ. We wish, then, to prove that RQ is at right angles to AO and of length $\alpha . AO$.

Now
$$\frac{PR}{AP} = \frac{PQ}{OP} = \alpha$$

and
$$\angle APO = \angle RPQ$$

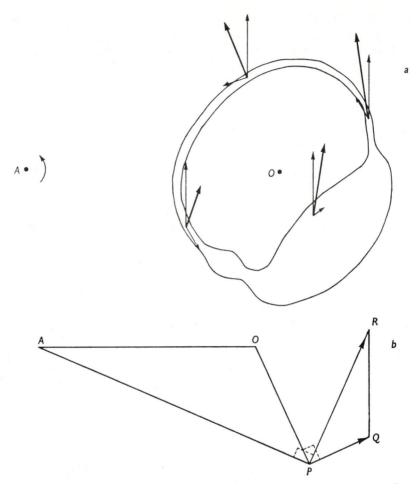

Fig. 16.9. Rotation of a semicircular canal about an eccentric axis. *a*: The accelerations produced by such a rotation (about the axis *A*) are shown by the thick arrows. These accelerations can be resolved into two components (thin arrows): an angular acceleration about the centre of the canal (*O*), and a linear acceleration at right angles to *AO*. *b*: Geometrical illustration of the resolution of these accelerations. Explanation in text.

therefore the triangles *AOP* and *QRP* are similar. Hence

$$\frac{RQ}{AO} = \alpha$$

or

$$RQ = \alpha \cdot AO$$

and, since ∠*APR* and ∠*OPQ* are right angles, *RQ* must be at right angles to *AO*. Notice that the positions of *A*, *P* and *O* are irrelevant to this proof.

So far we have considered the responses of a semicircular canal to rotation in the plane in which it lies, i.e. about an axis perpendicular to that plane. What happens if the axis of rotation is not perpendicular to the plane of the canal? This question has been theoretically investigated by Summers, Morgan and Reimann (1943); the following analysis is partially based on their account. An angular acceleration can be represented by a vector parallel to the axis of rotation whose length represents the magnitude of the acceleration and whose direction indicates the direction of rotation (it is conventional to draw this so that the rotation is clockwise when seen along the direction of the arrow). Now consider fig. 16.10, in which the line B is perpendicular to the plane in which the semicircular canal lies. The canal is subjected to an angular acceleration about an axis parallel to the line A, which lies at an angle θ with B. This angular acceleration is represented by the vector α. The canal is then effectively subjected to an angular acceleration about B represented by the vector β, which is obtained by dropping a perpendicular from the tip of the vector α onto B. It is clear that

$$\beta = \alpha \cos \theta.$$

Hence if θ is 90 degrees, i.e. if A lies in the plane of the semicircular canal, there can be no effective stimulation of the canal. This is the only set of positions of A for which this will occur; for all other positions there will be some response, and this will be maximal for a given angular acceleration when A coincides with B. Thus a single canal is subjected to the same effective stimulus when rotated about A with an angular acceleration α as it is when rotated about B with an angular acceleration $\alpha \cos \theta$. A single canal is therefore incapable of providing precise information as to the nature of the angular acceleration to which it is subjected.

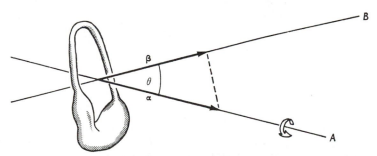

Fig. 16.10. Rotation of a semicircular canal about an axis (A) which is not perpendicular to the plane in which the canal lies. The vector α represents the magnitude and direction of the angular acceleration, and β is the projection of α onto the line B, which is perpendicular to the plane of the canal. β is then the effective acceleration to which the canal responds.

In order to provide such precise information, it is necessary to have three canals set in three planes which intersect in lines which are not parallel to one another, since there are three degrees of freedom in the orientation of an axis in space. Any particular angular acceleration can then be resolved as is shown in fig. 16.11, by dropping perpendiculars from the tip of its vector onto lines perpendicular to the planes in which the semicircular canals lie. It is in accordance with this that there are three semicircular canals in the labyrinths of all jawed vertebrates. It is not essential that the planes of these canals should be mutually at right angles to each other, but the sensitivity of measurement is maximal if they are.

In the jawless vertebrates the horizontal canal is absent, there being two vertical canals in each labyrinth in most cases, such as in the lamprey and the fossil *Cephalaspis*, and only one in the hagfish *Myxine*. Here, it is clear that a single labyrinth cannot determine the orientation

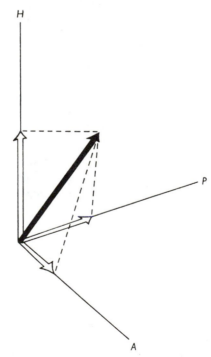

Fig. 16.11. How the semicircular canals resolve an angular acceleration about any axis in space into three components. The lines *H*, *P* and *A* represent perpendiculars to the planes of the horizontal, posterior vertical and anterior vertical canals respectively, and are therefore very approximately at right angles to one another. The black arrow is a vector indicating the imposed angular acceleration. Perpendiculars dropped from the tip of this arrow onto *H*, *P* and *A* give the white arrows, which are vectors showing the components of angular acceleration about these axes.

TABLE 16.1 *Responses of the six semicircular canals of* Raia *to angular displacements about the three primary axes.* ● *Excited*; ○ *unaffected*; ⊗ *inhibited.* (From Lowenstein and Sand, 1940*b*.)

Semicircular canal	longitudinal axis		Rotation about the transverse axis		vertical axis	
	Right[1]	Left	Forwards[2]	Backwards	Clockwise[3]	Anti-clockwise
Right ant. vert.	●	⊗	●	⊗	⊗	●
Left ant. vert.	⊗	●	●	⊗	●	⊗
Right post. vert.	●	⊗	⊗	●	●	⊗
Left post. vert.	⊗	●	⊗	●	⊗	●
Right horizontal	○	○	○	○	●	⊗
Left horizontal	○	○	○	○	⊗	●

[1] I.e. anticlockwise when viewed from the front.
[2] I.e. clockwise when viewed from the right.
[3] When viewed from above.

of the axis of rotation of the head. However, where there are two canals in each labyrinth, the four canals acting together could provide the information necessary for such a determination, provided that at least three of them are set in planes whose lines of intersection are not parallel to each other. It is interesting to note that this is in fact so: Stensiö (1927) describes the 'vertical' canals of *Cephalaspis* as being tilted outwards a little in each case.

Following their investigation on the horizontal canals of the ray, Lowenstein and Sand (1940*b*) measured the responses of all three canals on each side to rotation about the longitudinal, vertical and transverse axes. Their results are shown in table 16.1. Notice that the horizontal canals do not respond to rotation about the longitudinal or transverse axes, whereas the vertical canals respond to rotation about all three axes. The reason for this is simply that the longitudinal and transverse axes are parallel to the plane of the horizontal canal, whereas none of the three axes are parallel to the planes of the vertical canals.

An understanding of the precise workings of the system described in table 16.1 demands a knowledge of the orientation of the canals and involves an exercise in solid geometry. Fig. 16.12 represents the situation in the left labyrinth of the ray: the animal is in its normal horizontal position and faces north, and the system is viewed from a position about 45 degrees above the horizon to the south-west. The horizontal canal is in the horizontal plane and is represented by a perpendicular to this plane, i.e. by the vertical axis *H*. The canal is excited by rotation about *H* in the direction shown by the little curled arrow, and this is conventionally represented by the axis having an arrowhead on its upper end. The planes of the anterior and posterior vertical canals lie at about 45 degrees to the long axis of the animal, i.e. they run approximately north-west and south-west respectively; these planes are not precisely vertical – it is as if the tops of the two canals have been pushed together slightly. The perpendiculars to the planes of the vertical canals are represented by the lines A (anterior) and P (posterior) in fig. 16.12, and the arrows show the directions in which they must be rotated in order to be excited.

Fig. 16.13 shows what happens when the system is rotated about the vertical, longitudinal (north–south) and transverse (east–west) axes. Anticlockwise acceleration about the vertical axis (fig. 16.13*a*) produces an angular acceleration represented by the vector **h**, and so produces excitation of the horizontal canal. By dropping perpendiculars from the tip of **h** onto the lines A and P, we obtain the vectors **a** and **p** which represent the components of angular acceleration in the planes of the anterior and posterior canals respectively. The vector **p** is positive (since it points in the same direction as the arrowhead on line P in fig. 16.12), and so the left posterior vertical canal should be excited; reference to table 16.1 shows that this is so. The vector **a**, however, is negative (since it points in the opposite direction to the arrowhead on line A in fig. 16.12); thus the left anterior canal should be inhibited by

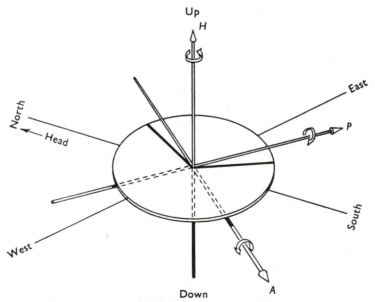

Fig. 16.12. Diagram showing the spatial arrangement of perpendiculars to the planes of the three semicircular canals in the left labyrinth of a ray. The disc is drawn to show the orientation of the horizontal plane of the fish. Further explanation in the text.

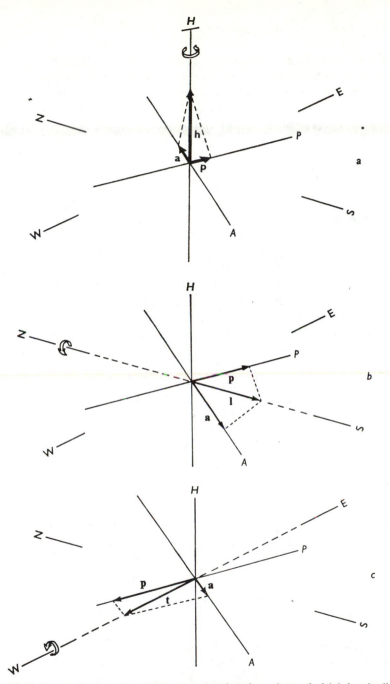

Fig. 16.13. Vector diagrams showing how acceleration about the vertical (*a*), longitudinal (*b*) and transverse (*c*) axes causes excitation or inhibition of the semicircular canals. The diagrams are based on fig. 16.12, so that *H*, *A* and *P* represent perpendiculars to the planes of the three canals. Vectors **h**, **l** and **t** represent the applied accelerations, and vector **h**, **a** and **p** represent the projections of these on *H*, *A* and *P* respectively. Further explanation in the text.

anticlockwise rotation about the vertical axis – as indeed it is. Figs.
16.13*b* and *c* show the components along *A* and *P* of angular accelerations about the longitudinal and transverse axes. Notice, in these cases,
that there is no component of the resulting vectors (**I** and **t** respectively)
along *H*; this is because *H* is at right angles to both these axes, or, to put
it another way, the horizontal canal lies precisely in the horizontal plane.

The otolith organs

In addition to the semicircular canals, the labyrinth contains three sacs
known as the utriculus, the sacculus and the lagena (fig. 16.5). The
receptor cells of these sacs are arranged in groups to produce areas of
sensory epithelia called *maculae*. The gelatinous material covering each
macula is filled with calcareous granules (*otoconia*); in bony fishes the
otoconia are fused to form *otoliths* (fig. 16.2*e*). There are typically three
otolith organs, the macula utriculi, the macula sacculi and the macula
lagenae, but the latter is absent in most mammals.

It seems that the essential stimulus for the receptor cells is a displacement of their sensory hairs, just as in the neuromasts of the lateral
line and the cristae of the semicircular canals. In this case such displacement is brought about by movements of the otoconial mass under
the influence of gravity or linear accelerations. Consider a flat macula
which is horizontal when the head is in its normal position. The otoconial mass exerts a downward force on the cilia of the receptor cells; in
the normal position these are upright and there is therefore no lateral
force applied to them. If now the head is tilted through an angle of α,
then the cilia will be subject to a lateral force proportional to $\sin \alpha$ (fig.

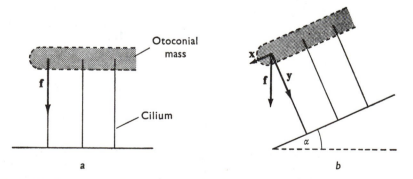

Fig. 16.14. The forces on the sensory cilia of a flat macula when it is horizontal (*a*), and
when it is tilted through an angle α (*b*). The vertical force **f** represents the weight of the
otoconial mass that is supported by one of the cilia. In *b* this force can be resolved into a
component acting along the axis of the cilium (**y**) and a force at right angles to this (**x**). It is
clear that $\mathbf{x} = \mathbf{f} \sin \alpha$.

16.14). A system such as this will give a sinusoidal relationship between the angle of tilt and the degree of sensory excitation whatever the initial orientation of the receptor cell is. Now let us see how this model system corresponds with actuality.

The equilibrium function of the otolith organs has been investigated electrophysiologically by Lowenstein and Roberts (1950), using single-fibre preparations from the ray, *Raia*. In almost all cases the sensory fibres showed a resting discharge in the normal position, which was increased or decreased by tilting the head. Many units appeared to act as true 'position receptors', with a more-or-less sinusoidal relation between the angle of rotation about one of the horizontal axes and the frequency of the sensory nerve impulses (fig. 16.15). In general, these units were of two types, those which had their positions of maximum activity in the 'side-up' and 'nose-up' positions, and those in which these positions were 'side-up' and 'nose-down'. In addition to these 'static' receptors, some units responded to changes in position irrespective of the direction of the change.

Lowenstein and Roberts (1951) later showed that fibres from certain parts of the otolith maculae in the ray are very sensitive to vibration. The otoliths are thus potentially hearing organs, and have been so developed in bony fishes and amphibians.

It is no easy matter to determine the functions of the individual otolith organs, and it is clear that there is some variation in this respect in the

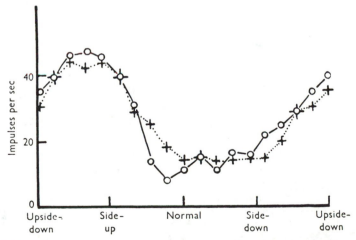

Fig. 16.15. Discharge frequencies in a single sensory fibre from the utriculus of *Raia* during slow rotation about the longitudinal axis. The continuous curve should be read from left to right, and the dotted curve from right to left. (From Lowenstein and Roberts, 1950.)

different classes of vertebrates (see Lowenstein, 1950). The utriculus is apparently always concerned with orientation to gravity. The sacculus is concerned with hearing in bony fishes, but this function is taken over by the cochlea in birds and mammals. The lagena seems to be concerned with equilibrium in elasmobranchs and amphibians.

The cochlea

The cochlea contains the receptor cells responsible for the perception of sound in mammals. We shall not examine the physiology of hearing in any detail here (see von Békésy, 1960; Whitfield, 1967); suffice it to give a rough outline of how the receptors are excited.

The mammalian cochlea consists of a tubular outgrowth of the membranous labyrinth, the *scala media*, surrounded by two tubular outgrowths of the osseous labyrinth, the *scala vestibuli* and the *scala tympani*. The structure of this complex, as seen in cross-section, is shown in fig. 16.16. The receptor cells, or 'hair cells' are mounted on the basilar membrane and their sensory hairs are inserted into the tectorial

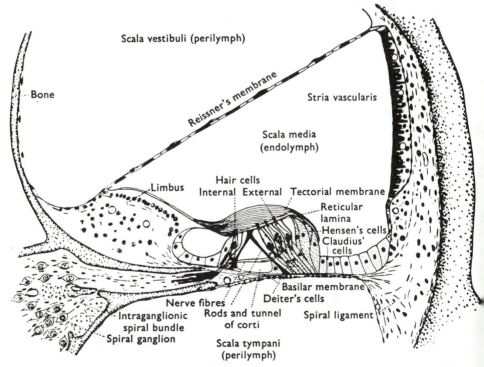

Fig. 16.16. Semi-diagrammatic transverse section of the organ of Corti in a guinea pig. (From Davis *et al.*, 1953.)

membrane (the whole complex forms the *organ of Corti*); hence we would expect these cells to be excited by movements of the basilar and tectorial membranes relative to each other.

Fig. 16.17 shows how the displacements produced by sound waves pass through the ear. Sound waves entering the external ear cause vibrations of the ear drum, which are transmitted through the air-filled middle ear by a chain of small bones (the auditory ossicles) called the *malleus*, *incus* and *stapes*. These ossicles form an 'impedance matching' device whereby the forces over the relatively large area of the ear drum (50 to 90 mm^2 in man) are concentrated on the relatively small (3.2 mm^2) area of the foot-plate of the stapes. This concentration of forces produces an increase in pressure sufficient to ensure that the sound vibrations can be successfully transmitted from the air outside to the liquid perilymph of the inner

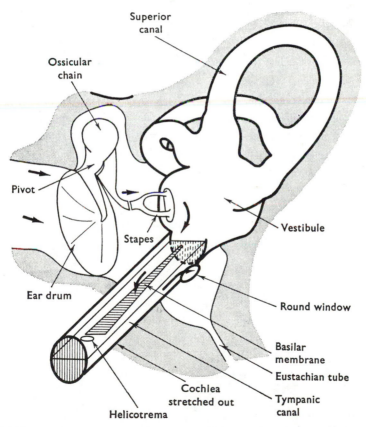

Fig. 16.17. Schematic diagram of the middle and inner ear in a mammal. The cochlea is shown uncoiled. The arrows show the displacements produced by an inward movement of the ear drum. (From von Békésy, 1962.)

ear. If there were no such mechanism, and the ear drum were directly backed by liquid, only a very small proportion of the sound energy would cross the air/liquid interface – the rest would be reflected.

The foot-plate of the stapes rests on a small membrane, the *oval window*, which covers a gap in the bone in which the inner ear is set. Behind this is the perilymph of the scala vestibuli, so that an inward movement of the stapes causes a displacement of fluid along the scala vestibuli. This displacement is accompanied by movement of the basilar membrane and a corresponding displacement of fluid in the scala tympani, which results in a bulging of the *round window*, a membrane between the scala tympani and the middle ear. The tectorial membrane moves up and down with the basilar membrane, but there is a lateral shearing action between them which displaces the ciliary processes of the hair cells and so provides the immediate stimulus for their excitation.

The mechanics of this system have been most beautifully worked out by von Békésy. In one of his experiments he measured the displacement of the different parts of the basilar membrane in response to vibration at the oval window by placing silver grains on it and observing their movement with the aid of a microscope and a stroboflash. He found that a travelling wave of the same frequency as the sound moves along the basilar membrane from base to apex, its amplitude varying with the distance along the membrane (fig. 16.18). The position at which the vibration of the basilar membrane is of maximum amplitude varies with

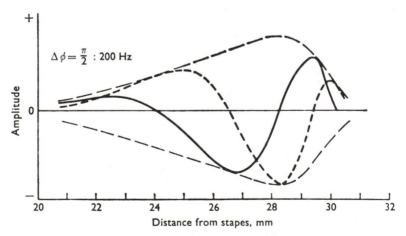

Fig. 16.18. The vibration of the basilar membrane in response to a 200 Hz tone. The full line shows the position of the membrane at one instant, and the broken line shows its position one quarter of a cycle later. The outer broken curves indicate the envelope of the movement. (From von Békésy, 1960.)

the sound frequency; it moves nearer to the stapes as the frequency is increased. The corollary of this observation is that any one point on the basilar membrane will vibrate in response to a number of different sound frequencies. This means that pitch discrimination depends not upon the excitation of particular receptors at very localized points along the length of the cochlea, as was once thought, but upon the differential excitation of relatively large numbers of receptors.

It has been suggested from time to time that von Békésy's experiments may not be relevant to the normal functioning of the ear since he

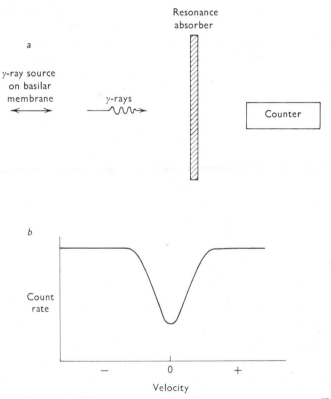

Fig. 16.19. How the Mössbaur effect can be used to measure velocities. The ^{57}Co atoms which form the gamma-ray source are part of the crystal lattice of a small piece of steel foil. The gamma-rays emitted are then all of nearly the same frequency (this is what Mössbauer discovered). The absorber is another piece of steel foil which contains ^{57}Fe atoms in its crystal lattice; these absorb gamma-rays over precisely the same narrow frequency band. If the gamma-ray source is moving away from or towards the absorber, the frequency of the gamma-rays hitting it will be lowered or raised, and so less absorption will occur. A gamma-ray counter on the far side of the resonance absorber will then produce a higher count rate than when the source was not moving. *a* shows how the components of the apparatus are arranged; *b* shows how the count rate varies with the velocity of the source.

had to use stimuli equivalent to very high sound pressures in order to produce visible movements of basilar membrane, and could not make measurements at high frequencies. Measurements of much lower amplitudes of movement have recently been made by Johnstone, Taylor and Boyle (1970), using the Mössbauer effect. The arrangement for their experiments is shown schematically in fig. 16.19. A small piece of foil impregnated with ^{57}Co, which produces gamma rays by radioactive emission, is placed on the basilar membrane. The gamma rays are measured by a proportional counter, but before meeting the counter they have to pass through a piece of stainless steel foil which is enriched with ^{57}Fe. The energy of the gamma rays is precisely that which is absorbed by the ^{57}Fe in the crystalline structure of the foil, and so only a proportion of them pass through the foil to be recorded by the counter. However, if the source is moving toward the absorbant, the frequency of the gamma rays 'seen' by the absorbant is raised; when it moves away, the frequency is lowered. (This is an example of the Doppler effect.) These changes in frequency mean that the energies of the gamma rays are no longer so precisely matched to the ^{57}Fe atoms, and so fewer of them are absorbed. Thus over a limited range (up to about ± 1 mm/sec) the output of the proportional counter increases with increasing velocity; and the velocity of the vibration is proportional to amplitude.

In fact the results obtained by Johnstone *et al.* (see the right-hand dotted curve in fig. 16.20) were in general similar to those of von Békésy. The curve from Mössbauer studies is a little sharper than von Békésy's curves, but this may simply be a feature of the higher frequencies (18 and 19 kHz) at which they were measured.

An electrode placed in the cochlea will record an oscillating potential of the same frequency as the sound stimulus when the ear is stimulated by a pure tone in the audible range. This is called the *microphonic potential* of the cochlea. Since it is largest in the immediate vicinity of the hair cells, it is very probable that it is produced by them and is therefore a receptor potential. In accordance with von Békésy's observations on the vibrations of the basilar membrane, it is found that cochlear microphonics are produced near the base of the cochlea in response to sound of all audible frequencies, but occur near the apex only in response to low frequencies.

The responses to sound of single nerve fibres in the auditory nerve have been investigated by Kiang, Sachs and Peake (1967) in cats and by Evans (1972) in guinea pigs, with similar results. Any particular fibre is most sensitive at one particular frequency (the 'characteristic frequency'): the threshold rises very rapidly at higher frequencies and rather less rapidly at lower frequencies. Different fibres have different characteristic frequencies (fig. 16.20 shows a selection), and we assume

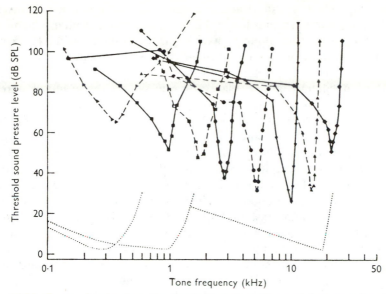

Fig. 16.20. Curves showing how the threshold varies with sound frequency for eight different fibres in the cochlear nerves of guinea pigs. Each of the dotted curves below shows the relative sound pressures necessary to produce a vibration of constant amplitude at a particular point on the basilar membrane, derived from measurements by von Békésy (the two lower frequency curves) and by Johnstone, Taylor and Boyle (the high frequency curve); these curves are positioned arbitrarily on the intensity scale. Notice that the curves of neural output are much more sharply tuned than are the mechanical response curves. (From Evans, 1972.)

that they are connected to hair cells at different places on the basilar membrane. It is particularly interesting that the 'sharpness' of these curves of threshold against frequency is much greater than that of the corresponding curve of basilar membrane movement, as is shown in fig. 16.20. This means that when an auditory fibre and its hair cell is at the peak of the basilar membrane displacement curve it is relatively more sensitive than when it is not; it is as if the auditory fibre 'knows' not only what the amplitude of basilar membrane movement curve is at its own hair cell, but also where the peak of the curve is. The mechanism of this remarkable phenomenon (discussed by Evans, 1975) remains a mystery.

The fine structure of acoustico–lateralis receptor cells

The receptor cells of the different components of the acoustico–lateralis system are remarkably uniform in structure. They are all set in an epithelial layer, have ciliary processes projecting from their distal surfaces, and are innervated proximally (fig. 16.1). Each cell (except those of the organ of Corti) possesses two types of ciliary processes: a group of *stereocilia*, which are short and apparently of uniform structure inside, and a *kinocilium*, which is longer and contains nine peripheral filaments and two central ones along its length, just as in motile cilia. The kinocilium is absent from the mature hair cells of the organ of Corti, but a basal region (the 'centriole') is still present.

The kinocilium is always placed on one side of the group of stereocilia, and possesses a 'basal foot' which projects from its basal body on the side away from the stereocilia; the sensory hair bundle is thus polarized in a particular direction. There is a strong correlation between the direction of this polarization and the direction in which the cilia must be bent in order to produce excitation of the hair cell. Fig. 16.21*a* shows the orientation of the hair cells in the labyrinth of a fish, *Lota*, the burbot (Wersäll, Flock and Lundquist, 1965); similar observations have been made on the ray (Lowenstein, Osborne and Wersäll, 1964). The cells in the cristae of any particular semicircular canal are all polarized in the same direction. Furthermore, it is evident from the pattern of orientation that the cells must be excited when the cilia are bent towards that side on which the kinocilium is situated: as mentioned previously, the receptor cells of the vertical canals (where the kinocilia are positioned on the side away from the utriculus) are excited by utriculofugal movements of the cupula, whereas those of the horizontal canal (where the polarization is in the opposite direction) are excited by utriculopetal movement.

The polarization of the sensory cell in the macula utriculi of *Lota* is shown in fig. 16.21*b*. Notice that there are two principle directions of polarization, set approximately at right angles to each other. This type of arrangement is just what is required to give information about the direction of tilting. It is instructive to compare this arrangement with the findings of Lowenstein and Roberts on the sensitivity of nerve fibres from the utricular macula of the ray. If the receptor cells are essentially unidirectional, one which is excited by both 'side-up' and 'nose-up' deflections will be directed contralaterally backwards at about 45° to the body axis, as are the cells in the anterior peripheral region of the macula utriculi in *Lota*. One which is excited by 'side-up' and 'nose-down' deflections will be directed contralaterally forwards, as are the cells in the posterior peripheral region in *Lota*. We would expect the majority of the cells in *Lota* to be excited by 'side-down' deflections (Flock, 1964) since they are ipsilaterally directed, but this question has yet to be investigated.

In the lateral line, the discharge frequency in any single sensory nerve fibre is increased by movement of the cupula in one direction and decreased by such movement in the opposite direction; each neuromast contains some fibres responsive to movements in one direction and

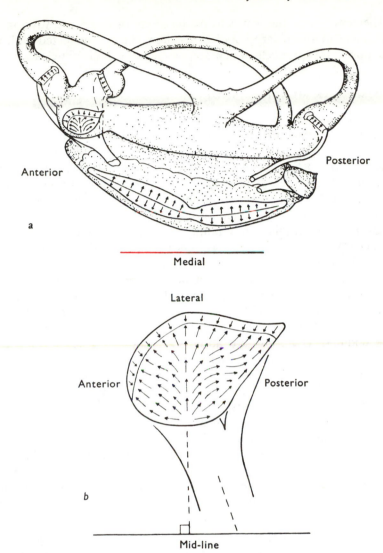

Fig. 16.21. *a*: The orientation of the hair cells in the right labyrinth of the fish *Lota vulgaris*: the labyrinth is viewed from above. The arrows point towards the side of the ciliary bundle on which the kinocilium is situated. *b*: Hair cell orientation in the macula utriculi, shown in more detail. (From Wersäll, Flock and Lundquist, 1965.)

some to movements in the other (p. 378). Flock (1965) has demon-strated the structural counter-part of these observations; adjacent hair cells in the neuromast are oriented with their kinocilia pointed in opposite directions, either up or down the canal. A similar situation occurs in free neuromasts (Dijkgraaf, 1963).

The centriole in the hair cells of the organ of Corti is always on the outer side of the ciliary bundle. Just how this is connected with the situation in receptor cells which do possess kinocilia is not clear.

MAMMALIAN MUSCLE SPINDLES

The afferent nerve fibres from mammalian limb muscles are of a number of different types (table 16.2), distinguishable according to their diameters and the nature of their sensory endings. There are two main types of receptor in the muscles: (i) the muscle spindles, which consist of small modified muscle fibres and are innervated by group Ia (primary) and group II (secondary) fibres and by motor fibres of the γ system, and (ii) the Golgi organs in the tendons, which are innervated by group Ib fibres. In addition there are a number of small diameter fibres (groups III and IV) which have free or encapsulated endings and may be responsive to pressure or pain.

This pattern of innervation can be illustrated by reference to the soleus muscle of the cat; Matthews (1964) combines the results of several workers to give the following figures. There are about 50 spindles, with the same number of primary endings and 50 to 70 secondary endings. There are also about 45 tendon organs, but individual group Ib fibres may innervate more than one tendon organ. A few group III and group IV afferents occur. The spindles contain a total of about 300 intrafusal muscle fibres, supplied by about 100 fusimotor nerve fibres of the γ system. Finally, there are about 25 000 extrafusal muscle fibres innervated by about 150 α motor nerve fibres.

Before examining the details of muscle spindle action, it would be as well to establish some general ideas as to how we might expect them to act under various conditions. Fig. 16.22 shows, in a schematic and much simplified fashion, the mechanical relations between the extrafusal muscle fibres, the muscle spindles and the Golgi tendon organs. Notice

TABLE 16.2. *The relation between function and diameter in the afferent fibres of mammalian muscle nerves.* (Based mainly on Hunt, 1954.)

Group	Diameter (μm)	Conduction velocity (m/sec)	Sensory endings
Ia	12–20	72–120	Primary endings on muscle spindles
Ib	12–20	72–120	Golgi tendon organs
II	4–12	24–27	Secondary endings on muscle spindles
III	1–4	6–24	Pressure/pain receptors
IV	Non-myelinated fibres		Pain

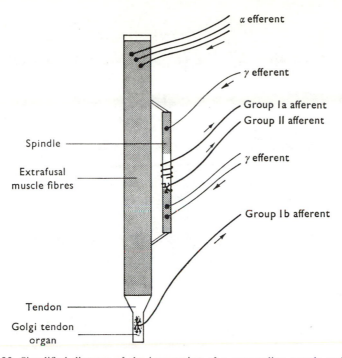

Fig. 16.22. Simplified diagram of the innervation of a mammalian muscle and muscle spindle. Contractile regions are shaded.

that the tendon organs are in series with the extrafusal muscle fibres, whereas the muscle spindles are in parallel with them: these positions are just right for the tendon organs to act as tension receptors and the spindles to act as length receptors. Our knowledge of the workings of mammalian muscle receptors begins with the work of B. H. C. Matthews (1933), and it is instructive to relate his results (some of which are shown in fig. 16.23) to the scheme shown in fig. 16.22. If the resting muscle is passively stretched, then both the muscle spindles and the tendon organs will be stimulated (figs. 16.23*a* and *b*), the former by the stretch itself and the latter by the consequent increase in passive tension. If the α fibres are stimulated so that the muscle contracts, the tendon organs will be stimulated by the increase in tension (fig. 16.23*c*) and the muscle spindle will shorten (even in an 'isometric' contraction there will be some shortening because of the compliance of the tendons) and so cease to fire (fig. 16.23*d*). Stimulation of the γ fibres in an otherwise resting muscle will stretch the equatorial regions of the intrafusal muscle fibres and so produce excitation of the muscle spindle afferents (fig. 15.23).

So much for the basic outlines of muscle spindle physiology. The work that has been done in recent years (see Matthews, 1972, 1977) has

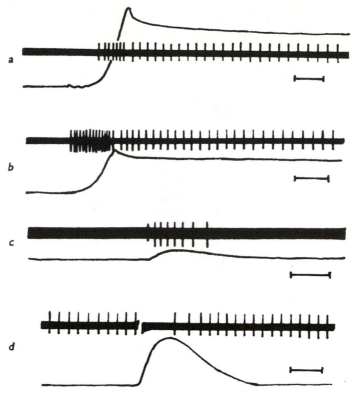

Fig. 16.23. Responses of sensory fibres in cat leg muscles to passive stretch (*a* and *b*) and to twitch contractions of the muscle (*c* and *d*). *a* and *c* show responses of tendon organ endings; *b* and *d* show responses of spindle endings. The thick lines show nervous activity, the thin traces show tension changes in the muscle. Time scales 0.1 sec. (From Matthews, 1933, redrawn.)

shown that muscle spindles are remarkably complex organs, whose study is no easy matter. As a result, much of the field has been beset with controversy, and it has been particularly difficult to relate the physiological properties of the spindles to their structure. But let us look at some of the details.

Structure

Anatomical investigations on mammalian muscle spindles have been carried out on a number of different species; what follows relates principally to cat limb muscles.

Each spindle contains five to ten intrafusal muscle fibres, which are held together centrally in a fluid-filled capsule. The fibres are generally of two distinct types, although some intermediate types may occur. The *nuclear bag* fibres are about 8 mm long and their equatorial regions are swollen and contain numbers of nuclei clustered together. The *nuclear chain* fibres are narrower than the nuclear bag fibres and about half their length; their equatorial regions are not swollen but contain a single string of nuclei. Myofibrils are almost absent from the equatorial regions in both cases, specially in the nuclear bag fibres. It has recently become clear that there are two types of nuclear bag fibres, one of them ('bag$_1$') shorter and thinner than the other ('bag$_2$'); there are also histochemical differences between the two types. Most spindles contain two nuclear bag fibres, one of each type, and four or five nuclear chain fibres.

Each spindle receives one group Ia afferent nerve fibre and from none to five group II fibres. The group Ia fibres end at the *primary endings*, where they wind round the equatorial regions of the intrafusal fibres to form 'annulospiral' terminations. All nuclear bag fibres have primary endings on them, and almost all nuclear chain fibres do. *Secondary endings*, derived from group II fibres, are found on either side of the primary endings on nuclear chain fibres; occasionally they also occur in a similar position on nuclear bag fibres. Most of the secondary endings in the cat are of the 'annulospiral' type, but there are some 'flower-spray' endings, in which the fibres split up into a number of branches which have slight thickenings at their ends.

The efferent nerve fibres supplying the muscle spindle are called fusimotor fibres. A number of these innervate extrafusal as well as

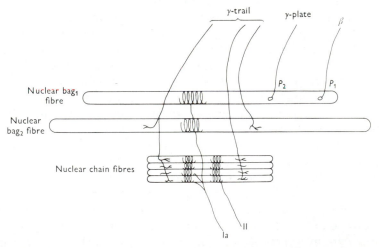

Fig. 16.24. Diagram showing the innervation of a typical mammalian muscle spindle. Sensory nerve axons (primary and secondary) are shown emerging below, fusimotor axons above. Based largely on Boyd *et al.* (1977). There may in addition be some innervation of bag$_1$ fibres by γ-trail axons (Barker, 1974).

intrafusal muscle fibres; they are known as β fibres and have terminals described as 'plate (p_1)' endings on the intrafusal muscle fibres. The γ fibres, which are exclusively fusimotor, have two types of terminal: 'plate (p_2)' and 'trail' endings. Plate endings form distinct end-plates with synaptic gutters, the p_2 endings being larger than the p_1 ones. Trail endings are more diffuse endings in which the fibre splits up into a number of fibre terminals, and synaptic gutters are absent.

There has been some controversy in the past about the details of fusimotor innervation (see, for example, Boyd, 1962 and Barker, 1962), but it seems that some sort of compromise view may now be emerging (fig. 16.24). The majority of β axons, and nearly all the γ-plate axons, innervate nuclear bag$_1$ fibres. γ-trail axons innervate nuclear chain and nuclear bag$_2$ fibres, and perhaps some nuclear bag$_1$ fibres as well.

Physiological properties

The general features of the responses of the afferent fibres from muscle spindles have already been dealt with. In this section we examine some of the details that have come to light in recent years.

A number of workers have investigated the differences between the responses of the primary and secondary endings to various mechanical stimuli in de-efferented spindles. We shall consider here some experiments by Bessou and Laporte (1962) which are notable in that comparisons were made between the responses of primary and secondary fibres from the same muscle spindle. The experiments were performed on spindles in the tenuissimus muscles of anaesthetized cats with the ventral roots to the muscle cut so as to eliminate the activity of the γ efferent system. Then the dorsal roots were dissected so that the activity of individual group Ia and group II fibres from these spindles could be recorded; the search for such fibres (identified by their response to stretching the muscle) was hastened by denervation of other muscles in the hind leg. Having prepared an afferent fibre from the muscle, the location of the spindle which it innervated was approximately determined by electrical stimulation at different points along the length of the muscle. Using this technique, action potentials appeared in the nerve fibre when the cathode of the stimulating electrodes was over the spindle (the tenuissimus was particularly suitable for this technique since its spindles are arranged in a chain along the length of the muscle). More precise localization was then obtained by applying very light pressures to the muscle via stiff bristle. The object of this procedure was to find a group Ia fibre and a group II fibre (so characterized by their conduction velocities) which came from the same muscle spindle; it was safe to assume that this was so if the regions responding to application of the bristle overlapped each other. Fig. 16.25 shows the response of such

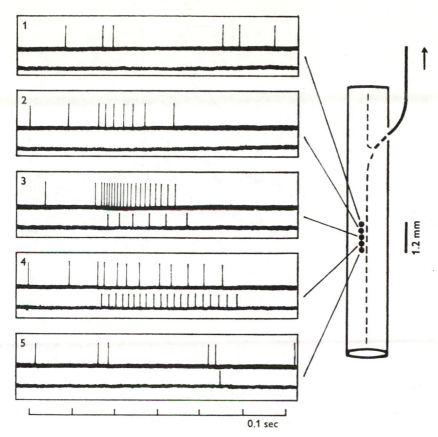

Fig. 16.25. Localization of a primary ending and a secondary ending from the same muscle spindle in a cat tenuissimus muscle. Records were obtained by lightly pressing a stiff hair onto the muscle at the points shown. Upper traces: responses of the primary ending, recorded in the dorsal root. Lower traces: responses of the secondary ending. (From Bessou and Laporte, 1962.)

a pair of afferent fibres to stimulation by the bristle at different points along the length of a spindle. Notice that the maximal responses from the two fibres occur at slightly different positions; this is just what we would expect to see, since the secondary endings on the spindle are placed in a different position from the primary endings.

Having obtained their preparation, it was then possible for Bessou and Laporte to investigate the properties of the primary and secondary endings in response to different types of stimuli. The steady-state responses from a typical spindle after stretching to different lengths is shown in fig. 16.26. Notice that, in this case, the secondary ending has a higher threshold than the primary ending (i.e. the primary ending is

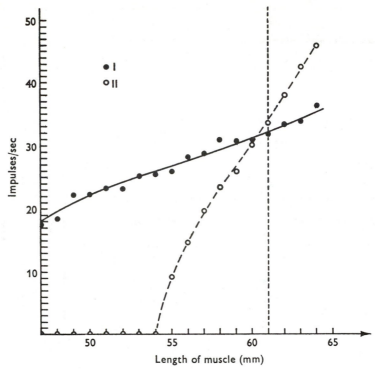

Fig. 16.26. The variation of steady impulse discharge frequency with length in the primary (black circles) and secondary (white circles) sensory fibres from a cat muscle spindle. Motor innervation to the muscle was cut. The dotted line shows the normal maximum length of the muscle in the body. (From Bessou and Laporte, 1962.)

active at shorter lengths) and that the secondary ending has a steeper relation between length and response.

The most marked difference between the primary and secondary endings was in their responses to *changes* in length. The primary endings always showed a fairly large increase in frequency of firing during extension, and a marked decrease in frequency, often amounting to a temporary absence of activity, during shortening. For the secondary endings these effects were very slight. This means that the secondary endings are primarily responsive only to the length of the muscle spindle, whereas the primary endings are responsive both to its length and its velocity of extension. These results, which agree with those obtained by others, are shown diagrammatically in fig. 16.27.

P. B. C. Matthews and his colleagues (Matthews, 1962; Crowe and Matthews, 1964; Brown, Crowe and Matthews, 1965) have shown that the γ efferent fibres in the cat are of two functionally distinct types, distinguished by their effects on the velocity-sensitive responses of the

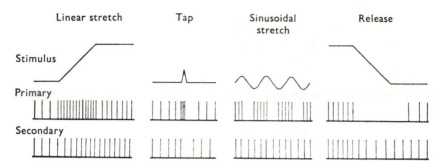

Fig. 16.27. Diagram summarizing the responses of primary and secondary endings to various length changes. (From Matthews, 1964.)

primary endings during extension. They define the 'dynamic index' of a response to stretching at a constant velocity as the difference between the frequency of firing of an afferent fibre just before the end of the period during which the muscle is being extended and that occurring at the final length half a second later. The two types of fusimotor fibres are then *dynamic fibres*, which increase the dynamic index of the primary afferent fibres, and the *static fibres*, which reduce it. In other words, dynamic fibres make the primary endings relatively more sensitive to the velocity of stretching, and the static fibres make them less sensitive. These features are shown in fig. 16.28.

Is it possible to find any anatomical basis for this physiological devision of fusimotor fibres into two types? An ingenious technique has recently been applied to the problem (Brown and Butler, 1973; Barker *et al.*, 1976). If a fusimotor axon is stimulated repetitively for some time, all the intrafusal fibres which it innervates will contract and so deplete their glycogen reserves, and this depletion can be detected by suitable histological staining. In this way it has been found that static γ axons innervate both bag and chain muscle fibres, whereas dynamic γ axons innervate predominantly bag_1 fibres. It looks as though static axons correspond to γ-trail axons and dynamic axons correspond to γ-plate and β axons.

Observations largely in agreement with these findings have been made by Boyd and his colleagues (1977) and by Bessou and Pagès (1975) using cinematography of isolated spindles. The main remaining point of difference appears to be whether or not the bag_1 fibres are ever innervated by static γ axons: Barker and his colleagues think they are whereas Boyd and his colleagues think they are not. Their conclusions on this matter have led Boyd and his colleagues to name the two types of nuclear bag fibres 'dynamic' (corresponding to bag_1) and 'static' (bag_2) nuclear bag fibres.

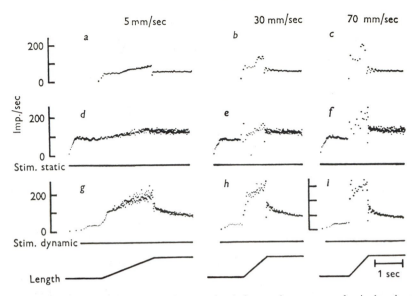

Fig. 16.28. Effect of two types of fusimotor stimulation on the response of a single primary ending, in a cat soleus muscle, to stretching the muscle at different velocities. Each action potential is shown as a dot whose vertical position is proportional to the instantaneous frequency (i.e. the reciprocal of the time since the preceding action potential). The time scale at the bottom right (1 sec) does not apply while the length is changing. *a–c*: No stimulation of fusimotor fibres. *d–f*: Stimulation of static fibre at 70/sec. *g–i*: Stimulation of dynamic fibre at 70/sec. (From Crowe and Matthews, 1964.)

This idea that the differences between the effects of static and dynamic γ axons arises from differences in the mechanical properties of the intrafusal muscle fibres which they innervate fits well with observations on the ultrastructure of the muscle fibres: bag_1 fibres have a much less well developed sarcoplasmic reticulum and T system than do bag_2 and chain fibres. The idea has an interesting history. Thus Matthews (1964) suggested that dynamic γ axons innervated bag fibres while static γ axons innervated chain fibres; this suggestion fitted well with Boyd's anatomical views at the time, but not with Barker's. The ensuing controversy was not resolved until the existence of the two types of nuclear bag fibres was established.

There is still a lot to be discovered about muscle spindles. In particular, it would be very pleasant to understand how the contractions of the intrafusal muscle fibres produce their static or dynamic effects on the primary afferent discharge.

CHEMORECEPTION IN INSECTS

Chemical stimuli play an important part in the lives of insects, as they do in many other animals, especially in the identification of food substances

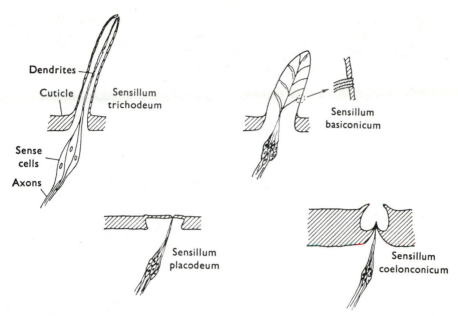

Fig. 16.29. Various types of chemoreceptor sensilla found in insects. Schematic.

and the location of mating partners. One of the great advantages of insects as material for the study of chemoreception is that single sense organs (sensilla) can be stimulated and recorded from individually in isolation from their neighbours.

The forms of some of these chemoreceptive sensilla are shown diagrammatically in fig. 16.29. Each consists of a modified region of cuticle through which the dendrites of a number of bipolar nerve cells are connected to the exterior. The cuticular part of the sensillum may be in the form of a hair (sensilla trichodea), a projecting peg (sensilla basiconica), a flat plate (sensilla placodea), or a short peg set in a small pit (sensilla coelonconica).

Behavioural experiments on taste

A remarkably large amount of information has been obtained from experiments involving measurement of the concentration of a substance required to elicit or inhibit a feeding response (Dethier, 1955, 1963). Suppose, for example, that we wish to measure the minimum concentration of sucrose which is detectable by a blowfly such as *Phormia* or *Calliphora*. The fly is suspended from a glass rod, starved, and then allowed to drink as much water as it wants. A drop of sucrose solution of known concentration is then brought into contact with one of the tarsi; if the sugar concentration is sufficient, the fly extends its proboscis, but there is no response if it is not. In this way it has been shown that the tarsal receptors of blowflies are most sensitive to sucrose and maltose, less sensitive to glucose, and insensitive to lactose.

The relative sensitivity to compounds which do not evoke the feeding response can be found by measuring the 'rejection thresholds' of the various compounds. In this technique, the concentration of the compound which is just necessary to prevent the feeding response to a sucrose solution is determined. Dethier and his colleagues measured the chemoreceptor sensitivity of blowflies to over 200 aliphatic organic compounds by this technique. A number of correlations between activity and molecular structure appeared, of which perhaps the most clear-cut was that the rejection thresholds were inversely related to the water solubility of the compounds involved. The simplest explanation of this phenomenon is that a response should depend on the stimulating substance becoming incorporated into the lipid membrane of the receptor cell.

Electrophysiological experiments on taste

Owing to the small diameter of the axons of the chemoreceptor sense cells, it has not been possible to record their responses to stimulation in

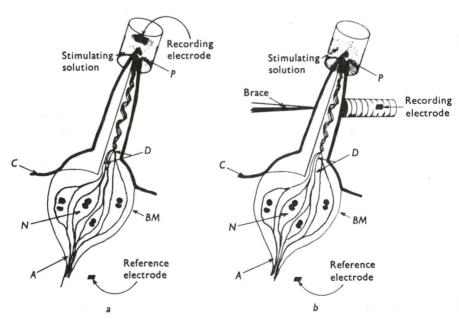

Fig. 16.30. Methods of stimulating and recording from individual taste receptor hairs in insects. *a* shows the technique in which the stimulating solution is held in the recording electrode, *b* shows the 'side wall cracking' technique. *A*, axons; *BM*, basement membrane; *C*, cuticle; *D*, dendrites; *N*, sensory nerve cell; *P*, pore in cuticle at tip of hair. (From Wolbarsht, 1965.)

the usual manner; two novel techniques had to be developed before electrophysiological investigations could be carried out. In the first of these, shown in fig. 16.30*a*, a pipette containing the stimulating solution is placed over the tip of the sensory hair, and also acts as the recording electrode (Hodgson, Lettvin and Roeder, 1955). This method can only be used with electrolytes in the stimulating solution. The second method, sometimes known as 'side wall cracking' (fig. 16.30*b*), is not subject to this limitation, since the recording electrode is applied to the base of the hair and gains contact with its interior by fracturing the cuticle; stimulating solutions are applied, as before, by means of a pipette applied to the tip of the hair.

A typical response to stimulation, as recorded by the 'side wall cracking' method, is shown in fig. 16.31. There is a negative receptor potential on which are superimposed a number of small, largely positive-going action potentials. The most probable interpretation of the mode of origin of these potentials, suggested by Hanson and Wolbarsht (1962), is as follows. First of all, the dendrite membrane becomes depolarized; this is recorded as a negative potential change by the extracellular electrode. This depolarization spreads down the dendrite to reach an impulse initiation site near the cell body. Action potentials then propagate down the sensory axon and (to some extent) up the dendrite. The local circuits associated with these action potentials cause the initial positive-going deflections seen on the record (compare fig. 4.16*c*) and the negative-going components occur when active regions pass under the recording electrode.

Electron micrographs show that there are four, or sometimes five, receptor cells associated with each sensillum trichodeum in *Phormia*. In

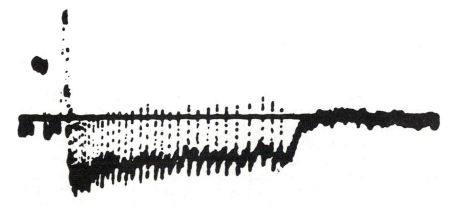

Fig. 16.31. Response of a labellar chemoreceptive hair of *Calliphora* to a 0.25 M sucrose solution, recorded by the side wall cracking technique. (From Morita and Yamashita, 1959.)

accordance with this, four physiologically distinct receptors have been found by electrophysiological methods. These are the 'salt' and 'sugar' receptors (Hodgson and Roeder, 1956), which respond to monovalent salts and certain sugars respectively, the 'water' receptor (Evans and Mellon, 1962), and a mechanoreceptor whose dendrite is attached to the base of the hair (Wolbarsht and Dethier, 1958). The function of the occasional fifth fibre is not yet known.

An interesting example of a very specific contact chemoreceptor has been dicovered by Rees (1969) in the beetle *Chrysolina brunsvicensis*, which feeds only on plants of the genus *Hypericum* (St. John's wort). These plants contain the compound hypericin, which is found nowhere else. Rees found that the tarsal chemoreceptors of the beetle are stimulated by hypericin, whereas those of related insects are not. The tarsal chemoreceptors thus provide a precise food-detector mechanism in this insect.

Olfaction

Much of our knowledge of the physiology of smell in insects is derived from recent experiments by Schneider and his colleagues (see Boeckh, Kaissling and Schneider, 1965; Schneider, 1966). They used two methods of measuring the receptor responses: (*a*) the mass response from the whole antenna (the electroantennogram), obtained with recording electrodes placed at the tip and base of an antenna, and (*b*) single unit responses recorded by means of a microelectrode inserted through the thin cuticle of a sensillum. Solutions of odoriferous substances were placed on small pieces of filter paper placed in the orifice of an air jet which was then directed at the antenna.

By measuring the responsiveness or otherwise of a large number of sensilla on the antennae of moths and honey-bees, it was found that the receptor cells fall into two main classes, called 'generalists' and 'specialists'. Odour 'generalists' are cells which respond to a number of different substances, some causing an increase in discharge frequency and others inhibiting it. Schneider describes the pattern of responsiveness of any particular receptor cell to a particular set of compounds as its 'reaction spectrum'. A number of such reaction spectra for the sensilla basiconica of a moth are shown in fig. 16.32. What is particularly interesting about these results is that no two cells have the same reaction spectrum. It seems probable that odours are identified by the patterns of responsiveness which they elicit in a large number of receptors. The system is an interesting example of the way in which a multi-channel sensory system can be used to sharpen the specificity of perception (see p. 371).

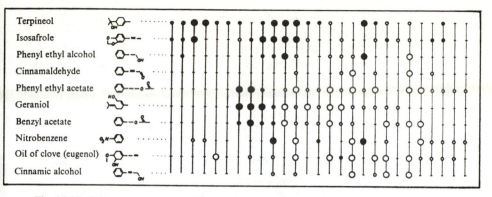

Fig. 16.32. Olfactory responses of 27 sensilla basiconica in the moth *Antherea pernyi* to 10 different compounds. Filled circles indicate excitation, open circles indicate inhibition; the sizes of these circles indicate the extent of the change in nerve impulse frequency. Small horizontal lines: no effect. (From Boeckh, Kaissling and Schneider, 1965.).

Odour 'specialists', the second type of receptor cell in Schneider's classification, are olfactory receptor cells which respond to a very limited number of compounds and which are usually present in large numbers, each cell having the same action spectrum. The antennae of male silkmoths (*Bombyx mori*), for example, contain a large number of sensilla trichodea which are specifically responsive to bombykol (hexadeca-dien-10-*trans*-12-*cis*-ol-1), a sex attractant substance that is produced by females of the species. Isomers of bombykol are somewhat less active than bombykol itself (as measured by the size of the electroantennogram which they produce), and other alcohols are very much less active. Males of another silkmoth, *Antherea pernyi*, possess antennal receptors which respond to the sex attractant of females of the species, but not to bombykol. Drone honey-bees possess receptors responsive to the 'queen substance' produced by queen bees. A male *Bombyx* possesses about 40 000 bombykol receptors, and responds to concentrations of bombykol as low as 200 molecules per cubic centimetre.

ELECTRORECEPTORS

Most sense cells are responsive to electric currents of sufficient magnitude, but this does not justify description of them as 'electroreceptors'. As Machin (1962) points out, a sense organ can only be described as an electroreceptor if it responds to electrical stimuli present in the environment and if the organism responds in a way appropriate to the detection of these electrical stimuli. Sense organs which satisfy these requirements are known only in certain fishes, where the ampullary organs of the lateral line system have in many cases been shown to be electroreceptors.

As mentioned in chapter 14, Lissmann (1951) suggested that fish with weak electric organs possess an object-location system. According to this theory, the activity of the electric organ sets up an electric field in the water, and the form of this field will be distorted if objects of different conductivity are placed in it, as is shown in fig. 16.33. The disturbance could be detected by the fish if it is able to measure the currents flowing across the skin at different points on its body surface, for which the presence of an array of electroreceptors must be necessary.

A number of experiments were adduced by Lissmann (1958) as evidence for sensitivity to weak electric fields in electric fish. Specimens of *Gymnarchus niloticus* responded to the closing of a switch between two wires dipped into the aquarium tank, and to movements of magnets or electrostatic charges outside the tank. Using the arrangement shown in fig. 16.34, he was able to show that *Gymnotus carapo* can be trained to differentiate between metal and plastic discs in total darkness, or between presence and absence of a magnet placed outside the tank and out of sight of the fish.

These experiments clearly establish that electric fish are sensitive to electrical changes in the environment, but they do not necessarily imply that the fish can detect objects by means of the distortion of the electric field produced by the electric organs, since metals immersed in water

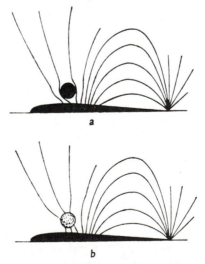

Fig. 16.33. The electric field produced by an electric fish in the presence of objects of low (*a*) and high (*b*) conductivity. (From Lissmann and Machin, 1958.)

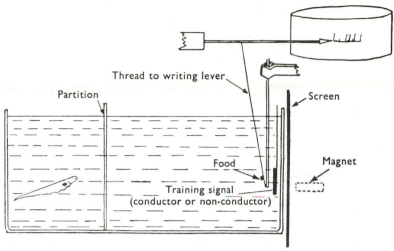

Fig. 16.34. Lissmann's arrangement for testing the ability of an electric fish to distinguish between objects of low and high conductivities. (From Lissmann, 1958.)

can give rise to small voltages and it might be these that the fish is responding to. The question was further investigated by Lissmann and Machin (1958) in some very elegant experiments on *Gymnarchus*. Their general technique was similar to that shown in fig. 16.34, except that the test objects were porous pots which were filled with materials of different conductivities (porous pot is opaque and is an effective diffusion barrier for short-term experiments, but does not offer any considerable resistance to electric current when it is immersed in a conducting solution). The fish could easily be trained to distinguish between pots filled with aquarium water and pots filled with distilled water. Such a fish would then respond differently to a pot filled with aquarium water according to whether or not it contained a glass rod more than 1.5 mm in diameter; the presence of the glass rod would, of course, alter the overall conductance of the pot.

Hagiwara and his colleagues have recorded the electrical activity in sensory nerve fibres innervating electroreceptors in various electric fish (Hagiwara and Morita, 1963; Hagiwara, Szabo and Enger, 1965). In each case the electroreceptors respond to the discharge of the electric organ, and the response is modified by the presence of objects of different conductivity from the surrounding water near to the receptor endings. An example of this effect is shown in fig. 16.35. The method of coding the intensity of the electric field by a single sense organ shows some variety. In those fishes in which the electric organ fires at a low frequency, each discharge is followed by a burst of action potentials in

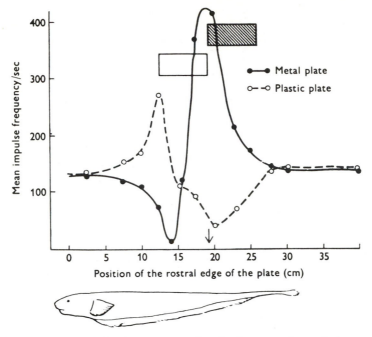

Fig. 16.35. The relation between nerve impulse frequency in an electroreceptor nerve fibre of *Sternarchus* and the position of metal and plastic plates placed close to the fish. The arrow indicates the location of the electroreceptor ending. The plates (the metal plate is shown cross-hatched and the plastic plate plain) are drawn to scale in the positions where they elicited the maximal response. (From Hagiwara, Szabo and Enger, 1965.)

the sensory axons, and coding is via the number of action potentials per burst. Three methods are known to be used in fishes whose electric organs discharge at a high frequency. In *Eigenmannia*, increase in stimulus intensity increases the probability of firing in response to each discharge. In *Sternarchus*, increase in stimulus intensity increases the frequency of firing in the sensory axons, but the individual action potentials are not elicited by particular discharges, so that the effective stimulus seems to be 'smoothed'; Machin and Lissmann (1960) reached a similar conclusion for *Gymnarchus*, on behavioural evidence. Finally, in *Sternopygus* an alteration in the stimulus intensity alters the latency of the sensory response to the electric organ discharge (Bullock and Chichibu, 1965).

The anatomy of the specialized lateral line organs present in the skin of various electric fish has been investigated by Szabo (1965). He finds that there are two types in gymnotids, named the ampullary and tuberous organs (fig. 16.36). The tuberous organs consist of a small group ('tuber') of sensory cells projecting into the epidermis from below

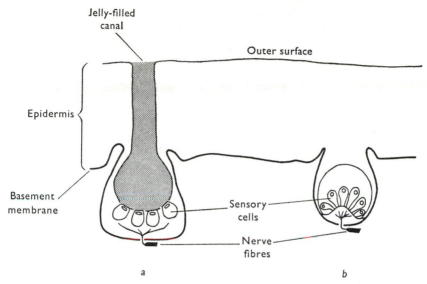

Fig. 16.36. Gymnotid electroreceptors. A diagrammatic section through the skin of a gymnotid fish showing an ampullary organ (*a*) and a tuberous organ (*b*). (Redrawn after Szabo, 1965, fig. 10.)

and innervated by a single nerve axon. The epidermal cells above the sensory cells are oriented vertically instead of horizontally; this arrangement probably reduces the transverse resistance of the skin at these points. The ampullary organs consist of groups of sensory cells at the bottom of jelly-fish canals which open to the exterior distally. Electrophysiological evidence indicates that the tuberous organs are almost certainly electroreceptors, and that the ampullary organs are probably not. A small specimen of *Sternarchus* possessed about 130 ampullary organs and over 2000 tuberous organs.

Cartilaginous fish possess a number of sense organs in the head, the ampullae of Lorenzini, which have long puzzled zoologists as to their use. Murray (1962) showed that they are particularly sensitive to electric currents, although they are also affected by changes in salinity and temperature and by mechanical stimulation. Evidence suggesting that they may be genuine electroreceptors has been provided by Dijkgraaf and Kalmijn (1963): dogfish and skates show behavioural responses to weak electric stimuli, and these responses are abolished after cutting the ampullary nerves.

What is the use of this extraordinary sensitivity? Kalmijn (1966, 1971) has found that plaice, flatfish upon which dogfish and skates feed, produce an appreciable electric field (up to $1000\,\mu\text{V/cm}$) from the action potentials of muscles used in respiration or movement. A dogfish can find a plaice even when it is held in an agar chamber under the sand, whereas it cannot find pieces of fish under the same conditions. So it seems as though these elasmobranchs are using their electroreceptors as highly sensitive detectors of live food buried in the sand.

17 The vertebrate eye

In this chapter we shall examine some aspects of the physiology of our most complex sense organ, the eye. The optical apparatus of the eye focuses an image of the visual field on the retina. The retina contains, in man, about 100 million receptor cells, which are connected in a rather complicated fashion to about a million fibres in the optic nerve. When light falls upon the receptor cells they are excited, and their excitation eventually leads to the production of action potentials in the optic nerve fibres.

The light sensitivity of the eye is primarily due to the existence in the receptor cells of a *visual pigment*, whose function is to absorb light and, in so doing, to change in some way so as to start the chain of events leading to excitation of the optic nerve fibres. As a reflection of this photochemical change in the pigment, we find that the pigment molecules are bleached by illumination, and have to be regenerated before they regain their photosensitivity.

The range of sensitivity of the eye is enormous: the brightest light in which we can see is about 10^{10} times the intensity of the dimmest. There are a number of mechanisms which enable this wide range to be perceived, which constitute the phenomena of visual adaptation. *Dark adaptation* is the increase in sensitivity which occurs when we pass from brightly lit to dim surroundings, and *light adaptation* is the reverse of this.

Visible light consists of electromagnetic radiation within a limited range of wavelengths. The visible spectrum, i.e. that range of electromagnetic radiation that can pass through the eye and cause a photochemical change in the visual pigments of the retina, covers wavelengths from about 400 nm (violet) to about 700 nm (deep red); within this range a number of different colours can be seen, as is shown in table 17.1.

Light energy is emitted and absorbed in discrete packets known as quanta or photons; there is no such thing as half a quantum. In a photochemical reaction, one molecule of pigment absorbs one quantum of light. The energy of a quantum varies with the wavelength of the light, and is given by $h\nu$, where ν is the frequency (the velocity of light

TABLE 17.1. *The colours of the visible spectrum. The wavelength ranges given in this table are only an approximation, since the naming of colours is a subjective phenomenon and differs for different observers.*

Colour	Wavelength (nm)
Red	Above 620
Orange	590–620
Yellow	570–590
Green	500–570
Blue	440–500
Violet	Below 440

divided by the wavelength) and h is Planck's constant, 6.63×10^{-27} erg sec.

THE STRUCTURE OF THE EYE

The photoreceptor cells of the eye, and the nerve cells which they innnervate, are found in a thin layer, the retina, which coats the inner surface of the eye. The rest of the eye consists of accessory structures concerned, directly or indirectly, with assisting the retina in the perception of the visual field.

The accessory structures

The gross structure of the human eye is shown in fig. 17.1. The outer coat of the eye is protective in function; it is called the *sclera*. The *cornea* is the transparent region of the sclera. Inside the sclera is a vascular layer, the *choroid*, which is usually pigmented. In some animals the choroid contains a reflecting layer, the *tapetum lucidum*, which serves to increase the effectiveness of the retina in catching light, but must necessarily lead to some loss of definition in the visual image. It is this layer which accounts for the reflecting properties of cat's eyes at night. In the bush-baby, the choroid contains a pigment which fluoresces in ultraviolet light and so enables ultraviolet light to be 'seen' by the retinal receptors (Dartnall *et al.*, 1965).

The interior of the eye contains the *lens*, the *aqueous humour* in front of it, and the *vitreous humour* behind it. The curved surface of the cornea acts as a lens, so that a parallel beam of light entering a human lensless eye is brought to a focus at about 31.2 mm behind the front face of the cornea. Since the length of the eye is about 24.4 mm, the cornea, aqueous and vitreous alone cannot produce a sharp image on the retina. The necessary extra focusing power is provided by the lens, which is of higher refractive index than the aqueous and vitreous humours. The

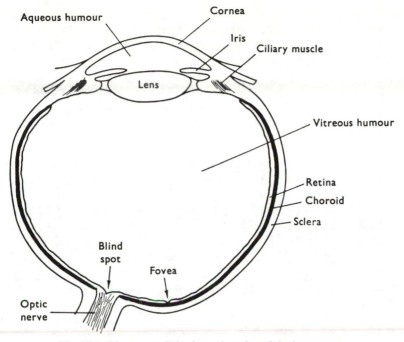

Fig. 17.1. Diagrammatic horizontal section of the human eye.

power of the lens can be increased by contraction of the ciliary muscle; this process, which is known as *accommodation*, allows the images of near objects to be focused on the retina. In many of the lower vertebrates (fishes, amphibia and snakes – see Walls, 1942), accommodation is produced not by altering the power of the lens but by moving it backwards and forwards.

In front of the lens is a diaphragm, the *iris*, the diameter of whose aperture (the *pupil*) can be varied by contraction of the iris muscles. Thus the activity of the iris can regulate the amount of light entering the eye; it is, in fact, reflexly controlled by the light intensity.

The direction in which the eye looks is controlled by the extraocular muscles, of which there are six.

The retina

The nervous structure of the retina is much more complex than that of most other peripheral sense organs. The developmental reason for this is that it is produced by an outgrowth of the brain in the embryo. The photoreceptors themselves are probably derived from flagellated cells forming the ependymal lining of the cavities of the brain, so that they are on the internal side of the mature retina, and light must first pass through the rest of the retina before reaching them.

The structure of the retina, as determined by light microscopy, is shown in fig. 17.2. Using a conventional non-selective stain, a number of layers can be distinguished. The interrelations of the cells in these layers can be observed by using a selective stain such as the Golgi silver impregnation method (see Polyak, 1941). The photoreceptor cells are of two types, described, from their shapes, as *rods* and *cones*. These

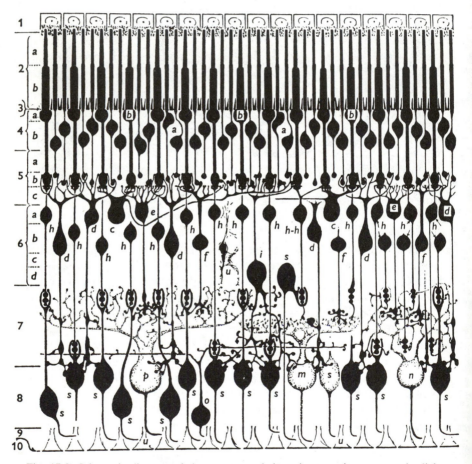

Fig. 17.2. Schematic diagram of the structure of the primate retina, as seen by light microscopy after using the Golgi silver staining method. The layers are as follows: 1, pigment epithelium; 2, receptor layer; 3, external limiting membrane; 4, outer nuclear layer; 5, outer plexiform layer; 6, inner nuclear layer; 7, inner plexiform layer; 8, ganglion cell layer; 9, nervous layer; 10, inner limiting membrane. Cell types: *a*, rods; *b*, cones; *c*, horizontal cells; *d, e, f, h*, bipolar cells; *i, l*, amacrine cells; *m, n, o, p, s*, ganglion cells; *u*, Müller fibres. (From Polyak, 1941.)

synapse with small interneurons, the *bipolar cells*, which themselves synapse with the *ganglion cells*. The axons of the ganglion cells form the optic nerve, and carry visual information from the retina into the brain. In addition to this sequential system, there are two lateral systems of neurons: the *horizontal cells*, which form interconnections between the receptor cells, and the *amacrine cells*, which synapse with each other, with the ganglion cells, and with the proximal ends of the bipolars. Filling the spaces between these various neurons are the *Müller fibres*, which are elongated glial cells.

These retinal cells are arranged in the retinal layers as follows. Inside the *pigment epithelium* is the *receptor layer*, which consists of the inner and outer segments of the rods and cones. Beneath the receptor layer is the *outer nuclear layer*, which contains the nuclei of the receptor cells. Between the receptor layer and the outer nuclear layer is the 'external limiting membrane'; electron micrographs show that this is not a true membrane but a level at which the receptor cells are closely attached to each other via thickenings of the cell membrane. The next layer is the *outer plexiform layer*, which contains the dendrites and synapses of the rods and cones, the bipolars and the horizontal cells. The *inner nuclear layer* contains the nuclei of the bipolars, horizontal cells and Müller fibres. The *inner plexiform layer* contains the synapses and dendritic processes of the bipolar cells, amacrine cells and ganglion cells. The *ganglion cell layer* contains the nuclei of the ganglion cells and, finally, the *nervous layer* contains the axons of the ganglion cells.

The rods and cones are not distributed evenly over the surface of the retina. In man (and in a number of other vertebrates, especially birds) there is a more or less central region which is specially modified for high visual acuity, known as the *fovea*. The fovea in man contains cones only; it is surrounded by a region in which some rods occur, the *parafovea*. In the extrafoveal retina the proportion of cones (in man) is very small.

There are about 100 million rods and about six million cones in the human eye. Since there are only about one million ganglion cells and optic nerve fibres, it follows that there must be some considerable convergence of the photoreceptors onto the ganglion cells. The anatomical basis of this convergence can be found in the synaptic contacts of the retinal cells. Fig. 17.3 summarizes the results of a number of investigations by electron miscroscopy (see, for example, Dowling, 1968, 1970). The presynaptic membrane of each receptor cell terminal is invaginated to form a pocket into which processes from bipolar and horizontal cells fit. These are called *ribbon synapses* since there is a dense ribbon or bar in the presynaptic cytoplasm, surrounded by an array of synaptic vesicles. Synapses of a more conventional structure are found between horizontal cells and bipolar cells. Some of the terminals

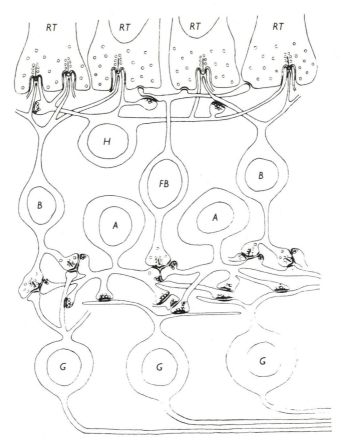

Fig. 17.3. Synaptic contacts in the vertebrate retina, as seen by electron microscopy, based on the frog. *RT*, synaptic terminals of the receptor cells (notice the synaptic ribbons and the invaginations into which the processes of the postsynaptic cells fit); *H*, horizontal cells; *B* and *FB*, bipolar cells (notice the synaptic ribbons); *A*, amacrine cells (notice the reciprocal synapses with bipolars and the serial synapses with other amacrines and ganglion cells); *G*, ganglion cells. (From Dowling, 1970.)

of the bipolar cells also show ribbon synapses, again where there are two types of postsynaptic cell: amacrine and ganglion cells in this case. Many of the synapses between bipolar cells and amacrines are *reciprocal*: presynaptic and postsynaptic areas may occur at different places on each of the membranes bounding the synaptic cleft between the same two cells. Amacrine cells also form conventional synapses with ganglion cells and with other amacrine cells. Some of the latter form *serial synapses* in which a terminal may be postsynaptic to one cell and presynaptic to another. All this structural complexity indicates that there must be an enormous amount of interaction between the various cells of the retina.

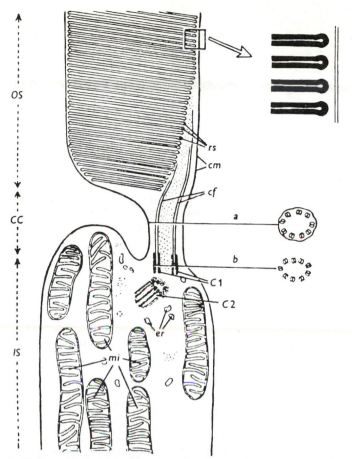

Fig. 17.4. Diagram to show the structure of a mammalian rod, as seen by electron microscopy. *OS*, outer segment; *CC*, connecting cilium; *IS*, inner segment; *C*1 and *C*2, centrioles. Transverse sections through the connecting cilium (*a*) and the centriole (*b*) are shown at the right. *rs*, rod sacs; *cf*, ciliary filaments; *cm*, cell membrane; *mi*, mitochondrion; *er*, endoplasmic reticulum. (From de Robertis, 1960.)

The physiological details of this interaction (p. 469) are only just beginning to be understood.

The structure of rods and cones is shown rather roughly in fig. 17.2, and part of a rod is shown in more detail in fig. 17.4. The outer segment, which contains the visual pigment, contains a stack of membranous discs (in rods) or infoldings of the cell membrane (in cones). The outer segment is connected to the inner segment, which contains numerous mitochondria, via a thin neck whose structure is very like that of a cilium. Below the inner segment is a region which contains the nucleus, and finally ends at the synaptic terminal.

THE DUPLICITY THEORY

The rods and cones have different functions as photoreceptors. The rods are used for vision at low light intensities and are not involved in colour vision. The cones are used at higher light intensities, and for colour vision. Visual acuity is higher for cone vision than for rod vision. These statements constitute the *duplicity theory*.

The duplicity theory was first propounded by Schultze in 1866. It is well known that, in very low light intensities such as occur on a dark night, the fovea is practically blind, and vision depends upon the extrafoveal regions of the retina; colour vision is absent under these conditions. Schultze pointed out that these features tie in with the distribution of the rods: there are no rods in the fovea. He went on to examine the retinae of a variety of different vertebrates, and showed that nocturnal animals tend to have a great preponderance of rods, and diurnal animals have a corresponding preponderance of cones.

The spectral sensitivity of rod (scotopic) vision differs from that of cone (photopic) vision. Scotopic vision is most sensitive to blue–green light and insensitive to red light, whereas photopic vision covers the whole of the visible spectrum, with the greatest sensitivity being in the yellow region. This movement of the region of maximum sensitivity to longer wavelengths, which accompanies the change from scotopic to photopic vision, is known as the 'Purkinje shift'.

A number of psychophysical experiments provide further evidence for the duplicity theory (see Hecht, 1937). A particularly useful experimental method is what is known as the 'fixation and flash' technique for determining visual thresholds. Since the threshold varies somewhat over different parts of the visual field, it is usually necessary in such experiments to ensure that a visual stimulus is always applied to the same part of the retina. This is done as follows. The subject 'fixates' on a small red light, so that its image falls on the fovea. The visual stimulus is then presented at some known angle to the line of fixation (fig. 17.5). In determining the threshold, the light intensity of the test field is adjusted

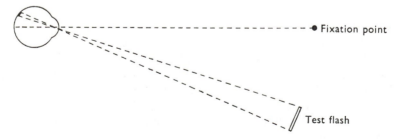

Fig. 17.5. The fixation and flash technique. (After Pirenne, 1962.)

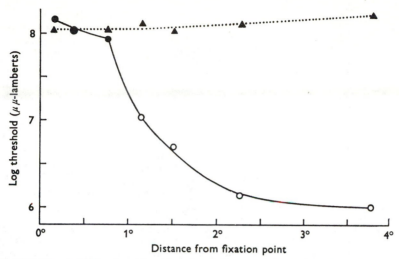

Fig. 17.6. Thresholds for deep blue (circles) and deep red (triangles) flashes of 10 minutes diameter at points near to the fovea. Black symbols show points at which the flash appeared coloured, open symbols where it did not. (From Pirenne, 1944.)

so that the subject can just see it (or, when flashes of short duration are used, when the subject sees the flash in 50 per cent of the times that it is presented).

Using this technique, Pirenne (1944) investigated the threshold of the dark-adapted eye to deep red and deep blue test fields of small diameter presented at small angles to the fixation line. His results are shown in fig. 17.6. The threshold for red light increases slightly as the field is presented further away from the fovea. The interpretation of this is that the cone density in the retina decreases correspondingly, and the rods are insensitive to red light. With blue light, however, to which the rods *are* sensitive, the threshold falls markedly with increasing eccentricities. This corresponds very neatly with the distribution of the rods, which are absent in the fovea and appear in increasing numbers at eccentricities greater than 0.5 degrees. Furthermore, the blue flash appears to the observer to be blue when seen by the fovea (cone vision), but appears colourless at the threshold in the extrafoveal region (rod vision), as is indicated by the filled and open circles in fig. 17.6.

During dark adaptation, the threshold falls progressively with time. Fig. 17.7 shows the result of one experiment on this phenomenon, in which the test field was a blue flash placed at an eccentricity of 7 degrees to the fixation line. The curve consists of two branches with a definite 'kink' at their intersection, and the flash appears blue above this kink and colourless below it. The interpretation of these results in terms of the duplicity theory is that the initial section of the curve is due to cone

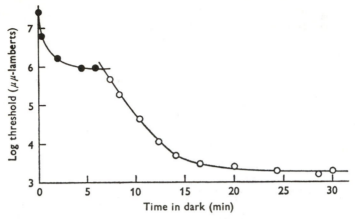

Fig. 17.7. The fall in threshold in the dark following a three-minute exposure to bright light, as tested by deep blue light flashes seen at an eccentricity of 7 degrees. For the first five points (black circles) the flash appeared blue or violet in colour, thereafter (open circles) it was apparently colourless. (From Hecht and Schlaer, 1938.)

vision, and the later section to rod vision. In accordance with this, the later section is absent if the test field is small in size and viewed by the fovea, or if it is deep red in colour; in each case we would not expect the rods to be excited.

VISUAL PIGMENTS *IN VITRO*

The properties of the visual pigments can be examined in two situations, either *in vitro*, after extraction from the eye, or *in vivo*, while still in the intact eye. As a half-way house between these two extremes, they can also be examined in isolated retinae or in fragments of the photoreceptor cells. The two methods are complementary to each other: *in vivo* studies must be interpreted in the light of the properties of the pigments, which can be determined much more precisely *in vitro*, and the relevance of *in vitro* studies can only be tested by reference to what happens in the intact eye.

Methods of investigating extracted pigments

In extracting a visual pigment from the retina of an animal, it is essential that the pigment should not be bleached by exposure to bright light. Hence it is usual to use dark-adapted animals, to dissect out the retina under darkroom conditions, using a dim deep red light, and to carry out the extraction procedures from then on in the dark. Visual pigments are not normally soluble in water, and have to be extracted from the retina by aqueous solutions of various substances; digitonin is the most commonly used extractant. Much purer solutions are obtained if the pigment is extracted from the outer segments of the rods and cones than if the whole retina is used; in particular, this method avoids contamination by haemoglobin from the retinal blood vessels. The outer segments are obtained by shaking the retina vigorously in 35–45 per cent sucrose solution followed by low speed centrifugation, which leaves the outer segments suspended in the supernatant fluid.

Visual pigments are characterized by measurement of the amount of light that they absorb at different wavelengths (see Dartnall, 1957, 1962). A convenient way of doing this is as follows. Light from a monochromator is passed through a cell (the sample cell) containing an extract of the visual pigment, and measured by means of a photocell. This is compared with the amount of light passed through a similar cell (the reference cell) containing the extractant solution but no visual pigment; the difference between the two measurements then gives the absorption by the pigment and by any impurities present in the extract.

Let the light intensity transmitted by the same cell be I_s, and let that transmitted by the reference cell be I_r. Then the amount of light absorbed by the pigment, I_p, is given by

$$I_p = I_r - I_s.$$

The percentage absorption by the pigment is therefore $100\,I_p/I_r$. If we plot the percentage absorption against wavelength, the curve obtained is known as the *absorption spectrum* of the pigment. Fig. 17.8a shows the absorption spectra of frog visual pigment at different pigment concentrations; in fig. 17.8b these curves have been replotted with their maxima made equal to 100 per cent. It can be seen that the absorption spectra become broader as the pigment concentration (or the thickness of the pigment solution) rises. The reason for this can be explained by an example as follows. Suppose that the amount of light transmitted is halved if we double the concentration of the pigment. Then, at a wavelength where the absorption is 90 per cent of the maximum, doubling the pigment concentration will raise the absorption to 95 per cent. But, at a wavelength where the absorption is only 40 per cent of the maximum, doubling the concentration will raise it to 70 per cent.

The fact that the absorption spectrum depends on the pigment concentration, even when expressed as a percentage of the maximum absorption, makes it difficult to compare results from solutions either of different strengths or in cells of different thicknesses. This difficulty is obviated by the use of *density spectra*.

Consider a very thin plane of pigment solution, of thickness dl. Of the light I incident on this plane, a portion dI will be absorbed. The fraction dI/I will be proportional to the thickness of the plane, dl, and the

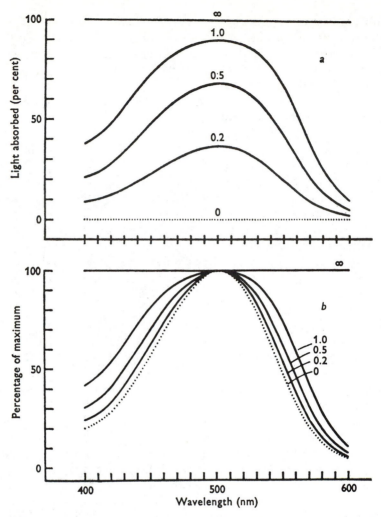

Fig. 17.8. Absorption spectra of solutions containing different concentrations of frog visual pigment. In *a* the ordinate is the percentage of light absorbed by the pigment ($100\ I_p/I_r$). In *b* these curves are replotted so that their maxima are made equal to 100 per cent; notice that the curves become broader as the concentration of pigment is increased. (From Dartnall, 1957.)

concentration of the pigment, *c*. Thus

$$\frac{\mathrm{d}I}{I} = \alpha_\lambda \cdot c\ \mathrm{d}l$$

where α_λ is a constant, the *extinction coefficient*, for any particular pigment at any particular wavelength λ. Integrating this equation

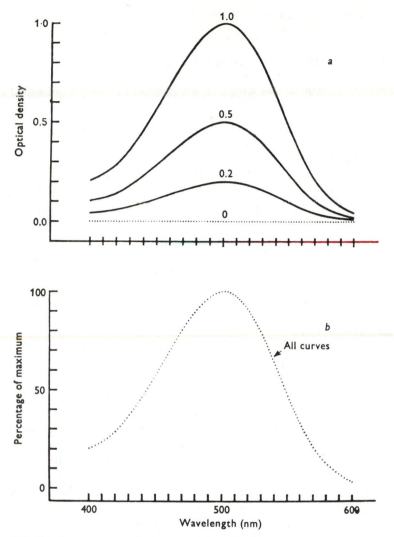

Fig. 17.9. Density spectra of solutions containing different concentrations of frog visual pigment. In *a* the ordinate is the optical density (D_λ) of the solution. In *b* these curves are replotted so that their maxima are made equal to 100 per cent; all concentrations now give identical curves. (From Dartnall, 1957.)

between the effective limits I_r and I_s, we get

$$\alpha_\lambda \,.\, c \,.\, l = \log_e I_r - \log_e I_s$$

where l is the width of the measuring cell. That is,

$$\log_e (I_r/I_s) = \alpha_\lambda \,.\, c \,.\, l.$$

When converted into decadic logarithms, this gives

$$D_\lambda = \log_{10}(I_r/I_s) = \alpha_\lambda \cdot c \cdot l/2.303.$$

This quantity D_λ is called the *optical density* of the solution. If we plot D_λ against λ, we obtain the density spectrum of the pigment, as is shown in fig. 17.9a. When density spectra for different concentrations of the same pigment are plotted as percentages of their maxima, all the curves coincide (fig. 17.9b). If we have two or more pigments in a solution, then the total optical density is the sum of the optical densities of the individual pigments. These two properties make the density spectrum a most useful tool for the characterization and comparison of different pigments. The absorption spectrum, on the other hand, is normally used when questions involving the concentration of a pigment arise.*

One of the difficulties of measuring the density spectra of visual pigments is that it is frequently very difficult to get rid of impurities in the extract, whose density spectra add to those of the visual pigment. This difficulty can be side-stepped by the measurement of the *difference spectrum*. It is a characteristic feature of visual pigments that they are bleached by exposure to bright lights. Hence it follows that the density spectrum of the products of bleaching differs from that of the unbleached pigment. The difference between the density spectra before and after bleaching constitutes the difference spectrum. Since any impurities which are not visual pigments will have the same density spectrum after bleaching as they did before, they do not contribute to the difference spectrum.

Finally, how are we to distinguish between an extract containing only one visual pigment and one containing two or more? The technique for dealing with this problem is known as partial bleaching, and was developed by Dartnall (1952). The method consists, essentially, in successively bleaching the extract with light of different wavelengths and measuring the difference spectrum after each bleach. If these difference spectra are different (when plotted on a percentage scale with their maxima made equal to 100 per cent), then more than one pigment is present. For example, suppose we have a retinal extract containing two visual pigments, with maximal absorptions (λ_{max}) at 500 nm and 540 nm. If we illuminate this with red light (λ equal to, say, 640 nm), the 540 pigment will absorb much more than the 500 pigment and will therefore be preferentially bleached (the absorption coefficients at this wavelength are likely to be, respectively, about 8 per cent and less than 1 per cent of their maxima). Hence the difference spectrum after this

---

* Unfortunately, as Dartnall (1962) points out, many authors use the term 'absorption spectrum' when referring to the density spectrum. The term 'extinction spectrum' is sometimes used as a synonym for the density spectrum.

first bleach will have its λ_{max} at about 540 nm. With a long enough exposure, nearly all the 540 pigment can be bleached. If we now illuminate the pigment with white light, or with light at any wavelength between about 560 and 420 nm, the 500 pigment will be bleached, and the difference spectrum for this second bleach will have its λ_{max} at about 500 nm. This shift in the difference spectrum indicates the presence of at least two visual pigments in the extract.

The human scotopic visual pigment: rhodopsin

Rhodopsin (visual purple) is the principal photosensitive pigment of many vertebrate retinae. It is found throughout the human retina except at the fovea, hence it is characteristic of the rods, and absent from the cones. Thus we would expect rhodopsin to be the visual pigment involved in scotopic vision, or, to put it more precisely, the primary photoreceptor event in human scotopic vision should be the capture of light by rhodopsin. The clearest evidence that this is so is provided by the results of an investigation by Crescitelli and Dartnall (1953), which we shall now consider.

Crescitelli and Dartnall extracted human rhodopsin from an eye which had to be removed because of the presence of a ciliary body melanoma. The patient was fully dark-adapted before the operation, which was then performed under deep red light. The density spectrum of the rhodopsin extract was measured before and after bleaching, with the results shown in fig. 17.10a.

If the extract did not contain any light-absorbing impurities, then the 'unbleached' curve in fig. 17.10a would be the density spectrum of human rhodopsin. But this is unlikely to be so, and indeed the 'unbleached' curve, by comparison with pure extracts of frog and cattle rhodopsin, looks as though it is derived from a mixture of rhodopsin and some impurities which absorb light increasingly at wavelengths below about 480 nm. Is it possible, then, to make a reasonable calculation as to what the density spectrum of pure human rhodopsin should be? Fig. 17.10b shows the difference spectrum of the retinal extract; this is effectively identical with that of rat rhodopsin, and very similar in shape to that of frog rhodopsin, but shifted along the wavelength axis so that its λ_{max} is about 5 nm less. This indicates that the density spectrum of human rhodopsin should be similar to that of frog rhodopsin but, again, with its λ_{max} about 5 nm less. Now the density spectrum of frog rhodopsin is well established; its λ_{max} is 502 nm, and the full spectrum is shown by the broken line in fig. 17.11. Dartnall (1953) produced a nomogram which could be used to predict the form of the density spectrum of any visual pigment, given the value of λ_{max}: since then this nomogram has been tested against a variety of visual pigments, and its use seems to be

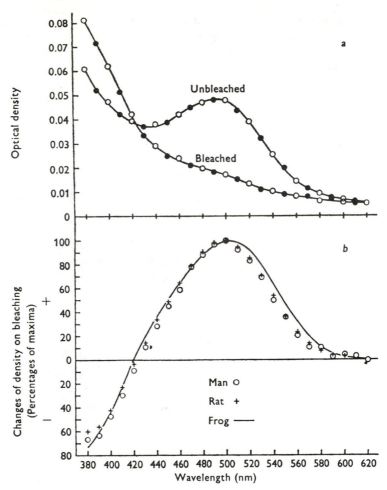

Fig. 17.10. Density and difference spectra of an extract containing human rhodopsin. Density spectra of the extract before and after bleaching are shown in *a*; readings were taken in the order 380 to 620 nm (open circles), then 610 to 390 nm (filled circles). The difference between the unbleached and bleached curves in *a* are plotted in *b* on a percentage scale (to give the difference spectrum), where similar curves for rat and frog rhodopsin are also shown. (From Crescitelli and Dartnall, 1953.)

fully justified in most cases. Using this nomogram to determine the density spectrum of a pigment whose λ_{max} was equal to 497 nm, Crescitelli and Dartnall obtained the curve shown by the full line of fig. 17.11. This curve, they suggested, represents the true density spectrum of human rhodopsin; it is difficult to believe that this conclusion is not justified.

The next step was to compare this curve with the scotopic sensitivity to light of different wavelengths. The scotopic visibility function was determined by Crawford (1949) from observations on fifty subjects under the age of thirty (this matter of the age of the subjects is rather important, since sensitivity to short wavelengths declines after thirty due to accumulation of yellow pigment in the lens). The subjects were dark-adapted, and looked at a field 20 degrees in diameter which was divided vertically into two halves. One half of the field was illuminated with white light of constant intensity, and the other half with mono-chromatic light of different wavelengths, whose intensity could be altered so as to make both halves of the field appear equally bright. The

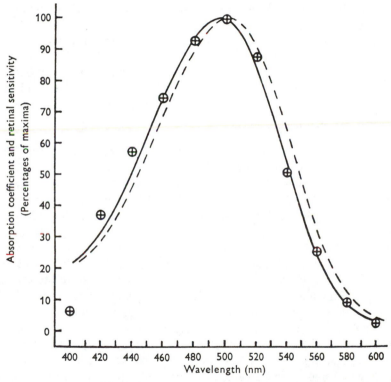

Fig. 17.11. Density spectrum of human rhodopsin and the scotopic visibility function. The broken curve shows the density spectrum of frog rhodopsin, as determined experimentally from very pure solutions. The continuous curve is the density spectrum (equal to the absorption spectrum at low pigment concentrations) of human rhodopsin, calculated from the difference spectrum in fig. 17.10*b* (see text). The points represent the scotopic sensitivity of the eye at different wavelengths as determined by Crawford, corrected for the absorption of short wavelengths between the cornea and the retina and expressed as an equal quantal content spectrum. (From Crescitelli and Dartnall, 1953.)

actual intensities were about fifteen times threshold, well within the scotopic range.

Crawford's results were expressed as relative sensitivities, derived from the reciprocals of the light intensities needed to produce a certain sensation of brightness. Light intensity is usually measured in units equivalent to a certain amount of energy per unit time, hence Crawford's results were expressed as an 'equal energy spectrum'. But the absorption of light by a pigment takes place in quanta, whose energy is inversely proportional to the wavelength. Hence, in order for Crawford's results to be compared with the rhodopsin density spectrum, they must be recalculated in terms of an 'equal quantal content spectrum'. A second correction is necessary, to allow for the absorption of light between the cornea and the retina, which is more pronounced at short wavelengths. Values for this preretinal absorption were obtained from the results of Ludwigh and McCarthy (1938). The points shown in fig. 17.11 are Crawford's results as plotted by Crescitelli and Dartnall after these two corrections had been made. It is evident that the agreement between the scotopic sensitivity function and the absorption of light by rhodopsin is very precise, and we can therefore conclude that rhodopsin is the visual pigment used in human scotopic vision.*

The chemistry of rhodopsin

In order to understand the role of rhodopsin in the visual process it is necessary to know something about its chemical structure and the changes it undergoes on illumination. We shall here give only the outlines of what is known about rhodopsin chemistry; fuller accounts are given in a number of recent articles (e.g. Dartnall, 1962; Bridges, 1967).

When a solution of rhodopsin is exposed to bright light at room temperature, it is bleached, and the rhodopsin splits to form two components: retinaldehyde (or, as it used to be known as, retinene$_1$) and a lipoprotein called opsin. Retinaldehyde is the aldehyde of vitamin A_1 (retinol); it can exist in a number of different sterioisomeric forms, some of which are shown in fig. 17.12. Rhodopsin can be reconstituted by mixing opsin with 11-*cis* retinaldehyde (Hubbard and Wald, 1952); of the other isomers, only 9-*cis* retinaldehyde forms a photosensitive pigment (which has been named 'isorhodopsin'), and its difference spec-

---

* There is a further correction that is not taken into account in fig. 17.11. The absorption of light by the retinal rhodopsin is given by its absorption spectrum, not by its density spectrum; the two are only equal when the net percentage of light absorbed is very low. Crescitelli and Dartnall estimated that the retinal absorption was only 3.5 per cent, in which case the correction would be minute. Other estimates, however, suggest that this figure is too low – about 55 per cent absorption seems more likely (Alpern and Pugh, 1974) – so that the curve in fig. 17.11 should be broadened somewhat.

Vitamin A_1 (retinol)

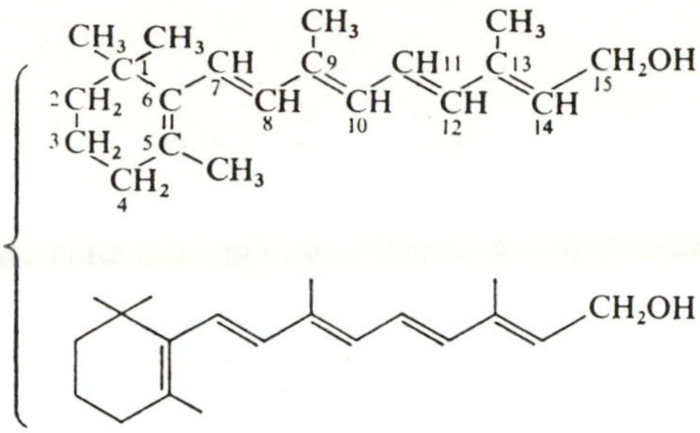

All-*trans* retinaldehyde

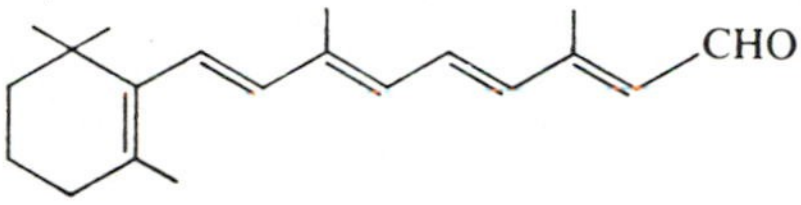

11-*cis* retinaldehyde
(neo-*b*)

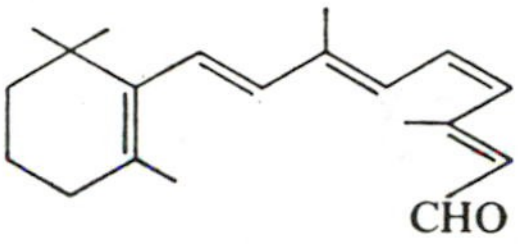

9-*cis* retinaldehyde
(iso-*a*)

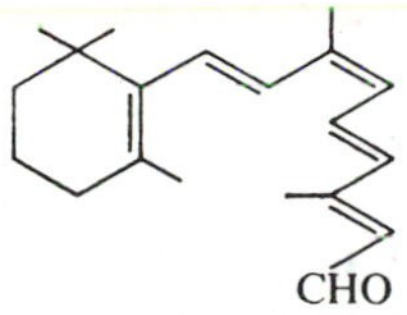

Fig. 17.12. Structural formulae of vitamin A_1, shown in full and in shorthand notation, and of three stereoisomers of retinaldehyde, shown in shorthand notation.

trum is different from that of natural rhodopsin. The retinaldehyde released after exposure of rhodopsin to light, on the other hand, is in the all-*trans* configuration. Rhodopsin can also be broken down in the dark by heating so as to denature the protein part of the molecule, but in this case the retinaldehyde liberated is largely in the 11-*cis* form (Hubbard, 1959). From these various results we can conclude that the retinaldehyde in rhodopsin is in the 11-*cis* form, but that it is converted into the all-*trans* form by the action of light.

The breakdown of rhodopsin to all-*trans* retinaldehyde and opsin after exposure to light takes place in a series of stages (fig. 17.13). The direct action of light appears to be simply to change the retinaldehyde from the 11-*cis* to the all-*trans* form; as a result of this a transient intermediate is produced, prelumirhodopsin, which can only be observed at temperatures below −140 °C. At higher temperatures there

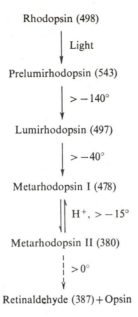

Fig. 17.13. The breakdown of rhodopsin after exposure to light. There are a number of stages between metarhodopsin II and retinaldehyde, but these are not included here as there is no general agreement as to their precise nature. (Simplified after Bridges, 1967.)

follows a series of changes that are apparently alterations in the internal structure of the opsin molecule and its means of attachment to the retinaldehyde. These changes appear to be fairly minor in nature during the lumirhodopsin and metarhodopsin I stages; the evidence for this view is that metarhodopsin I is produced within microseconds of illumination at room temperatures, that totally dry rhodopsin is converted to metarhodopsin I by the action of light, and the overall change in the absorption spectrum up to this stage is not very great. The next stage, to metarhodopsin II, appears to involve a more drastic alteration in the structure of the molecule, since it takes longer (the half-time for the change is about 1 msec at room temperature), it requires hydrogen ions and it involves a large change in absorption spectrum which is readily visible as 'bleaching'.

It has been known for some time that, in the later stages of the changes produced by light, the retinaldehyde is attached to the protein by means of a Schiff base link (Morton and Pitt, 1955), probably via the ε-amino group of a lysine residue (Bownds, 1967). There has been some controversy about whether or not this linkage also occurs in rhodopsin and the earlier stages as well (Poincelot *et al.*, 1969; Hall and Bacharach, 1970).

In solution at temperatures above 0 °C, metarhodopsin II gradually breaks down to give retinaldehyde and opsin. In the intact retina retinol (vitamin A_1) is formed instead, probably by the action of the enzyme retinol dehydrogenase. The nature of the intermediate steps in this process is not yet fully agreed upon (see Bridges, 1967); what is clear is that the reactions are far too slow to be of any direct importance in the excitation of the visual receptors.

Other visual pigments

We have already seen that the density spectrum of human rhodopsin is slightly different from that of frog rhodopsin, a circumstance which indicates that the two rhodopsins are different pigments. In fact, there are a large number of different pigments, produced by combination of retinaldehyde with different opsins. Different authors use different nomenclatures for these pigments. Wald describes all retinaldehyde pigments extracted from rods as 'rhodopsins' and all those extracted from cones as 'iodopsins'. In Dartnall's system the pigment is named from its λ_{max} value, with a subscript to indicate that it contains retinaldehyde (retinene$_1$); thus frog rhodopsin is 'visual pigment 502_1', and human rhodopsin is 'visual pigment 497_1'.

In addition to these retinaldehyde pigments, there is another set of pigments which contain 3-dehydroretinaldehyde, which used to be known as retinene$_2$. 3-dehydroretinaldehyde is the aldehyde of vitamin A_2, which differs from vitamin A_1 in possessing a double bond between carbon atoms 3 and 4 (fig. 17.14). Wald describes these pigments as 'porphyropsins' when they are extracted from rods; when 3-dehydroretinaldehyde is combined with cone opsin (from iodopsin) an artificial pigment, 'cyanopsin', is produced. In Dartnall's nomenclature, 3-dehydroretinaldehyde (retinene$_2$) pigments are indicated by the subscript; thus the porphyropsin of tadpoles is 'visual pigment 523_2'.

Vitamin A_2

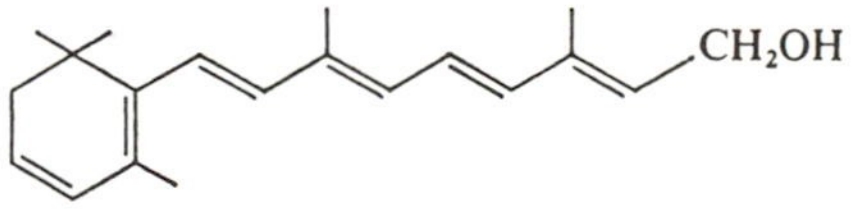

All-*trans*
3-dehydroretinaldehyde

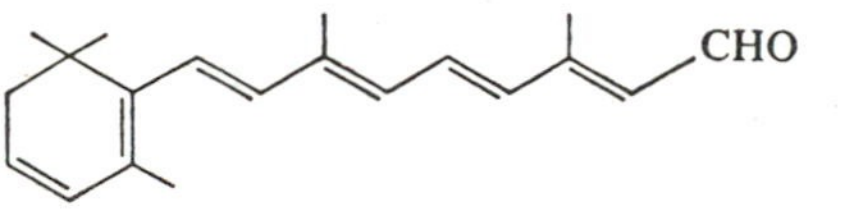

Fig. 17.14. Vitamin A_2 and all-*trans* 3-dehydroretinaldehyde.

VISUAL PIGMENTS *IN SITU*

Retinal densitometry

We shall now consider some of the applications of a very elegant technique, known as retinal densitometry, which is used to measure the concentrations and characteristics of visual pigments in the living eye. The principle of the method is to shine lights of two wavelengths, of which one is absorbed by the pigment and the other not, into the eye and measure the intensity of the light reflected back from the choroid layer behind the retina. Since this light passes through the retina twice, some of it will be absorbed by the visual pigment, and so the absorptive properties of the pigment can be determined from comparison of the intensities of the reflected beams of the two lights.

There are various ways in which the technique can be carried out; that used by Campbell and Rushton (1955) in their measurements on rhodopsin in the human eye may serve as an example. The apparatus is shown diagrammatically in fig. 17.15. White light from the source S_1 is focused on the wheel W by the lens L_1; the neutral density filter F_1 serves to control its intensity. The wheel W is divided into two semicircular halves, one of which is an orange filter and the other a blue–green filter; it is rotated rapidly, so that the light passing through it is alternately orange and blue–green. Blue–green light will be absorbed by rhodopsin, whereas orange will not. The light passes through further

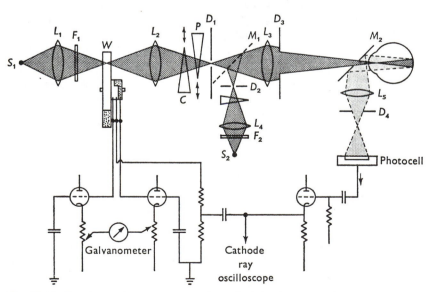

Fig. 17.15. One form of apparatus used for retinal densitometry. Explanation in text. (Redrawn after Campbell and Rushton, 1965.)

lenses and diaphragms, and is eventually focused on the pupil by the lens L_3, so that a circular patch of light falls on the retina. The position on the retina of this patch is determined by fixation of a small red light by the other eye. The light reflected from the back of the eye is then reflected by the Mirror M_2, through the lens L_5 and the diaphragm D_4 onto the photomultiplier cell, which produces an electrical signal proportional to the intensity of the light falling on it.

The intensity of the coloured beams can be altered by moving the two wedges placed just after L_2; one of these (C) is blue–green, and so can be used to alter the intensity of the orange beam, and the other (P) is purple, and therefore controls the intensity of the blue–green beam. The light source S_2 is used for bleaching the retina with high intensities, and is incorporated into the main optical system by placing the mirror M_1 in position.

The output from the photomultiplier is fed via a high-pass filter and cathode follower, to be displayed on an oscilloscope, producing an oscillatory trace whose amplitude is proportional to the difference in the intensities of the blue–green and orange light emerging from the eye. The apparatus is used on the null principle: the coloured wedges are moved until there is no difference in the intensities of the reflections of the two colours from the back of the eye. The point at which this occurs can be approximately estimated by observation of the oscilloscope trace, but a more accurate estimation is obtained as follows. The photomultiplier signal is passed through a commutator on the axle of the wheel W, and from there to the grids of two cathode followers, which are also connected to capacitors. The commutator is constructed so that the signal is fed to one side of the system when the blue–green light is transmitted by the wheel, and to the other side when orange light is transmitted. The capacitors smooth out the irregularities in the signal, so that the outputs of the two cathode followers are proportional respectively to the intensities of the two reflected light colours. A galvanometer measures the difference between these two outputs; when it reads zero, the light intensities are the same.

The net result of this system is that the difference between the absorption of blue–green light and that of orange light can be determined from the position of the purple wedge when this is adjusted so that the galvanometer reading is zero. Since rhodopsin absorbs blue–green light but not orange light, it is to be expected that this difference will be proportional to the amount of rhodopsin present in the area of retina whose reflection is measured.

Fig. 17.16 shows the results of an experiment with this apparatus. The eye was dark-adapted and the reflection measured from a patch about 15 degrees away from the fovea. The white circles show the effect on the

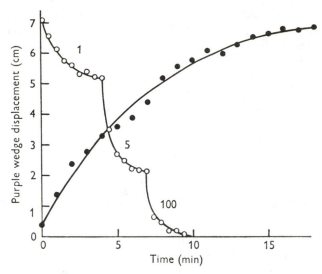

Fig. 17.16. Measurement of rhodopsin concentration in the human eye during bleaching under three levels of illumination (white circles) and its subsequent regeneration in the dark (black circles). The displacement of the purple wedge is proportional to the optical density of the retinal rhodopsin. Further explanation in the text. (From Campbell and Rushton, 1965.)

rhodopsin concentration of bleaching with white lights. When the bleaching light intensity was not too high, the rhodopsin concentration reached a steady level at which its rate of bleaching was counter-balanced by regeneration. With increasing bleaching light intensities, this steady level occurred at lower concentrations. The black circles show the recovery of rhodopsin concentration in the dark. The time course of this curve is very similar to that of the scotopic dark adaptation curve when this is plotted in the normal way as log threshold against time. In other words, the *logarithm* of the scotopic sensitivity is pro-portional to the concentration of rhodopsin in the patch of retina which is illuminated; we shall examine the implications of this situation later.

Campbell and Rushton performed various other experiments to check that the difference between the reflection of the blue–green and orange beams really was due to absorption of the blue–green light by rhodop-sin. Rhodopsin is a rod pigment, and therefore the differential absorp-tion in different parts of the retina should be proportional to the rod concentration. Fig. 17.17 shows that this is so (notice, particularly, that there is no absorption at the fovea or at the blind spot). The effectiveness of bleaching of lights of different wavelengths should be related to the scotopic sensitivity curve. This, again, was found to be so: lights which looked as though they were equally bright produced equal amounts of

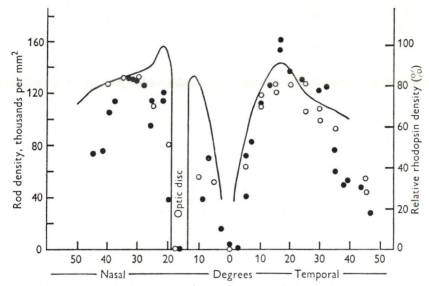

Fig. 17.17. Relative rhodopsin concentrations in the human eye at different positions in the horizontal plane, as measured by retinal densitometry. White and black circles indicate the change in absorption following bleaching (two different methods) with a constant quantity of light. The curve shows the number of rods per mm², as determined by Østerberg. (From Campbell and Rushton, 1965.)

bleaching when brightened by the same factor. Using apparatus of a somewhat different design, Rushton (1956) was able to measure the reflection from the back of the eye at different wavelengths. If such measurements were made before and after bleaching, the difference spectrum of the absorbing substance could be obtained; fig. 17.18 shows that this is very similar to that of rhodopsin in solution.*

Rushton has also applied the retinal densitometry technique to the investigation of foveal cone pigments. Colour vision in the normal fovea is dichromatic, which suggests that there are two different pigments present. In the type of colour blindness known as protanopia, it is impossible to distinguish between red and green, and the fovea is apparently monochromatic. Thus the fovea of the protanope offers a very useful means of investigating a particular cone pigment in isolation.

Fig. 17.19 shows the results of an experiment by Rushton (1963) using retinal densitometry on the fovea of the protanope. The measuring light was 540 nm (green), and the comparison beam 700 nm (deep red). The plan of the experiment was very similar to that of the earlier

---

* The *in vivo* and *in vitro* difference spectra are not quite identical. This suggests that the products of bleaching are different in the two cases. It may be that an intermediate product is more stable in the intact rods than it is in solution.

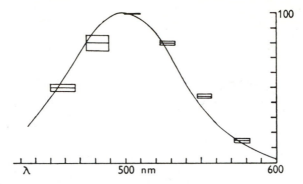

Fig. 17.18. Difference spectrum of human rhodopsin in the eye, as determined by retinal densitometry measurements at six different wavebands. The width of each box gives the waveband of the measuring light, and its height is twice the standard error of the mean. The curve is the difference spectrum for extracted human rhodopsin as determined by Crescitelli and Dartnall. (From Rushton, 1956.)

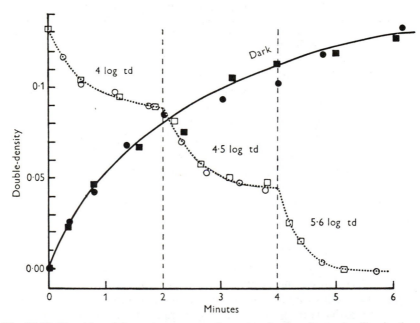

Fig. 17.19. Bleaching of the protanope cone pigment under three increasing illuminations (white symbols) and its subsequent regeneration in the dark (black symbols), as measured by retinal densitometry. Circles and squares are repetitions of the same experiment. Retinal illuminations are measured in trolands (td), one troland being the illumination produced by an external surface of luminance 1 candela/m^2 seen through a pupil of area 1 mm^2. (Rushton, 1963.)

rhodopsin experiment shown in fig. 17.16. It can be seen that the results are almost identical, except that the rate of regeneration of the cone pigment is about three times as rapid as that of rhodopsin. Furthermore, the intensity of the bleaching light required to bleach 50 per cent of the pigment is much the same both for the cone pigment and for rhodopsin. This means that the ability of a cone to catch light is similar to that of a rod; the difference in the thresholds for rod and cone vision must therefore lie almost entirely in differences in the excitation process in the two types of photoreceptor, or in the amount of convergence of the rods and cones upon ganglion cells, or both.

The homogeneity of the cone pigment was tested by applying the method of partial bleaching. It was found that the difference spectrum of the pigment was identical after bleaching with red, blue–green and white lights (fig. 17.20), indicating that there is indeed only one pigment

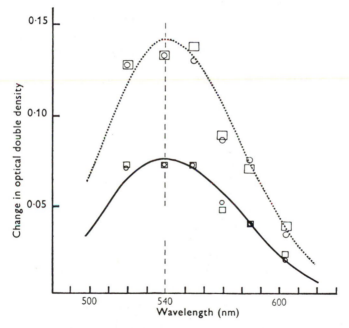

Fig. 17.20. Results of a partial bleaching experiment on the protanope cone pigment, using retinal densitometry. The ordinate shows the change in optical double density (the optical density of the pigment is apparently doubled because the measuring light passes through it twice) following bleaching. Small circles: bleaching by red light, followed by further bleaching by white light (large circles). Small squares: bleaching by blue–green light, followed by further bleaching by white light (large squares). The white lights were strong enough to bleach fully, and the coloured lights were strong enough to bleach 50 per cent. The curves show the spectral sensitivity of the protanope (as determined by Pitt), suitably scaled. (From Rushton, 1963.)

present. Rushton called this pigment 'chlorolabe'. The absorption spectrum of chlorolabe was determined by measuring the amount of light needed to produce equal amounts of bleaching at different wavelengths, and it was further found that this corresponded with the foveal spectral sensitivity of the subject.

Rushton (1962) has evidence for the existence of a second foveal pigment in normal subjects (it is absent from protanopes), known as 'erythrolabe', which is red-sensitive.

Microspectrophotometry

It would obviously be of considerable interest to measure the light-absorbing properties of individual rods and cones. A method whereby this can be done has been developed in recent years, known as microspectrophotometry. It is not an easy technique: a very small beam of light has to be focused on the receptor cells of isolated retinae, and difficulties arise from the small amounts of light that must be used in order to avoid excessive bleaching of the pigments. Fig. 17.21 shows some results obtained by Marks (1965) from individual goldfish cones; the absorption spectra fall into three groups, suggesting that there is one of three separate pigments in each cone.

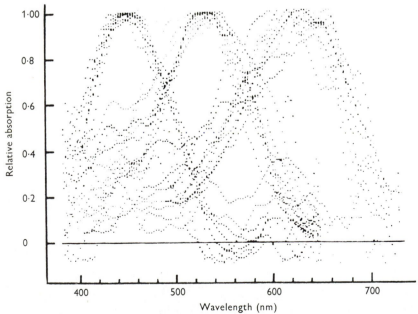

Fig. 17.21. Difference spectra of single goldfish cones, determined by microspectrophotometry. They clearly fall into three different groups, corresponding presumably to three different visual pigments. (From Marks, 1965.)

Microspectrophotometry has been applied to the cones of the parafoveal regions of human and macaque monkey retinae by Marks, Dobelle and MacNichol (1964). Their results indicate that there are probably three different types of cones, each containing a different visual pigment, with λ_{max} values at 445 nm (blue-sensitive), 535 nm (green-sensitive, probably equivalent to Rushton's chlorolabe) and 570 nm (red-sensitive). Similar results have been obtained by Brown and Wald (1964), who also measured the difference spectrum of single rods, and found it to be similar to that of rhodopsin in solution but with a slight shift towards the red (as was also found by Rushton, 1956, using retinal densitometry).

The organization of rhodopsin in rods

If the molecules of a light-absorbing substance are not arranged at random, but are oriented in one particular plane or one particular direction, then light waves vibrating in a particular plane are preferentially absorbed. Thus we can determine whether the visual pigment molecules in a photoreceptor cell are so organized by observing the absorption of light polarized in different directions. This was first done by Schmidt (1938), using the outer segments of frog rods. He found that when light was shone along the axis of the rods, they appeared to be red (indicating that light was absorbed) whatever the plane of polarization. However, with illumination from the side, the rods appeared to be red when the plane of polarization was parallel to the transverse plane of the outer segment, but colourless (indicating that no light was absorbed) when it was parallel to the longitudinal axis (fig. 17.22). These results have been confirmed and extended by Denton (1959) and by Wald,

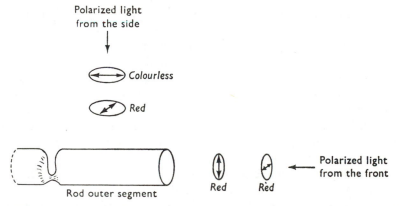

Fig. 17.22. The absorption of polarized light by an intact retina. No absorption occurs (so that the rods appear colourless, instead of red) when the transverse components of the light waves are parallel to the longitudinal axes of the rods.

Brown and Gibbons (1962). They imply that the rhodopsin molecules (in particular, the retinene moieties) are oriented in transverse planes across the outer segment, but that there is no preferential direction of orientation within these planes. The most obvious way in which this could be done is by incorporation of the rhodopsin molecules in or on the membranes of the sacs in the outer segment.

Such an arrangement must increase the efficiency of light absorption by the rod pigment, since light arrives 'end-on' at the rods in the intact eye; any molecules oriented so as to absorb light vibrating in a plane parallel to the longitudinal axis would be ineffective in vision. The absorption of unidirectional light by rhodopsin solutions (in which the molecules are, of course, oriented at random) is therefore less than that of rhodopsin in the rods, by a factor of 2/3.

THE ABSOLUTE THRESHOLD

What is the minimum amount of light that can be detected by the fully dark-adapted eye? Because of the quantal nature of light, this question can be rephrased as: what is the minimum number of quanta required for visual excitation? This problem has been investigated, in a most elegant set of experiments, by Hecht, Schlaer and Pirenne (1942), whose work we shall now examine.

The principle of this work was to ensure that conditions were optimal for the perception of light by human subjects, to measure the intensity of light at the visual threshold, and finally to calculate the number of quanta absorbed by the visual pigment in the retina under these conditions. In order to secure optimum conditions for perception, a number of precautions were taken. Subjects were fully dark-adapted by remaining in complete darkness for at least half an hour. Using the 'fixation and flash' technique, the test flash was positioned so that its image fell 20 degrees to the left (in the left eye) of the fovea, a position which is in the region of maximum sensitivity of the retina. The diameter of the test flash was 10 minutes, which is the angle at which the product of area and threshold intensity is least. The product of threshold intensity times length of flash is constant at times up to about 10 msec, but increases with longer flashes; hence flashes of 1 msec duration were used. The wavelength of the light used was 510 nm, which is near the peak of the scotopic sensitivity curve.

The arrangement of the apparatus is shown in fig. 17.23. The eye looks through the artificial pupil *P*, fixates the red light point *FP*, and sees the test field formed by the lens *FL* and the diaphragm *D*. The light for this field comes from the lamp *L* through the neutral filter *F* and the wedge *W* (which together control the intensity), and through the double

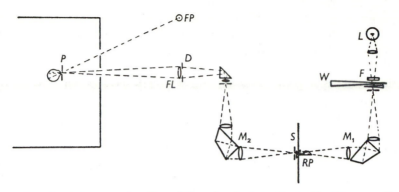

Fig. 17.23. Optical system for determining the absolute threshold of human vision. Explanation in text. (From Hecht, Schlaer and Pirenne, 1942.)

monochromator M_1M_2 (which selects the wavelength). The shutter S opens for 1 msec, and is controlled by the subject. Careful calibrations of the light intensity at P were made, so that the light energy incident on the cornea during a flash could be calculated for different neutral filters and different settings of the wedge W.

For each subject, a series of flashes of different intensities was presented many times, and the frequency of seeing the flash was determined for each intensity. Graphs of frequency of seeing against intensity were then drawn, and the threshold was regarded as that intensity at which the flash could be seen in 60 per cent of the trials. Seven subjects were tested, some of them more than once. Their thresholds, measured as the light energy incident on the cornea, were in the range 2.1 to 5.7×10^{-10} ergs.

As we have seen, the energy content, $h\nu$, of a quantum of light is dependent upon its wavelength. For light with a wavelength of 510 nm, $h\nu$ is 3.89×10^{-12} ergs. Hence the number of quanta incident on the cornea at the absolute threshold was in the range 54 to 148.

These values do not represent the number of quanta absorbed by the visual pigment, since some light is lost between the cornea and the retina, and not all of the light incident upon the retina is absorbed by it. Measurements by other workers had shown that about 4 per cent of the light incident upon the cornea is reflected from its surface, and that (at a wavelength of 510 nm) about 50 per cent of the light entering the eye is absorbed by the lens and ocular media before reaching the cornea. The main problem here, then, is the measurement of the proportion of the light incident upon the retina which is absorbed by the visual pigment. Hecht and his colleagues solved this problem in a rather ingenious manner. We have seen that the shape of the absorption spectrum of a pigment depends on its concentration (and therefore on the proportion of light absorbed), being broader at higher concentrations (see fig. 17.8b). It follows that the proportion of light absorbed by the rhodopsin in the retina can be obtained by comparing the shape of the scotopic luminosity curve with the shapes of the absorption spectra of different concentrations of rhodopsin in solution. The results suggested that between 5 and 20 per cent of the light incident on the retina is absorbed by the visual pigment; for the purposes of calculation, the 20 per cent value was used. (More recent measurements suggest that this figure was still too low; but we shall return to this point later.)

Putting these figures together, we find that, of the light incident at the cornea at a wavelength of 510 nm, 96 per cent enters the eye, 48 per cent reaches the retina, and about 9.6 per cent is absorbed by the visual pigment. Hence, at the threshold of vision, when 54 to 148 quanta are incident on the cornea, 5 to 14 quanta are absorbed by the visual pigment.

This figure is of such importance that it is most desirable to have an independent check on its accuracy. Such a check was obtained by Hecht and his colleagues from measurements of 'frequency-of-seeing' curves and consideration of the statistical properties of light flashes containing small numbers of quanta. Any flash of light of constant *average* quantal content (measured at the retina) will in fact produce fluctuating numbers of quanta; the actual number of quanta absorbed by the retina will also fluctuate, and the relative frequencies of the quantal contents of the flashes will be given by the terms of a Poisson distribution (compare (6.3)). Thus, if n is the number of quanta which it is necessary for the retina to absorb in order to be able to see a flash, and a is the average quantal content per flash in a series of flashes of 'constant' intensity, then the probability P_n that any one flash will yield exactly the necessary number of quanta is given by

$$P_n = e^{-a} \cdot \frac{a^n}{n!}. \tag{17.1}$$

From this equation, it is possible to calculate the probabilities of n or more quanta occurring for different values of a and n; the results of this calculation are shown in fig. 17.24.

The next step was to determine 'frequency-of-seeing' curves at flash intensities in the region of the absolute threshold. The method used was identical with that for determining the absolute threshold in the first instance. Three observers were used, and the results are shown in fig. 17.25. The abscissae in these curves represent the amounts of light incident upon the cornea, whereas that of the theoretical curves in fig. 17.24 is equivalent to the amount of light absorbed by the visual pigment. Hence the theoretical curves must be fitted to the experimental

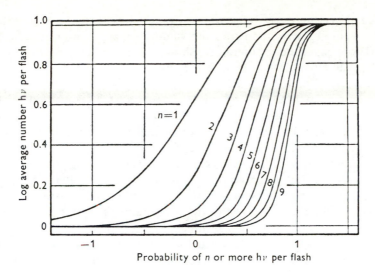

Fig. 17.24. Solutions of (17.1) for different values of *n*. (From Hecht, Schlaer and Pirenne, 1942.)

curves by moving them along the *x* axis. When this was done, it was clear that the value of *n* must be in the range 5 to 8. (This argument assumes that the variation in frequency-of-seeing is entirely due to the fluctuations in the stimulus; if there is also some fluctuation in the stimulus/response ratio, then the frequency-of-seeing curves would be

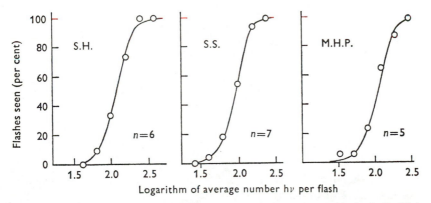

Fig. 17.25. The relation between average quantal content (at the cornea) of a flash of light and the frequency with which it was seen, for three different observers. The curves are drawn as in fig. 17.26 for *n* = 5, 6 and 7, but are moved along the abscissa in order to fit the points (this procedure is permissible, since the abscissa measures the number of quanta incident at the cornea, whereas *n* is the number absorbed by the retinal rods). (From Hecht, Schlaer and Pirenne, 1942.)

·stretched over a greater range of average quantal contents, so that their slopes would be less, and hence the values of n obtained would be too low. These experiments, therefore, show that the true value of n cannot be less than 5 to 8; it may be a little more.)

The general conclusion of this investigation is that, under optimal conditions, a minimum of about 6 quanta must be absorbed by the retina if a light source is to be detected. What does this mean in terms of the excitation of the retinal cells? Let us first consider the requirements for excitation of the rods. We can suppose either (i) that the absorption of one quantum is sufficient to excite a rod, or (ii) that two or more quanta need to be absorbed by a rod before it is excited. We could eliminate the second hypothesis if we could show that the probability of any of the illuminated rods absorbing the requisite number of quanta from a flash of light at the visual threshold is less than the probability of seeing the flash. Our problem can therefore be stated as follows: if n quanta are randomly distributed over N rods, what is the chance of one or more of these rods receiving two or more quanta? The mean number of quanta absorbed per rod is n/N. If the actual numbers of quanta absorbed by each rod are distributed according to a Poisson distribution, then the probability P_x that any particular rod will absorb x quanta is

$$P_x = e^{-n/N} \cdot \frac{(n/N)^x}{x!}.$$

Thus when $x = 0$,

$$P_0 = e^{-n/N}$$

and when $x = 1$,

$$P_1 = e^{-n/N} \cdot \frac{n}{N}.$$

Hence the probability that any particular rod will absorb either 1 or 0 quanta is

$$P_0 + P_1 = e^{-n/N}\left(1 + \frac{n}{N}\right).$$

Therefore the probability that *every* rod in the group of N rods will absorb either 1 or 0 quanta is

$$(P_0 + P_1)^N = e^{-n}\left(1 + \frac{n}{N}\right)^N.$$

Hence the probability, Q, that one or more rods will absorb more than one quantum is given by

$$Q = 1 - e^{-n}\left(1 + \frac{n}{N}\right)^{N}. \tag{17.2}$$

We have now to decide what values of n and N are to be inserted in (17.2) for the conditions obtaining at the absolute threshold. The power of the average human eye is about 60 dioptres, which is equivalent to a focal length of 16.67 mm. Thus the image of a 10 minute circular field is $(16.67 \times \sin 10')$ mm in diameter, i.e. 48.5 μm. Its area is therefore 1.84×10^{-3} mm^2. According to Østerberg (1935), there are about 150 000 rods/mm^2 in the region of the retina on which the experiments of Hecht, Schlaer and Pirenne were performed, so that a 10 minute field will contain about 273 rods. However, the optical image of a 10 minute field will be rather larger than this, due to diffraction effects at the pupil; Pirenne (1962) suggests that the actual number of rods in the illuminated field should be about 500. However, let us take a very conservative value of N, 250. Using a similarly conservative value for n, 10, (17.2) becomes

$$Q = 1 - e^{-10}(1.04)^{250}$$

$$= 0.178.$$

Thus, if the two-quanta-per-rod hypothesis were correct, we would expect only 17.8 per cent of the flashes to be seen when the flash intensity is at the visual threshold. But, in fact, 60 per cent of them are seen, since this is how the visual threshold has been defined. We must conclude that a visual sensation arises if a small number of rods absorb a single quantum each.

A complication

Hecht, Schlaer and Pirenne assumed that 20 per cent of the light incident on the retina is absorbed by the photopigment. More direct measurements have since been made which suggest that a much higher figure should be used. Thus Alpern and Pugh (1974), using retinal densitometry conclude that about 55 per cent of the light incident on the retina is absorbed, and Dobelle, Marks and MacNichol (1969) using microspectrophotometry conclude that the figure is at least 50 per cent. How does this affect Hecht, Schlaer and Pirenne's conclusions? Of the light incident at the cornea, about 24 per cent is absorbed by the visual pigment. So at the threshold of vision, when 54 to 148 quanta are incident on the cornea, 13 to 35 are absorbed by the visual pigment. This makes the conclusion that one quantum suffices to excite a rod less well founded. Thus, in (17.2), when n is 13, Q is 0.28, when n is 20, Q is 0.53 and when n is 25, Q is 0.69.

However, as Brindley (1970) points out, measurements on thresholds for larger fields are quite conclusive in this respect. Thus Stiles (1939), again using the fixation-and-flash method, measured the threshold for detection of a square of side 1.04° at 5° from the fovea, with a light of 510 nm exposed for 63 msec. His results indicate that such a stimulus had to contain 122 quanta (at the cornea) in order for 50 per cent of the flashes to be seen. So about 59 quanta will reach the retina and about 33 of them will be absorbed by the visual pigment. At 5° from the fovea there are about 90 000 rods per mm^2, and so the number of rods in the illuminated field is just over 8000. So, applying (17.2) with $n = 33$ and $N = 8000$, we get

$$Q = 1 - e^{-33}\left(1 + \frac{33}{8000}\right)^{8000}$$

$$= 0.066$$

Which is clearly very much less than 0.50, the probability of seeing the flash. Even with $n = 59$ (corresponding to absorption of all the incident quanta by the retina), $Q = 0.195$, which is still a satisfactory figure for the one-quantum-per-rod hypothesis.

The quantal laws of visual excitation

We have seen that there must be some considerable convergence of signals from the rods onto the ganglion cells. We may therefore regard each ganglion cell, with its associated bipolars, as a 'summation pool' which collects the signals from a large number of rods. The conclusions from the experiments on the absolute threshold may then be restated as what we may call 'the quantal laws of visual excitation': (i) the absorption of one quantum suffices to excite a rod, and (ii) a small number of rods need to be excited in order to excite one summation pool and so lead to a visual sensation.

THE ELECTRICAL ACTIVITY OF THE RETINA

It is easy to show that electrical activity arises in the retina in response to illumination. The precise nature of this activity is not so easy to elucidate; the recorded events depend, as we shall see, very much on the positioning of the recording electrodes.

The electroretinogram

The electroretinogram is a mass electrical response of the retina recorded with fairly large external electrodes, one placed on the cornea and the other at some indifferent point in the body or (when using

Fig. 17.26. The electroretinogram of a dark-adapted decerebrate cat, to show the *a*, *b*, *c* and *d*-waves. The eye was illuminated for the duration of the thick line below the record, on which time intervals of 0.5 sec are marked. (From Granit, 1933.)

excised eyes) behind the optic bulb. An enormous amount of work has been done on this response since its discovery in 1865 (see Granit, 1947, 1962), but we shall here just outline the main features of the phenomenon.

Fig. 17.26 shows an electroretinogram recorded by Granit (1933) from a decerebrate cat. The first part of the response is the small, cornea-negative *a*-wave. This followed by a rapid cornea-positive response, the *b*-wave, which is finally followed by a much slower response, the *c*-wave. On the cessation of illumination, there is a small cornea-positive deflection, the *d*-wave. In vertebrates other than mammals, the *d*-wave is usually much larger, and is comparable in size with the *b*-wave.

The components of the electroretinogram can also be detected when records of local activity in the retina are made with a glass micropipette electrode. Using this technique, Brown and Wiesel (1961) determined the amplitude of these various components at different levels in the retina, in an attempt to localize their origin. They found that the *a* and *c*-waves were associated with the receptor layer; there was some evidence that the *a*-wave arises from activity in the rods and cones, whereas the *c*-wave arises from the cells of the pigment epithelium. The *b*-wave was maximal when the electrode was in the inner nuclear region; it appears to be due to activity in the glial cells (Müller cells, fig. 17.2*u*) which occur at that level (Miller and Dowling, 1970).

The early receptor potential

The early stages of the response which constitutes the electroretinogram have recently been investigated by a number of workers; since these potentials arise in the receptor cells they can be described as receptor potentials. Using intense brief flashes as the light stimulus, Brown and Murakami (1964) found that the early stages of the locally recorded electroretinogram consisted of two parts (fig. 17.27), an initial *early receptor potential* with very brief latency, followed by a larger *late receptor potential*. The rising phase of the late receptor potential corresponds to the *a*-wave of the electroretinogram. Cone (1964) showed

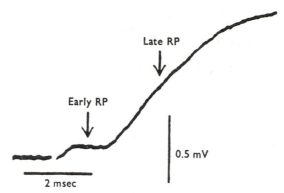

Fig. 17.27. The early and late receptor potentials, as recorded by a tungsten micro-electrode in the extrafoveal retina of a monkey. (From Brown and Murakami, 1964.)

that the early receptor potential in the rat could be seen in electro-retinograms recorded by conventional methods, that its action spectrum was similar to that of the visual pigment, and that its size was linearly proportional to the amount of pigment bleached by the flash. If the retina is cooled, it can be seen that the early receptor potential is biphasic, consisting of an initial cornea-positive wave, R_1, followed by a cornea-negative wave, R_2 (fig. 17.28).

Many different observations suggest that the early receptor potential is not brought about by changes in membrane conductance (see Pak, 1968; Arden, 1969). It is resistant to anoxia and to treatment with a wide variety of different chemical agents which affect most other bio-electric potentials, such as potassium chloride solutions, acids and alkalies, fixatives and so on. R_1 and R_2 have different sensitivities in this respect; R_1 is not much affected by cooling and can be recorded at temperatures down to $-35\,°C$, whereas R_2 is greatly diminished by cooling and is scarcely evident below $0\,°C$. R_2 is abolished by fixation with glutaraldehyde, whereas R_1 is not. R_2 is increased by sodium bisulphite and reduced by oxygen, whereas R_1 is unaffected by these agents.

It seems probable that the R_1 component of the early receptor potential is caused by charge displacement in the visual pigment molecules during the conformational changes occurring in the early reactions of the sequence rhodopsin to metarhodopsin I (or analogous sequences in other visual pigments). R_2 may be produced in the change from meta-rhodopsin I to metarhodopsin II (Cone, 1967) or may reflect rapid movements of membrane charges (Arden *et al.*, 1968). A finding which may be relevant here is that most of the early receptor potential is produced by the cones in frogs and primates (Goldstein, 1968, 1969;

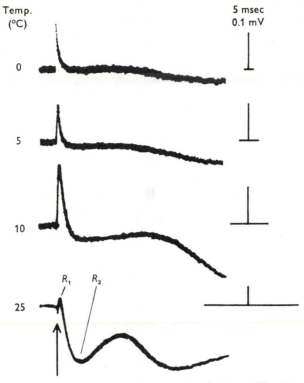

Fig. 17.28. Responses of isolated rat eyes to intense light flashes at different temperatures. R_1 and R_2 are the two phases of the early receptor potential; this is followed by a further cornea-negative (downward) deflection, the late receptor potential. Notice that the time and voltage scales differ for the different records. (From Arden, 1969.)

Goldstein and Berson, 1969). Goldstein suggests that this may be linked with the fact that rod visual pigments are in enclosed sacs which are not in contact with the cell membrane, whereas cone pigments are arranged on infoldings of the cell membrane.

Photoreceptor potentials

Rods and cones are rather small cells and it is technically not easy to make intracellular recordings from them. Tomita and his colleagues have been able to insert high resistance microelectrodes into single cones of the carp by jolting the retina upwards towards the electrode in short steps of about 1 μm (Tomita, 1965, 1970, 1972; Tomita *et al.*, 1967). The microelectrode will then record the membrane potential of the inner segment of the cone; outer segments are too small to give satisfactory records, but of course electrical changes in the outer segment cell membrane will be recorded, with some attenuation, in the

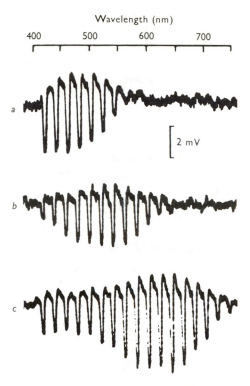

Fig. 17.29. Intracellular recordings from cones of the carp, showing the hyperpolarizations produced by light. In each case the cone was illuminated with a series of flashes of monochromatic light in steps of 20 nm across the spectrum. The scale at the top gives the light wavelength. These records show the three types of spectral sensitivity found, with the maximal responses in the blue (*a*), green (*b*) and red (*c*) regions of the spectrum. (From Tomita *et al.*, 1967.)

inner segment. They found that illumination of the cones produced a hyperpolarization, as is shown in fig. 17.29.

Investigations on other preparations (gecko and frog rods, mudpuppy and turtle cones) have produced similar results. These hyperpolarizations produced by vertebrate photoreceptors on illumination are in marked contrast to the response of most other sensory cells (including most invertebrate photoreceptors), in which stimulation causes a depolarization of the cell membrane. Further, it is found that during the hyperpolarization the membrane resistance increases (Toyoda, Nosaki and Tomita, 1969; Baylor and Fuortes 1970), suggesting that illumination reduces the conductance of the cell membrane to some ion or ions. Fig. 17.30 shows some of the evidence for this conclusion: the voltage change produced by brief current pulses increases on illumination.

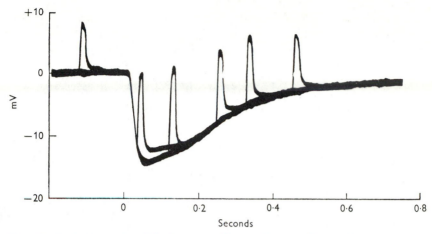

Fig. 17.30. Evidence that light increases the membrane resistance in turtle cones. Superimposed tracings of six responses to brief flashes of light at zero time. For each trace a brief pulse of depolarizing current, of the same strength but occurring at different times, was applied through the microelectrode. The voltage changes produced by this pulse are larger during the hyperpolarization following the flash of light. (From Baylor and Fuortes, 1970.)

Current flow in rods

The current flow associated with the late receptor potential and also in the dark has been investigated by Hagins and his colleagues (Penn and Hagins, 1969; Hagins, Penn and Yoshikami, 1970; Hagins, 1972). They used slices of rat retina which they were able to keep alive mounted on a microscope stage. In order to examine these slices without bleaching their rhodopsin, they were illuminated by infrared light and visualized by means of image converters. Microelectrodes could be accurately placed at different depths in the retina so that the extracellular potentials at these depths could be measured. They found that, in the dark, the potential difference between the measuring electrode and a reference electrode placed at the tips of the rod outer segments increased with increasing penetration into the rod layer.

If the resistance of the extracellular space were constant, then the longitudinal current flow would simply be the rate of change of voltage with distance (compare (4.2)) and therefore could be directly obtained from fig. 17.31. However it is not, and must therefore be measured. To do this, two microelectrodes with tips separated by a depth of 10 μm, were driven through the retina, and the potential (ΔV) between them recorded. At the same time current pulses were passed across the whole retina: the size of ensuing voltage pulses is proportional to the resistance (ΔR) between the two electrodes. The longitudinal flow between the

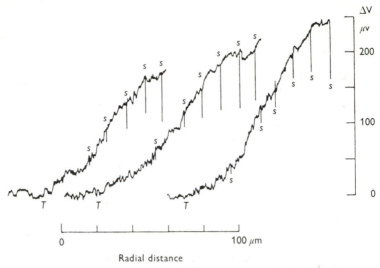

Fig. 17.31. Extracellular potentials recorded from rat retina in the dark. One electrode was positioned at the tips of the rod outer segments, the other had penetrated the retina from that side to a depth of about 100 μm and was being withdrawn while the potential between the two electrodes was recorded. Three separate records are shown; T marks the tips of the outer segments. Brief light flashes of constant intensity were delivered at s, producing the photovoltages shown as vertical lines. (From Hagins, Penn and Yoshikami, 1970.)

two electrodes (I) is then given by Ohm's law:

$$I = \frac{\Delta V}{\Delta R}.$$

The results (fig. 17.32) show that the longitudinal dark current increases steadily as the electrode traverses the layer of rod outer segments and then begins to fall as they enter the layer beneath them. This means that there is a steady inflow of current into the rod outer segments, supplied by a corresponding outflow from the inner segments and nuclear regions. In similar experiments on frog rods, which are rather larger than rat rods, Zuckerman (1973) found that there is also some inward current in the synaptic region, as is shown in fig. 17.33.

The size of this dark current is quite considerable; the maximum longitudinal current is just over 400 μA/cm^2 in the rat rods of fig. 17.32. If there are 10^7 rods/cm^2, this corresponds to 40 pA per rod, which means that 40 pC, equivalent to about 4×10^{-16} moles of univalent ions, flow into each outer segment every second. The outer segments are about 25 μm long by 1.7 μm in diameter, and so have a volume of about 56 μm^3. About 60 per cent of their content is aqueous, so if we assume that the cationic concentration in this phase is 140 mM,

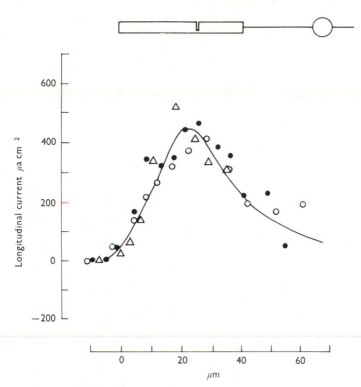

Fig. 17.32. The longitudinal (radial) distribution of the dark current in the rod retina. The three symbols show results from three separate measurements. The curve is calculated on the assumption that the whole of the rod outer segment acts as a sink for the current, whose source is the inner segment and nuclear region. (From Hagins, Penn and Yoshikami, 1970; redrawn.)

each outer segment contains about 4.7×10^{-15} moles of univalent ions. Clearly the action of the dark current is sufficient to replace the ions in the outer segment every 12 seconds!* It is also instructive to consider the longitudinal current inside the rod, which must be equal and opposite to the external current. At its maximum this will be 40 pA/rod, which gives a current density of about 18 pA/μm^2 for a rod 1.7 μm in diameter. In the ciliary segment, diameter about 0.3 μm, the current density must be a staggering 560 pA/μm^2. Compare these figures with the longitudinal current in a myelinated nerve fibre, which may reach about 2000 pA during an action potential (fig. 4.24): in an axon 10 μm in diameter the current density is then 25 pA/μm^2.

* This calculation was inspired by a similar one by Hagins, Penn and Yoshikami (1970) in which they calculate that the turnover time for the cations in the whole rod cell is about 47 seconds.

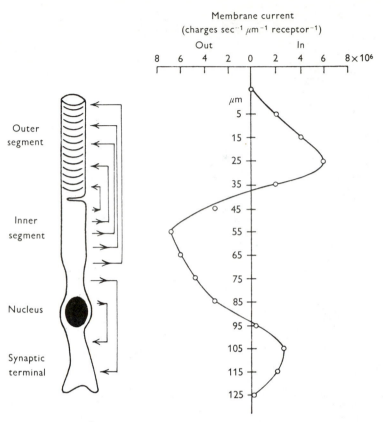

Fig. 17.33. Current flow in a frog rod in the dark. The membrane current is calculated from measurements of longitudinal current (similar in general to those of fig. 17.32) by application of (4.3). (From Zuckerman, 1973, with lines of current flow added.)

Hagins and his colleagues found that illumination of the retina causes a reduction in the potential recorded by the extracellular measuring electrode, by an amount which was roughly proportional to the size of the dark potential measured at that point (as shown at the points marked '*s*' in fig. 17.31). This decrease in potential indicates a reduction in the magnitude of the dark current. (This reduction in the dark current is sometimes known as the 'photocurrent'; but this is a concept which can cause confusion if not handled carefully.) The dark current is completely eliminated by a flash of light which allows 200 or more photons to be absorbed per rod, and is reduced to half by 30 to 50 photons per rod (Penn and Hagins, 1972). The relation between light absorption and response is linear at levels up to about 30 photons per rod, so we can see that absorption of one photon reduces the dark current by about 0.5 pA per rod. The response to a very brief flash reaches a peak after about 0.1 sec and decays to zero within a second, and hence the reduction in dark current corresponds to about 0.1 to 0.2 pC of electric charge or 1 to 2×10^{-18} moles of univalent ions. Since each mole contains 6×10^{23} ions, we can say that the absorption of one photon by a rod causes about a million ions not to flow through the rod membrane.

What is the mechanism of this million-fold amplification process, and what is the cause of the current flow through which it acts? Let us consider the nature of the current flow first. It had been known for some time that the ERG and the receptor potential with which it begins are dependent upon the presence of sodium ions (Furukawa and Hanawa, 1955; Sillman, Ito and Tomita, 1969), and more recently it has been shown that frog rods hyperpolarize and become unresponsive to light in sodium-free solution (Brown and Pinto, 1974). It seems reasonable to suggest, then, that the dark current is caused by an inflow of sodium ions into the outer segments, which is reduced by the action of light. This would account for the hyperpolarization and increase in resistance of rods and cones in the light.

Some nice evidence in favour of this idea has been provided by Korenbrot and Cone (1972) from measurements on the osmotic behaviour of rod outer segments. Suspension of these can be obtained by shaking the retina gently in Ringer solution, and their volume can easily be measured from photomicrographs. When the outer segments were put into a strong solution of potassium chloride (3.5 times the isotonic concentration) in the dark, their volume shrank to about 70 per cent of its initial value within about 3 seconds, and then remained constant. This suggests that there is a rapid exit of water from the outer segments in the first 3 seconds, after which they are in osmotic equilibrium; hence their plasma membranes must be impermeable to potassium ions. Illumination had no effect on this response. When they were put from Ringer into a similar strong solution of sodium chloride in the dark, however, their volume fell equally rapidly to 70 per cent and then began to rise again, reaching 80 per cent after 10 seconds. This suggests that after the initial rapid exit of water there is a slower entry of sodium ions which allows water to flow back into the outer segments. Illumination prevents the later increase in volume and therefore makes the plasma membrane impermeable to sodium ions. All these results are in delightful agreement with expectation.

If sodium ions are continually flowing into the outer segment in the dark, there must be some region in the rod cell where they are extruded. Ouabain, an inhibitor of the sodium pump in many cells, greatly reduces the dark current within a minute or so of its application to the retina

(Yoshikami and Hagins, 1973; Zuckerman, 1973), but Korenbrot and Cone found that it had no effect on the osmotic behaviour of isolated outer segments. So it looks as though there is a metabolically driven sodium extrusion pump in the plasma membrane of the inner segment (where there are many mitochondria) and perhaps in the nuclear region also. This pump serves to maintain the ionic gradient of sodium ions which provides the driving force for the flow of current into the outer segment through the open sodium channels in its plasma membrane.

Now light, it would seem, causes the closing of these outer segment sodium channels. Is it possible that each rhodopsin molecule is associated with a sodium channel which closes when that rhodopsin molecule absorbs a photon? There are a number of reasons why this cannot be so. Firstly, the response of the rod to light saturates, that is to say all the sodium channels are closed, when only 200 or so of its 30 million rhodopsin molecules have absorbed photons. Secondly the rhodopsin molecules are set in the disc membranes, not in the plasma membrane where the sodium channels are. And finally the change in current flow seems to be rather too large to be accommodated by a single channel. Suppose, for example, that each sodium channel carries the same current as one in a squid axon at about zero membrane potential. Current flow in the axon is then about 1 mA/cm^2 (fig. 5.12) or 0.02 pA/channel if there are 500 channels per μm^2. A change in flow in the rod of 0.5 pA, produced by absorption of one photon, would thus demand the closing of 25 channels.

This conclusion that the absorption of one photon leads to closure of a number of channels has led to the idea that there is some 'internal transmitter' which is released in the outer segment cytoplasm by the action of light on rhodopsin, and it is this transmitter which closes the sodium channels in the plasma membrane (see, for example, Fuortes and Hodgkin, 1964; Baylor and Fuortes, 1970; Hagins 1972; Cone 1973). Yoshikami and Hagins (1973) suggest that the internal transmitter is calcium. In support of this idea, they find that the dark current is reduced as the calcium ion concentration is raised from 10^{-5} to 2×10^{-2} M, (they suggest that high calcium ion concentrations raise the internal calcium level, so blocking the sodium channels), and that the rods become insensitive to light at very low external calcium ion concentrations ($< 10^{-8}$ M).

More direct evidence has since been provided by Brown, Coles and Pinto (1977). They injected calcium ions into the outer segments of toad rods and found that brief hyperpolarizations resulted, similar in time course to receptor potentials. When EGTA (p. 254), which chelates calcium ions, was injected, a depolarization ensued and the responses to light flashes were reduced.

Hagins (1972) suggests that calcium ions are released from a rod disc when one of its rhodopsin molecules absorbs a photon; perhaps each rhodopsin molecule is also a calcium channel. In cones, it is suggested, a similar photon-induced calcium leak occurs across the plasma membrane where the visual pigment molecules are located.

The ganglion cells

Hartline (1938) found that the ganglion cells of the frog could be divided into three types according to their response to illumination. 'On' units responded to the onset of illumination, 'off' units responded to the cessation of illumination, and 'on–off' units responded to both onset and cessation.

Studies on the structure of the retina show that, as we have seen, a large number of photoreceptor cells may converge, via the bipolars, on any one ganglion cell. Thus a single ganglion cell can respond to illumination anywhere within a certain area of the retina: this area is known as the *receptive field* of the ganglion cell. The receptive fields of cat retinal ganglion cells were investigated by Kuffler (1953), using very small spots of light as stimuli. He found that many cells responded with 'on' discharges to illumination of the centre of the field, but with 'off' discharges when the periphery was illuminated, and with 'on–off' discharges in an intermediate region (fig. 17.34); this type of cell is called an 'on-centre' unit. Other cells showed the reverse type of organization ('off-centre' units), giving 'off' responses to illumination at the centre of the field and 'on' responses to illumination of its periphery. Since the receptive fields of the ganglion cells show considerable overlap, a spot of light which produces an 'on' response in a ganglion cell immediately under it may produce 'off' responses in ganglion cells some distance away.

The organization of the receptive fields differs according to the state of adaptation of the eye. Kuffler suggested that they increase in size during dark adaptation, but, in a later investigation (Barlow, Fitzhugh and Kuffler, 1957), it was found that the inhibitory effects in the periphery of the field disappear during dark adaptation.

The size of a receptive field can be roughly estimated by measuring the threshold of a ganglion cell with different areas of illumination. The threshold falls with increasing area as more receptors within the field become excited, but when the area covers the whole of the receptive field further increase cannot lead to further lowering of the threshold. Using this technique, Wiesel (1960) found that the receptive fields in a cat's retina range from 0.125 mm (0.5 degrees) to 2 mm (8 degrees) in diameter. The smaller fields were found mainly in the central region of the retina, the larger ones in the periphery. Similar results were

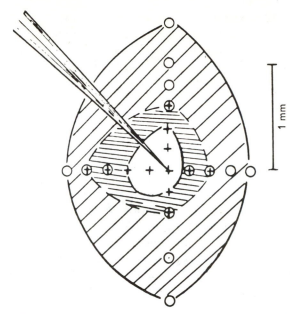

Fig. 17.34. The distribution of discharge patterns within the receptive field of a ganglion cell in a cat retina. The cell is located at the tip of the electrode. The field was explored with a spot of light 0.2 mm in diameter. 'On' responses (crosses and unshaded zone) occurred in the centre of the field, 'off' responses (circles and diagonal shading) in its periphery. 'On–off' responses (horizontal shading) occurred in an intermediate zone. (From Kuffler, 1953.)

obtained by Hubel and Wiesel (1960) in the spider monkey, the smallest field detected here being only 4 minutes in diameter.

Some ganglion cells have more specialized response patterns than those so far described. Barlow, Hill and Levick (1964), investigating the receptive fields of rabbit ganglion cells, discovered that many cells were sensitive to the direction of movement of an object across them. These cells gave a much greater discharge of impulses for motion in one direction than they did for motion in the reverse direction, whatever the nature of the moving stimulus.

In frogs the situation is even more complicated, with ganglion cells acting as 'edge-detectors', 'moving-edge-detectors' and so on (Maturana, Lettvin, McCulloch and Pitts, 1960). Clearly there is quite a variety of types of receptive field in the ganglion cells of different vertebrates (see Levick, 1972).

Is it possible for us to account for this variety? Dowling and Werblin (1969) suggest that the more complicated receptive fields are found where amacrine cells are interposed in the pathway between bipolars and ganglion cells. Perhaps such complexity is not necessary when the visual information from the retina is processed by a well-developed visual cortex such as occurs in cats and primates, which have relatively simple ganglion cell receptive fields.

From photoreceptor to ganglion cell

The rods and cones are connected to the ganglion cells via the bipolar cells (and in some cases via amacrine cells as well), with lateral connections via horizontal cells and amacrine cells (figs. 17.2, 17.3). The ganglion cells send their axons into the brain and their activity is readily measurable as a series of action potentials.

Responses of *bipolar cells* have been investigated by Werblin and Dowling (1969) in *Necturus* and by Kaneko (1970) in the goldfish. In studies of this type it is necessary to mark the cell from which recordings have been made by afterwards injecting a dye (Niagara Sky Blue or, better, Procion Yellow) by electrophoresis from the electrode. The retina is then fixed, sectioned and examined histologically so as to determine which sort of retinal cell had been impaled and recorded from. The results showed that bipolar cells respond to light by means of graded hyperpolarizations or depolarizations; they do not produce action potentials. Further, they show a feature known as centre-surround antagonism: stimulation of the centre of their receptive fields produces one type of response whereas stimulation of a peripheral area surrounding this produces its opposite. In fig. 17.35, for example,

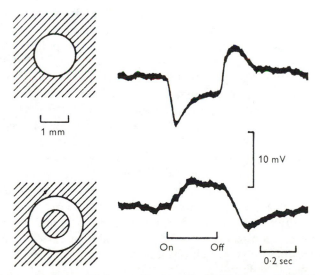

Fig. 17.35. Intracellular records from a bipolar cell which is hyperpolarized by a spot of white light (upper trace) and depolarized by an annulus (lower trace). (From Kaneko, 1970.)

stimulation of the centre of its receptive field by a spot of light produces a hyperpolarization of the bipolar cell, whereas stimulation of the periphery by an annulus produces a depolarization. In about half of the cells the responses were the opposite of this, in which illumination of the centre caused depolarization and illumination of the periphery caused hyperpolarization.

The size of the central area corresponds approximately to that of the receptive fields of the bipolar cells, and so it would seem that these central responses are caused by activity in photoreceptors to which the bipolars are directly connected. The dimensions of the surround fit reasonably well with the dendritic fields of horizontal cells, and it therefore seems likely that photoreceptors in the peripheral field of a bipolar cell produce their action on it via horizontal cells.

The *horizontal cells* themselves respond to illumination with graded hyperpolarizing potentials. Their receptive fields are fairly extensive and do not show centre-surround antagonism. Their responses are relatively easily recorded and were for a long time named '*S*-potentials' (although it is possible that the full range of *S*-potentials, described for example, by MacNichol and Svaetichin, 1958, includes potentials from retinal glial cells as well).

Now let us consider the means whereby these responses in horizontal and bipolar cells are produced by the *photoreceptors*. Light, we have seen, causes hyperpolarization of the rods and cones, and it is pertinent firstly to enquire whether this hyperpolarization is an essential element in the subsequent transmission of information. Baylor and Fettiplace (1976) have performed some useful experiments on turtle retinae which show quite clearly that it is. They found that hyperpolarization of the cone by electric current causes responses in the ganglion cell similar to those produced by light, and that depolarization of the cone can prevent the ganglion cell response to light from occurring.

The synaptic terminals of the rods and cones contain synaptic vesicles and are separated by the usual synaptic cleft from the bipolar and horizontal cells, so it seems very likely that these are chemically-transmitting synapses. There are two possible ways in which they might act: either (i) the terminals are inactive in the dark and respond to hyperpolarization (produced by light) by releasing a transmitter or (ii) they are continually releasing transmitter in the dark and respond to hyperpolarization by ceasing to do so. Evidence for the second possibility has been provided, among others, by Dowling and Ripps (1973): they found that magnesium ions, which inhibit transmission at most chemical synapses, cause hyperpolarization of skate horizontal cells in the dark and abolish their responses to light. So it looks as though the transmitter substance causes depolarization of the horizontal cells and some of the bipolars, and hyperpolarization of other bipolars.

The *amacrine cells* show transient depolarizing responses, including what are apparently all-or-nothing action potentials, at the onset or cessation of light. Ganglion cell responses are also depolarizing, with all-or-nothing action potentials, as we have seen. They may or may not be more responsive to changes in light intensity than to steady levels, perhaps depending on how far they are influenced by amacrine cells rather than by bipolars.

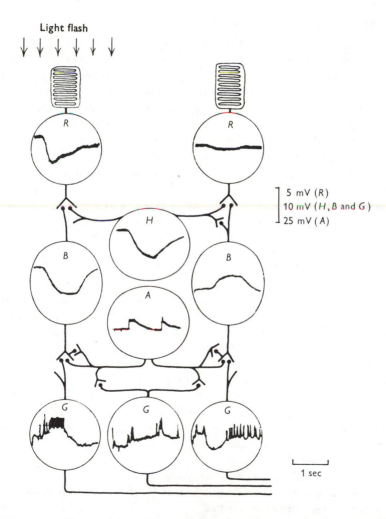

Fig. 17.36. Diagram to show the intracellular responses of the various retinal cells and their connections with each other. The receptor on the left is illuminated with a brief flash of light imposed on a dim background which illuminates both receptors. *R*, receptors; *H*, horizontal cell; *B*, bipolar cells; *A*, amacrine cells; *G*, ganglion cells. The responses were actually obtained from the retina of *Necturus*. (From Dowling, 1970.)

Dowling (1970) has neatly summarized the electrical activity of the various retinal cells in the diagram shown here as fig. 17.36. Three important generalizations emerge: (1) light produces hyperpolarization in the initial stages of the sequence (i.e. in receptors, horizontals and many of the bipolars), but depolarization in the later stages, (2) lateral connections occur at two levels, via the horizontals and the amacrines, and (3) only the amacrines and the ganglion cells produce action potentials.

Receptor coupling

So far we have assumed that single rods and cones are independent units, each one only responding when light is absorbed in its own outer segment. However it has recently become clear that the receptors may be to some extent electrically coupled to each other, so that a change in potential in one cell can spread into neighbouring cells. Baylor, Fuortes and O'Bryan (1971) measured the membrane potentials of turtle cones illuminated with small or large spots of light of the same intensity. The small (4μm radius) spot would be sufficient in size to illuminate the whole of the cone, the large one ($70\ \mu$m radius would also illuminate many neighbouring cones. They found that the resulting hyperpolarizations were larger with the large spots than with the small ones. A single cone would produce detectable hyperpolarizations in response to illumination of other cones up to about $40\ \mu$m away. They also passed current through one cone and recorded potential changes in a neighbouring one.

Further evidence has been provided by some very interesting experiments by Fain (1975; see also Fain, Gold and Dowling, 1976) on the responses of toad rods to diffuse flashes of dim light. He found that the response of an individual rod to a series of flashes was much less variable than would be predicted by the Poisson distribution. For example in one series, when the flashes were of an intensity to bleach an average of 1.4 rhodopsin molecules per receptor, the Poisson distribution (eq. 6.3) predicts that about 25 per cent of the flashes would have bleached no rhodopsin molecules in that rod, 35 per cent would have bleached one molecule, 24 per cent two and 16 per cent three or more. And yet the hyperpolarizations recorded from the rod all fell in the narrow range of 440 to 660 μV. Fain concluded that a rod is able to respond even when no rhodopsin molecules are bleached in its own outer segment, because it receives signals from other rods in which rhodopsin molecules are bleached.

The structural basis of coupling lies in the existence of gap junctions which exist between the terminals of neighbouring receptors. The functional reasons for this phenomenon of receptor coupling are as yet not at all clear. Perhaps it serves to increase the sensitivity of the system, but it would seem that this must be at the expense of visual acuity.

DARK ADAPTATION

Bright lights cause (i) bleaching of the visual pigments, and (ii) an increase of threshold when subsequently tested in the dark. The threshold then falls progressively if the subject remains in the dark (dark adaptation) and, at the same time, visual pigment is regenerated. Thus it might seem, at first sight, that the photosensitivity of the eye should be directly and linearly proportional to the quantity of visual pigment in its photoreceptors. But it has become clear, in recent years, that this is not so. Dowling (1960) measured the time course of dark adaptation in rats, using the electroretinogram as a measure of visual excitation, and also extracted the rhodopsin from the eyes at different times during dark

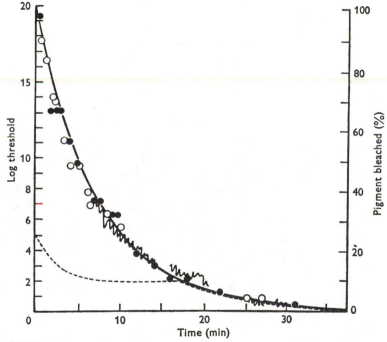

Fig. 17.37. Dark adaptation and the regeneration of rhodopsin. The open circles show the time course of rhodopsin regeneration in the normal eye following a full bleach, as measured by retinal densitometry. The scale on the right shows the percentage of rhodopsin in the bleached condition. Black circles show the same for a rod monochromat. The lines show the threshold ($\log_{10}$ scale, on the left) during dark adaptation: irregular line, traced by a rod monochromat after full bleaching exposure; dotted line, cone and rod branches in the normal eye. The continuous curve is an exponential with a time constant of 7.5 min. (From Rushton, 1965, by permission of the Royal Society.)

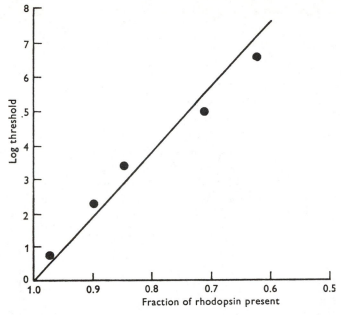

Fig. 17.38. The relation between log threshold and the fraction of rhodopsin present in a rod monochromat. (From Rushton, 1961.)

adaptation. He found that the *logarithm* of the electroretinogram threshold fell in proportion to the rise in rhodopsin concentration.

We have seen that a similar logarithmic relation between threshold and rhodopsin concentration is indicated by the experiments of Campbell and Rushton, using retinal densitometry on the human eye (fig. 17.16). More precise evidence has since been obtained by Rushton (1961) from experiments on the dark adaptation of a subject deficient in cone vision. Rushton's results are shown in fig. 17.37, from which it can be seen that the rod threshold fell to less than a millionth of its initial value while the percentage of unbleached pigment rose from 60 to 98 per cent. Fig. 17.38 shows the relation between log threshold and the rhodopsin concentration.

COLOUR VISION

By 'colour vision', we mean the ability to distinguish lights of different wavelengths. A glance at a paint specification chart will immediately show that we are able to recognize an enormous variety of different colours. Although the spectrum is conventionally divided into six or seven different colours (table 17.1), lights of wavelengths as little as 1 or

2 nm apart can be clearly distinguished from each other. What is the physiological basis of this ability?

Colour vision is trichromatic. The meaning of this statement, and the main evidence for it, has been explained by Brindley (1970) as follows: 'Given any four lights, whether spectroscopically pure or not, it is always possible to place two of them in one half of a photometric field and two in the other half, or else three in one half and one in the other, and by adjusting the intensities of three of the four lights to make the two halves of the field appear indistinguishable to the eye. This property of human colour discrimination, whereby the adjustment of three independent continuous controls makes possible an exact match, though two are generally insufficient, is known as trichromacy.'

Trichromacy was first hinted at by Newton, and became generally accepted during the eighteenth centry. But at this time it was thought to be a property of light, and it needed the genius of Thomas Young (1802) to suggest that trichromacy arose from the properties of the eye; that, to put it in modern terms, colour vision is mediated by three sensory channels which have different spectral sensitivities. Young's suggestion was adopted by Helmholtz (1866) in his great work on physiological optics, so that the idea that colour vision is brought about by three mechanisms with differing spectral sensitivities is frequently known as 'the Young–Helmholtz theory'.

The study of colour vision has given rise to a wide variety of theories as to its mechanism, some of them being more creditable to their authors' powers of imagination than to their critical faculties. There is not the space to examine these theories here. Suffice it to give a particular modern formulation of the Young–Helmholtz theory: that colour vision is mediated by three types of cones, each containing a different visual pigment. As we have seen, the results of recent experiments with microspectrophotometry are exactly in accordance with this view. It is not always, in Science, that the paths of understanding run so straight.

References

ABBOTT, B. C. and RITCHIE, J. M. (1951). Early tension relaxation during a muscle twitch. *J. Physiol., Lond.* **113**, 330–5.

ABBOTT, B. C. and WILKIE, D. R. (1953). The relation between velocity of shortening and the tension-length curve of skeletal muscle. *J. Physiol., Lond.* **120**, 214–23.

ADRIAN, E. D. (1926). The impulses produced by sensory nerve endings. I. *J. Physiol., Lond.* **61**, 47–72.

ADRIAN, E. D. and ZOTTERMAN, Y. (1926). The impulses produced by sensory nerve endings. II. The response of a single end-organ. *J. Physiol., Lond.* **61**, 151–71.

ADRIAN, R. H., CHANDLER, W. K. and RABOWSKI, R. F. (1976). Charge movement and mechanical repriming in skeletal muscle. *J. Physiol., Lond.* **254**, 361–88.

AHLQUIST, R. P. (1948). A study of the adrenotropic receptors. *Am. J. Physiol.* **153**, 586–600.

AIDLEY, D. J. (1965a). The effect of calcium ions on potassium contracture in a locust leg muscle. *J. Physiol., Lond.* **177**, 94–102.

AIDLEY, D. J. (1965b). Transient changes in isotonic shortening velocity of frog rectus abdominis muscles in potassium contracture. *Proc. R. Soc.* B **163**, 214–23.

AIDLEY, D. J. (1967). The excitation of insect skeletal muscles. *Adv. Insect Physiol.* **4**, 1–31.

AIDLEY, D. J. (1969). Sound production in a Brazilian cicada. *J. exp. Biol.* **51**, 325–37.

AIDLEY, D. J. and WHITE, D. C. S. (1969). Mechanical properties of glycerinated fibres from the tymbal muscles of a Brazilian cicada. *J. Physiol., Lond.* **205**, 179–92.

ALPERN, M. and PUGH, E. N., JNR. (1974). The density and photosensitivity of human rhodopsin in the living retina. *J. Physiol., Lond.* **237**, 341–70.

ALTAMARINO, M., COATES, C. W. and GRUNDFEST, H. (1955). Mechanisms of direct and neural excitability in electroplaques of electric eel. *J. gen. Physiol.* **38**, 319–60.

ANDERSON, C. R. and STEVENS, C. F. (1973). Voltage clamp analysis of acetylcholine produced end-plate current fluctuations at frog neuromuscular junction. *J. Physiol., Lond.* **235**, 655–91.

APRISON, M. H. and WERMAN, R. (1965). The distribution of glycine in cat spinal cord and roots. *Life Sci.* **4**, 2075–83.

ARAKI, T., ECCLES, J. C. and ITO, M. (1960). Correlation of the inhibitory postsynaptic potential of motoneurones with the latency and time course of inhibition of monosynaptic reflexes. *J. Physiol., Lond.* **154**, 354–77.

ARAKI, T. and TERZUOLO, C. A. (1962). Membrane currents in spinal motoneurones associated with the action potential and synaptic activity. *J. Neurophysiol.* **25**, 772–89.

ARDEN, G. B. (1969). The excitation of photoreceptors. *Progr. Biophys.* **19**, 371–421.

ARDEN, G. B., BRIDGES, C. D. B., IKEDA, H. and SIEGEL, I. M. (1968). Mode of generation of the early receptor potential. *Vision Res.* **8**, 3–24.

ARMETT, C. J., GRAY, J. A. B., HUNSPERGER, R. and LAL, S. (1962). The transmission of information in primary receptor neurones and second order neurones of a phasic system. *J. Physiol., Lond.* **164**, 395–421.

ARMSTRONG, C., HUXLEY, A. F. and JULIAN, F. J. (1966). Oscillatory responses in frog skeletal muscle fibres. *J. Physiol., Lond.* **186**, 26–27P.

ARMSTRONG, C. M. (1974). Ionic pores, gates, and gating currents. *Quart. Rev. Biophys.* **7**, 179–210.

ARMSTRONG, C. M. and BEZANILLA, F. M. (1973). Current related to the movement of the gating particles of the sodium channels. *Nature, Lond.* **242**, 459–61.

ARMSTRONG, C. M. and BEZANILLA, F. M. (1974). Charge movement associated with the opening and closing of the activation gates of the Na channels. *J. gen. Physiol.* **63**, 675–89.

ARMSTRONG, C. M., BEZANILLA, F. M. and ROJAS, E. (1973). Destruction of sodium conductance inactivation in squid axon perfused with pronase. *J. gen. Physiol.* **62**, 375–91.

ASHHURST, D. E. (1967). The fibrillar flight muscles of giant water bugs. An electron microscope study. *J. Cell Sci.* **2**, 435–44.

ASHLEY, C. C. and RIDGEWAY, E. B. (1968). Simultaneous recording of membrane potential, calcium transient and tension in single muscle fibres. *Nature, Lond.* **219**, 1168–9.

ASHLEY, C. C. and RIDGEWAY, E. B. (1970). On the relationships between membrane potential, calcium transient and tension in single barnacle muscle fibres. *J. Physiol., Lond.* **209**, 105–30.

ATWATER, I., BEZANILLA, F. and ROJAS, E. (1969). Sodium influxes in internally perfused squid giant axon during voltage clamp. *J. Physiol., Lond.* **201**, 657–64.

AUBERT, X. (1956). La relation entre la force et la vitesse d'allongement et de raccourcissement du muscle strié. *Archs int. Physiol. Biochim.* **64**, 121.

AXELSSON, J. and THESLEFF, S. (1959). A study of supersensitivity in denervated mammalian skeletal muscle. *J. Physiol., Lond.* **147**, 178–93.

BAKER, P. F. (1976). The regulation of intracellular calcium. *Symp. Soc. exp. Biol.* **30**, 67–88.

BAKER, P. F. and CRAWFORD, A. C. (1972). Mobility and transport of magnesium in squid giant axons. *J. Physiol., Lond.* **227**, 855–74.

BAKER, P. F., HODGKIN, A. L. and SHAW, T. I. (1962a). Replacement of the axoplasm of giant nerve fibres with artificial solutions. *J. Physiol., Lond.* **164**, 330–54.

BAKER, P. F., HODGKIN, A. L. and SHAW, T. I. (1962b). The effects of changes in internal ionic concentrations on the electrical properties of perfused giant axons. *J. Physiol., Lond.* **164**, 355–74.

BAKER, P. F., HODGKIN, A. L. and RIDGEWAY, E. B. (1971). Depolarization and calcium entry in squid giant axons. *J. Physiol., Lond.* **218**, 709–55.

BAKER, P. F. and WILLIS, J. S. (1972). Binding of the cardiac glycoside ouabain to intact cells. *J. Physiol., Lond.* **224**, 441–62.

BARKER, D. (1962). The structure and distribution of muscle receptors. In *Symposium on Muscle Receptors* (ed. D. Barker). Hong Kong: University Press.

BARKER, D. (1974). The morphology of muscle receptors. In *Handbook of Sensory Physiology* (ed. H. Autrum *et al.*) **3**(2), 1–190. Berlin: Springer.

BARKER, D., EMONET-DÉNAND, F., HARKER, D. W., JAMI, L. and LAPORTE, Y. (1976). Distribution of fusimotor axons to intrafusal muscle fibres in cat tenuissimus spindles as determined by the glycogen-depletion method. *J. Physiol., Lond.* **261**, 49–69.

BARLOW, H. B., FITZHUGH, R. and KUFFLER, S. W. (1957). Change of organization in the receptive fields of the cat's retina during dark adaptation. *J. Physiol., Lond.* **137**, 338–54.

BARLOW, H. B., HILL, R. M. and LEVICK, W. R. (1964). Retinal ganglion cells responding selectively to direction and speed of image motion in the rabbit. *J. Physiol., Lond.* **173**, 377–407.

BARNARD, E. A., WIECKOWSKI, J. and CHIU, T. H. (1971). Cholinergic receptor molecules and cholinesterase molecules at mouse skeletal muscle junction. *Nature, Lond.* **234**, 207–09.

BARRON, D. H. and MATTHEWS, B. H. C. (1938). The interpretation of potential changes in the spinal cord. *J. Physiol., Lond.* **92**, 276–321.

BAYLOR, D. A. and FETTIPLACE, R. (1976). Transmission of signals from photoreceptors to ganglion cells in the eye of the turtle. *Cold Spring Harb. Symp. quant. Biol.* **40**, 529–36.

BAYLOR, D. A. and FUORTES, M. G. F. (1970). Electrical responses of single cones in the retina of the turtle. *J. Physiol., Lond.* **207**, 77–92.

BAYLOR, D. A., FUORTES, M. G. F. and O'BRYAN, P. M. (1971). Receptive fields of cones in the retina of the turtle. *J. Physiol., Lond.* **214**, 265–94.

BEAM, K. G. and GREENGARD, P. (1976). Cyclic nucleotides, protein phosphorylation and synaptic function. *Cold Spring Harb. Symp. quant. Biol.* **40**, 157–68.

BEELER, G. W. and REUTER, H. (1970). Membrane calcium current in ventricular myocardial fibres. *J. Physiol., Lond.* **207**, 191–209.

BENDALL, J. R. (1953). Further observations on a factor (the 'Marsh' factor) effecting relaxation of ATP-shortened muscle fibre models, and the effect of Ca and Mg ions on it. *J. Physiol., Lond.* **121**, 232–54.

BENNETT, M. R. (1972). *Autonomic Neuromuscular Transmission.* Cambridge University Press.

BENNETT, M. V. L. (1960). Electrical connections between supermedullary neurons. *Fedn Proc. Fedn Am. physiol. Socs* **19**, 282.

BENNETT, M. V. L. (1966) Physiology of electrotonic junctions. *Ann. N.Y. Acad. Sci.* **137**, 509–39.

BENNETT, M. V. L. (1971). Electric Organs. In *Fish Physiology*, (ed. W. S. Hoar and D. J. Randall), **5**, pp. 347–492. New York: Academic Press.

BENNETT, M. V. L. (1972). A comparison of electrically and chemically mediated transmission. In *Structure and Function of Synapses* (ed. G. D. Pappas and D. P. Purpura), pp. 257–302. New York: Raven Press.

BENNETT, M. V. L., ALJURE, E., NAKAJIMA, Y. and PAPPAS, G. D. (1963). Electrotonic junctions between teleost spinal neurones: electrophysiology and ultrastructure. *Science* **141**, 262–4.

BENNETT, M. V. L. and GRUNDFEST, H. (1959). Electrophysiology of electric organ in *Gymnotus carapo*. *J. gen. Physiol.* **42**, 1067–104.

BENNETT, M. V. L. and GRUNDFEST, H. (1961a). The electrophysiology of electric organs of marine electric fishes. II. The electroplaques of main and accessory organs of *Narcine brasiliensis*. *J. gen. Physiol.* **44**, 805–18.

BENNETT, M. V. L. and GRUNDFEST, H. (1961b). The electrophysiology of electric organs of marine electric fishes. III. The electroplaques of the stargazer. *Astroscopus y-graceum*. *J. gen. Physiol.* **44**, 819–42.

BENNETT, M. V. L. and GRUNDFEST, H. (1961c). Studies on the morphology and electrophysiology of electric organs. III. Electrophysiology of electric organs in mormyrids. In *Bioelectrogenesis* (ed. C. Chagas and A. Paes de Carvalho), pp. 113–14. Amsterdam: Elsevier.

BENNETT, M. V. L., WURZEL, M. and GRUNDFEST, H. (1961). The electrophysiology of electric organs of marine electric fishes. I. Properties of electroplaques of *Torpedo nobiliana. J. gen. Physiol.* **44**, 757–804.

BERL, S., PUSZKIN, S. and NICKLAS, W. J. (1973). Actomyosin-like protein in brain. *Science* **179**, 441–6.

BERNARD, C. (1857). *Lecons sur les Effects des Substances Toxiques et Medicamenteuses.* Paris.

BERNSTEIN, J. (1902). Untersuchungen zur Thermodynamik der bioelektrischen Ströme. *Pflügers Arch. ges. Physiol.* **92**, 521.

BESSOU, P. and LAPORTE, Y. (1962). Responses from primary and secondary endings of the same neuromuscular spindle of the tenuissimus muscle of the cat. In *Symposium on Muscle Receptors* (ed. D. Barker). Hong Kong: University Press.

BESSOU, P. and PAGÈS, B. (1975). Cinematographic analysis of contractile events produced in intrafusal muscle fibres by stimulation of static and dynamic fusimotor axons. *J. Physiol., Lond.* **252**, 397–427.

BEZANILLA, F., ROJAS, E. and TAYLOR, R. E. (1970). Sodium and potassium conductance changes during a membrane action potential. *J. Physiol., Lond.* **211**, 729–51.

BIANCHI, C. P. (1961). Calcium movements in striated muscle during contraction and contracture. In *Biophysics of Physiological and Pharmacological Actions* (ed. A. M. Shanes). Washington: American Association for the Advancement of Science.

BIANCHI, C. P. and SHANES, A. M. (1959). Calcium influx in skeletal muscle at rest, during activity, and during potassium contracture. *J. gen. Physiol.* **42**, 803–15.

BINSTOCK, L. and LECAR, H. (1969). Ammonium ion currents in the squid giant axon. *J. gen. Physiol.* **53**, 342–61.

BIRKS, R., HUXLEY, H. E. and KATZ, B. (1960). The fine structure of the neuromuscular junction of the frog. *J. Physiol., Lond.* **150**, 134–44.

BIRKS, R. and MACINTOSH, F. C. (1961). Acetylcholine metabolism of a sympathetic ganglion. *Can. J. Biochem. Physiol.* **39**, 787–827.

BOECKH, J., KAISSLING, K.-E. and SCHNEIDER, D. (1965). Insect olfactory receptors. *Cold Spring Harb. Symp. quant. Biol.* **30**, 263–80.

BOISTEL, J. and FATT, P. (1958). Membrane permeability change during transmitter action in crustacean muscle. *J. Physiol., Lond.* **144**, 176–91.

BONE, Q. (1964). Patterns of muscular innervation in the lower chordates. *Int. Rev. Neurobiol.* **6**, 99–147.

BOWNDS, D. (1967). Site of attachment of retinal in rhodopsin. *Nature, Lond.* **216**, 1178–81.

BOYD, I. A. (1962). The structure and innervation of the nuclear bag muscle fibre system and nuclear chain muscle fibre system in mammalian muscle spindles. *Phil. Trans. R. Soc.* B **245**, 81–136.

BOYD, I. A., GLADDEN, M. H., MCWILLIAM, P. N. and WARD, J. (1977). Control of dynamic and static nuclear bag fibres and nuclear chain fibres by gamma and beta axons in isolated cat muscle spindles. *J. Physiol., Lond.* **265**, 133–62.

BOYLE, P. J. and CONWAY, E. J. (1941). Potassium accumulation in muscle and associated changes. *J. Physiol., Lond.* **100**, 1–63.

BRADLEY, K., EASTON, D. M. and ECCLES, J. C. (1953). An investigation of primary or direct inhibition. *J. Physiol., Lond.* **122**, 474–88.

BRADLEY, P. B. and WOLSTENCROFT, J. H. (1965). Actions of drugs on single neurones in the brain-stem. *Br. med. Bull.* **21**, 15–18.

BRETSCHER, M. S. (1973). Membrane structure: some general principles. *Science* **181**, 622–9.

BRIDGES, C. D. B. (1967). Biochemistry of the visual process. In *Comprehensive Biochemistry* (ed. M. Florkin and E. H. Stotz), **27**, 31–78. Amsterdam: Elsevier.

BRINDLEY, G. S. (1970). *Physiology of the Retina and the Visual Pathway* 2nd edn. London: Arnold.

BROCK, L. G., COOMBS, J. S. and ECCLES, J. C. (1952). The recording of potentials from motoneurones with an intracellular electrode. *J. Physiol., Lond.* **117**, 431–60.

BROCK, L. G., ECCLES, R. M. and KEYNES, R. D. (1953). The discharge of individual electroplates in *Raia clavata*. *J. Physiol., Lond.* **122**, 4–6P.

BROCKES, J. P., BERG, D. W. and HALL, Z. W. (1976). The biochemical properties and regulation of acetylcholine receptors in normal and denervated muscle. *Cold Spring Harb. Symp. quant. Biol.* **40**, 253–62.

BROWN, D. E. S. (1936). The effect of rapid compression upon events in the isometric contraction of skeletal muscle. *J. cell. comp. Physiol.* **4**, 257–81.

BROWN, G. L. (1965). The release and fate of transmitter liberated by adrenergic nerves. *Proc. R. Soc.* B **162**, 1–19.

BROWN, G. L., DALE, H. H. and FELDBURG, W. (1936). Reactions of the normal mammalian muscle to acetylcholine and to eserine. *J. Physiol., Lond.* **87**, 394–424.

BROWN, J. E. and PINTO, L. H. (1974). Ionic mechanism for the photoreceptor potential of the retina of *Bufo marinus*. *J. Physiol., Lond.* **236**, 575–91.

BROWN, J. E., COLES, J. A. and PINTO, L. H. (1977). Effects of injections of calcium and EGTA into the outer segments of retinal rods of *Bufo marinus*. *J. Physiol., Lond.* **269**, 707–22.

BROWN, K. T. and MURAKAMI, M. (1964). A new receptor potential of the monkey retina with no detectable latency. *Nature, Lond.* **201**, 626–8.

BROWN, K. T., WATANABE, K. and MURAKAMI, M. (1965). The early and late receptor potentials of monkey cones and rods. *Cold Spring Harb. Symp. quant. Biol.* **30**, 457–82.

BROWN, K. T. and WIESEL, T. N. (1961). Localization of origins of electroretinogram components by intraretinal recording in the intact cat eye. *J. Physiol., Lond.* **158**, 257–80.

BROWN, M. C. and BUTLER, R. G. (1973). Studies on the site of termination of static and dynamic fusimotor fibres within spindles of the tenuissimus muscles of the cat. *J. Physiol., Lond.* **233**, 553–73.

BROWN, M. C., CROWE, A. and MATTHEWS, P. B. C. (1965). Observations on the fusimotor fibres of the tibialis posterior muscle of the cat. *J. Physiol., Lond.* **177**, 140–59.

BROWN, P. K. and WALD, G. (1964). Visual pigments in single rods and cones of the human retina. *Science* **144**, 45–52.

BÜLBRING, E. (1955). Correlation between membrane potential, spike discharge and tension in smooth muscle. *J. Physiol., Lond.* **128**, 200–21.

BÜLBRING, E., BURNSTOCK, G. and HOLMAN, M. (1958). Excitation and conduction in the smooth muscle of the isolated taenia coli of the guinea pig. *J. Physiol., Lond.* **142**, 420–37.

BULLARD, B., LUKE, B. and WINKELMAN, L. (1973). The paramyosin of insect flight muscle. *J. molec. Biol.* **75**, 359–67.

BULLER, A. J. (1975). *The Contractile Behaviour of Mammalian Skeletal Muscle* (Oxford Biology Reader No. 36). London: Oxford University Press.

BULLOCK, T. H. (1973). See the world through a new sense: electroreception in fish. *Amer. Sci.* **61**, 316–25.

BULLOCK, T. H. and CHICHIBU, S. (1965). Further analysis of sensory coding in electroreceptors of electric fish. *Proc. natn. Acad. Sci. U.S.A.* **54**, 422–9.

BULLOCK, T. H. and DIEKE, F. P. J. (1956). Properties of an infra-red receptor. *J. Physiol., Lond.* **134**, 47–87.

BULLOCK, T. H. and HAGIWARA, S. (1957). Intracellular recording from the giant synapse of the squid. *J. gen. Physiol.* **40**, 565–77.

BULLOCK, T. H. and HORRIDGE, G. A. (1965). *Structure and Function in the Nervous Systems of Invertebrates.* San Francisco: W. H. Freeman & Co.

BURKHARDT, D. (1958). Die Sinnesorgane des Skeletmuskels und die nervöse Steuerung der Muskeltätigkeit. *Ergebn. Biol.* **20**, 27–66.

BURNSTOCK, G. (1972). Purinergic nerves. *Pharmacol. Rev.* **24**, 509–81.

BURNSTOCK, G. and COSTA, M. (1975). *Adrenergic neurons.* London: Chapman and Hall.

CAIN, D. F., INFANTE, A. A. and DAVIES, R. E. (1962). Chemistry of muscle contraction. Adenosine triphosphate and phosphoryl creatine as energy supplies for single contractions of working muscle. *Nature, Lond.* **196**, 214–17.

CALDWELL, P. C., HODGKIN, A. L., KEYNES, R. D. and SHAW, T. I. (1960). The effects of injecting 'energy-rich' phosphate compounds on the active transport of ions in the giant axons of *Loligo. J. Physiol., Lond.* **152**, 561–90.

CALDWELL, P. C. and WALSTER, G. E. (1963). Studies on the microinjection of various substances into crab muscle fibres. *J. Physiol., Lond.* **169**, 353–72.

CAMPBELL, F. W. and RUSHTON, W. A. H. (1955). Measurement of the scotopic pigment in the living human eye. *J. Physiol., Lond.* **130**, 131–47.

CARLSON, F. D. (1963). The mechanochemistry of muscular contraction, a critical revaluation of *in vivo* studies. *Progr. Biophys.* **13**, 261–314.

CARLSON, F. D., HARDY, D. J. and WILKIE, D. R. (1963). Total energy production and phosphocreatine hydrolysis in the isotonic twitch. *J. gen. Physiol.* **46**, 851–91.

CARLSON, E. D. and WILKIE, D. R. (1974). *Muscle Physiology.* New Jersey: Prentice-Hall.

CHANGEUX, J.-P., BENEDETTI, L., BOURGEOIS, J.-P., BRISSON, A., CARTAUD, J., DEVAUX, P., GRÜNHAGEN, H., MOREAU, M., POPOT, J.-L., SOBEL, A. and WEBER, M. (1976). Some structural properties of the cholinergic receptor protein in its membrane environment relevant to its function as a pharmacological receptor. *Cold Spring Harb. Symp. quant. Biol.* **40**, 211–30.

CHANDLER, W. K., HODGKIN, A. L. and MEVES, H. (1965). The effect of changing the internal solution on sodium inactivation and related phenomena in giant axons. *J. Physiol., Lond.* **180**, 821–36.

CHANDLER, W. K. and MEVES, H. (1965). Voltage clamp measurements on internally perfused giant axons. *J. Physiol., Lond.* **180**, 788–820.

CHANDLER, W. K., RAKOWSKI, R. F. and SCHNEIDER, M. F. (1976). Effects of glycerol treatment and maintained depolarization on charge movement in skeletal muscle. *J. Physiol., Lond.* **254**, 285–316.

CIVAN, M. M. and PODOLSKY, R. J. (1966). Contraction kinetics of striated muscle fibres following quick changes in load. *J. Physiol., Lond.* **184**, 511–34.

CLOSE, R. I. (1972). Dynamic properties of mammalian skeletal muscle. *Physiol. Rev.* **52**, 129–97.

COHEN, M. J. (1964). The peripheral organization of sensory systems. In *Neural Theory and Modelling* (ed. R. F. Reiss). pp. 273–92. Stanford University Press.

COHEN, M. J., HAGIWARA, S. and ZOTTERMAN, Y. (1955). The response spectrum of taste fibres in the cat: a single fibre analysis. *Acta physiol. scand.* **33**, 316–32.

COLE, K. S. and CURTIS, H. J. (1939). Electric impedance of the squid giant axon during activity. *J. gen. Physiol.* **22**, 649.

CONE, R. A. (1964). The early receptor potential of the vertebrate retina. *Nature, Lond.* **204**, 736–9.

CONE, R. A. (1967). Early receptor potential: photoreversible charge displacement in rhodopsin. *Science* **155**, 1128–31.

CONE, R. A. (1973). The internal transmitter model for visual excitation: some quantitative implications. In *Biochemistry and Physiology of Visual Pigments* (ed. H. Langer), pp. 275–82. Berlin: Springer.

CONWAY, E. J. (1957). Nature and significance of concentration relations of potassium and sodium ions in skeletal muscle. *Physiol. Rev.* **37**, 84–132.

COOMBS, J. S., CURTIS, D. R. and ECCLES, J. C. (1957a). The interpretation of spike potentials of motoneurones. *J. Physiol., Lond.* **139**, 198–231.

COOMBS, J. S., CURTIS, D. R. and ECCLES, J. C. (1957b). The generation of impulses in motoneurones. *J. Physiol., Lond.* **139**, 232–49.

COOMBS, J. S., ECCLES, J. C. and FATT, P. (1955a). The electrical properties of the motoneurone membrane. *J. Physiol., Lond.* **130**, 291–325.

COOMBS, J. S., ECCLES, J. C. and FATT, P. (1955b). The specific ionic conductances and the ionic movements across the motoneuronal membrane that produce the inhibitory postsynaptic potential. *J. Physiol., Lond.* **130**, 326–73.

COOMBS, J. S., ECCLES, J. C. and FATT, P. (1955c). Excitatory synaptic action in motoneurones. *J. Physiol., Lond.* **130**, 374–95.

COOMBS, J. S., ECCLES, J. C. and FATT, P. (1955*d*). The inhibitory suppression of reflex discharges from motoneurones. *J. Physiol., Lond.* **130**, 396–413.

COSTANTIN, L. L. (1970). The role of sodium current in the radial spread of contraction in frog muscle fibres. *J. gen. Physiol.* **55**, 703–15.

COUCEIRO, A. and DE ALMEIDA, D. F. (1961). The electrogenic tissue of some Gymnotidae. In *Bioelectrogenesis* (ed. C. Chagas and A. Paes de Carvalho), pp. 3–13. Amsterdam: Elsevier.

COUTEAUX, R. (1960). Motor end-plate structure. In *The Structure and Function of Muscle*, vol. 1 (ed. G. H. Bourne). New York: Academic Press.

CRAIG, R. and OFFER, G. (1976). Location of C protein in rabbit skeletal muscle. *Proc. R. Soc.* B **192**, 451–61.

CRAWFORD, B. H. (1949). The scotopic visibility function. *Proc. phys. Soc.* B **62**, 321–34.

CRESCITELLI, F. and DARTNALL, H. J. A. (1953). Human visual purple. *Nature, Lond.* **172**, 195–6.

CROWE, A. and MATTHEWS, P. B. C. (1964). The effects of stimulation of static and dynamic fusimotor fibres on the response to stretching of the primary endings of the muscle spindles. *J. Physiol., Lond.* **174**, 109–31.

CURTIN, N. A. and DAVIES, R. E. (1973). ATP breakdown following activation of muscle. In *The Structure and Function of Muscle* 2nd edn. vol. III (ed. G. H. Bourne), pp. 471–515. New York: Academic Press.

CURTIS, D. R. (1959). Pharmacological investigations upon inhibition of spinal neurones. *J. Physiol., Lond.* **145**, 175–92.

CURTIS, D. R. (1965). Actions of drugs on single neurons in the spinal cord and thalamus. *Brit. med. Bull.* **21**, 1–9.

CURTIS, D. R. (1969). The pharmacology of spinal postsynaptic inhibition. *Progress in Brain Research* **31**, 171–89.

CURTIS, D. R. and ECCLES, J. C. (1959). The time courses of excitatory and inhibitory synaptic actions. *J. Physiol., Lond.* **145**, 529–46.

CURTIS, D. R. and ECCLES, R. M. (1958*a*). The excitation of Renshaw cells by pharmacological agents applied electrophoretically. *J. Physiol., Lond.* **141**, 435–45.

CURTIS, D. R. and ECCLES, R. M. (1958*b*). The effect of diffusional barriers upon the pharmacology of cells within the central nervous system. *J. Physiol., Lond.* **141**, 446–63.

CURTIS, D. R., PHILLIS, J. W. and WATKINS, J. C. (1960). The chemical excitation of spinal neurones by certain acidic amino acids. *J. Physiol., Lond.* **150**, 656–82.

CURTIS, D. R. and RYALL, R. W. (1964). Nicotinic and muscarinic receptors of Renshaw cells. *Nature, Lond.* **203**, 652.

CURTIS, H. J. and COLE, K. S. (1942). Membrane resting and action potentials in giant fibres of squid nerve. *J. cell. comp. Physiol.* **19**, 135–44.

DALE, H. H. (1914). The action of certain esters and ethers of choline, and their relation to muscarine. *J. Pharmac. exp. Ther.* **6**, 147–90.

DALE, H. H. and DUDLEY, H. W. (1929). The presence of histamine and acetylcholine in the spleen of the ox and the horse. *J. Physiol., Lond.* **68**, 97–123.

DALE, H. H., FELDBURG, W. and VOGT, M. (1936). Release of acetylcholine at voluntary motor nerve endings. *J. Physiol., Lond.* **86**, 353–80.

DARTNALL, H. J. A. (1952). Visual pigment 467, a photosensitive pigment present in tench retinae. *J. Physiol., Lond.* **116**, 259–89.

DARTNALL, H. J. A. (1957). *The Visual Pigments.* London: Methuen & Co. Ltd.

DARTNALL, H. J. A. (1962). The photobiology of the visual process. In *The Eye* (Davson, 1962), vol. 2, pp. 321–533.

DARTNALL, H. J. A., ARDEN, G. B., IKEDA, H., LUCK, C. P., ROSENBERG, M. E., PEDLER, C. M. H. and TANSLEY, K. (1965). Anatomical electrophysiological and pigmentary aspects of vision in the bush baby: an interpretative study. *Vision Res.* **5**, 399–424.

DARWIN, C. (1859). *The Origin of Species.* London: John Murray.

DAVIDSON, N. (1976). *Neurotransmitter Amino Acids.* London: Academic Press.

DAVIES, R. E. (1963). A molecular theory of muscle contraction: calcium-dependent contractions with hydrogen bond formation plus ATP-dependent extensions of part of the myosin–actin cross-bridges. *Nature, Lond.* **199**, 1068–74.

DAVIES, R. E. (1964). Adenosine triphosphate breakdown during single muscle contractions. *Proc. R. Soc.* B **160**, 480–5.

DAVIS, H. (1961). Some principles of sensory receptor action. *Physiol. Rev.* **41**, 391–416.

DAVIS, H. *et al.* (1953). Acoustic trauma in the guinea pig. *J. acoust. Soc. Am.* **25**, 1180–9.

DAVSON, H. (ed.) (1962). *The Eye.* New York: Academic Press.

DAVSON, H. and DANIELLI, J. F. (1943). *The Permeability of natural Membranes.* London: Cambridge University Press.

DEL CASTILLO, J. and ENGBAEK, L. (1954). The nature of the neuromuscular block produced by magnesium. *J. Physiol., Lond.* **124**, 370–84.

DEL CASTILLO, J. and KATZ, B. (1954*a*). The membrane change produced by the neuromuscular transmitter. *J. Physiol., Lond.* **125**, 546–65.

DEL CASTILLO, J. and KATZ, B. (1954*b*). Quantal components of the end-plate potential. *J. Physiol., Lond.* **124**, 560–73.

DEL CASTILLO, J. and KATZ, B. (1954*c*). Statistical factors involved in neuromuscular facilitation and depression. *J. Physiol., Lond.* **124**, 574–85.

DEL CASTILLO, J. and KATZ, B. (1955). On the localization of acetylcholine receptors. *J. Physiol., Lond.* **128**, 157–81.

DEL CASTILLO, J. and KATZ, B. (1956). Biophysical aspects of neuromuscular transmission. *Progr. Biophys.* **6**, 121–70.

DEL CASTILLO, J. and MOORE, J. W. (1959). On increasing the velocity of a nerve impulse. *J. Physiol., Lond.* **148**, 665.

DE LORENZO, A. J. D. (1963). Studies on the ultrastructure and histophysiology of cell membranes, nerve fibres and synaptic junctions in chemoreceptors. In *Olfaction and Taste* (ed. Y. Zotterman). Oxford: Pergamon Press.

DENTON, E. J. (1959). The contributions of the oriented photosensitive and other molecules to the absorption of whole retina. *Proc. R. Soc.* B **150**, 78–94.

DE ROBERTIS, E. (1960). Some observations on the ultrastructure and morphogenesis of photoreceptors. *J. gen. Physiol.* **43**, suppl. 1–13.

DE ROBERTIS, E. and BENNETT, H. S. (1954). Submicroscopic vesicular component in the synapse. *Fedn Proc. Fedn Am. Socs exp. Biol.* **13**, 38.

DE ROBERTIS, E. and FRANCHI, C. M. (1956). Electron microscope observations on synaptic vesicles in synapses of the retinal rods and cones. *J. biophys. biochem. Cytol.* **2**, 307–18.

DEROSIER, D. J. and KLUG, A. (1968). Reconstruction of three dimensional structures from electron micrographs. *Nature, Lond.* **217**, 130–4.

DESMEDT, J. E. (1958). Myasthenic-like features of neuromuscular transmission in myasthenic patients: 'post-tetanic exhaustion'. *Nature, Lond.* **179**, 156–7.

DETHIER, V. G. (1953). Summation and inhibition following contralateral stimulation of the tarsal chemoreceptors of the blowfly. *Biol. Bull. mar. biol. Lab., Woods Hole* **105**, 257–68.

DETHIER, V. G. (1955). The physiology and histology of the contact chemoreceptors of the blowfly. *Quart. rev. Biol.* **30**, 348–71.

DETHIER, V. G. (1963). *The Physiology of Insect Senses*. London: Methuen.

DIAMOND, J., GRAY, J. A. B. and INMAN, D. R. (1958). The relation between receptor potentials and the concentration of sodium ions. *J. Physiol., Lond.* **142**, 382–94.

DIJKGRAAF, S. (1934). Untersuchungen über die Funktion der Seitenorgane an Fischen. *Z. vergl. Physiol.* **20**, 162–214.

DIJKGRAAF, S. (1952). Bau und Funktionen der Seitenorgane und des Ohrlabyrinths bei Fischen. *Experientia* **8**, 205–16.

DIJKGRAAF, S. (1963). The functioning and significance of the lateral line organs. *Biol. Rev.* **38**, 51–105.

DIJKGRAAF, S. and KALMIJN, A. J. (1963). Untersuchungen über die Funktion der Lorenzinischen Ampullen an Haifischen. *Z. vergl. Physiol.* **47**, 438–56.

DIONNE, V. E. and STEVENS, C. F. (1975). Voltage dependence of agonist effectiveness at the frog neuromuscular junction: resolution of a paradox. *J. Physiol., Lond.* **251**, 245–70.

DOBELLE, W. H., MARKS, W. B. and MACNICHOL, E. F., JNR. (1969). Visual pigment density in single primate foveal cones. *Science* **166**, 1508–10.

DODGE, F. A. and RAHAMIMOFF, R. (1967). On the relationship between calcium concentration and the amplitude of the end-plate potential. *J. Physiol., Lond.* **189**, 90–2P.

DOHLMAN, G. (1935). Some practical and theoretical points in labyrinthology. *Proc. R. Soc. Med.* **28**, 1371–80.

DOWDALL, M. J., BOYNE, A. F. and WHITTAKER, V. P. (1974). Adenosine triphosphate: a constituent of cholinergic synaptic vesicles. *Biochem. J.* **140**, 1–12.

DOWLING, J. E. (1960). Chemistry of visual adaptation in the rat. *Nature, Lond.* **188**, 114–18.

DOWLING, J. E. (1968). Synaptic organization of the frog retina: an electron miscroscopic analysis comparing the retinas of frogs and primates. *Proc. R. Soc.* B **170**, 205–28.

DOWLING, J. E. (1970). Organization of vertebrate retinas. *Investigative Ophthalmol.* **9**, 655–80.

DOWLING, J. E. and RIPPS, H. (1973). Effect of magnesium on horizontal cell activity in skate retina. *Nature, Lond.* **242**, 101–3.

DOWLING, J. E. and WERBLIN, F. S. (1969). Organisation of the retina of the mudpuppy, *Necturus maculosus*. I. Synaptic structure. *J. Neurophysiol.* **32**, 315–38.

DRAPER, M. H. and WEIDMANN, S. (1951). Cardiac resting and action potentials recorded with an intracellular electrode. *J. Physiol., Lond.* **115**, 74–94.

DREYER, F. and PEPER, K. (1975). Density and dose–response curve of acetylcholine receptors in frog neuromuscular junction. *Nature, Lond.* **253**, 641–3.

DU BOIS REYMOND, E. (1877). *Gesammelte Abhandl.d.allgem. Muskel und Nervenphysik* **2**, 700.

DUDEL, J. and KUFFLER, S. W. (1961). Presynaptic inhibition at the crayfish neuromuscular junction. *J. Physiol., Lond.* **155**, 543–62.

EATON, R. C., BOMBARDIERI, R. A. and MEYER, D. L. (1977). The Mauther-initiated startle response in teleost fish. *J. exp. Biol.* **66**, 65–81.

EBASHI, S. and ENDO, M. (1968). Calcium ions and muscle contraction. *Progr. Biophys.* **18**, 123–83.

EBASHI, S., ENDO, M. and OHTSUKI, I. (1969). Control of muscle contraction. *Quart. Rev. Biophys.* **2**, 351–84.

ECCLES, J. C. (1957). *The Physiology of Nerve Cells.* London: Oxford University Press.

ECCLES, J. C. (1964). *The Physiology of Synapses.* Berlin: Springer-Verlag.

ECCLES, J. C., ECCLES, R. M. and FATT, P. (1956). Pharmacological investigations on a central synapse operated by acetylcholine. *J. Physiol., Lond.* **131**, 154–69.

ECCLES, J. C., ECCLES, R. M. and ITO, M. (1964). Effects produced on inhibitory postsynaptic potentials by the coupled injections of cations and anions into motoneurons. *Proc. R. Soc.* B **160**, 197–210.

ECCLES, J. C., ECCLES, R. M. and MAGNI, F. (1961). Central inhibitory action attributable to presynaptic depolarization produced by muscle afferent volleys. *J. Physiol., Lond.* **159**, 147–66.

ECCLES, J. C., MAGNI, F. and WILLIS, W. D. (1962). Depolarization of central terminals of group I afferent fibres from muscle. *J. Physiol., Lond.* **160**, 62–93.

EDMAN, K. A. P. and SCHILD, H. O. (1962). The need for calcium in the contractile responses induced by acetylcholine and potassium in the rat uterus. *J. Physiol., Lond.* **161**, 424–41.

EINTHOVEN, W. (1924). The string galvanometer and measurement of the action currents of the heart. Nobel Lecture. Republished in 1965 In *Nobel lectures, Physiology or Medicine 1921–41.* Amsterdam: Elsevier.

ELLIOTT, G. F., LOWY, J. and MILLMAN, B. M. (1965). X-ray diffraction from living striated muscle during contraction. *Nature, Lond.,* **206**, 1357–8.

ELLIOTT, T. R. (1904). On the action of adrenaline. *J. Physiol., Lond.* **31**, 20P.

ENGSTRÖM, H. and WERSÄLL, J. (1958). The ultrastructural organization of the organ of Corti and the vestibular sensory epithelia. *Exp. Cell Res.* suppl. 5, 460.

ERLANGER, J. and GASSER, H. S. (1937). *Electrical Signs of Nervous Activity.* Philadelphia: University of Pennsylvania Press.

EVANS, D. R. and MELLON, DE F. (1962). Electrophysiological studies of a water receptor associated with the taste sensilla of the blowfly. *J. gen. Physiol.* **45**, 487–500.

EVANS, E. F. (1972). The frequency response and other properties of single fibres in the guinea-pig cochlear nerve. *J. Physiol., Lond.* **226**, 263–87.

EVANS, E. F. (1975). Cochlear nucleus. In *Handbook of Sensory Physiology* (ed. H. Autrum *et al.*) **5**(2), 1–108. Berlin: Springer.

EYZAGUIRRE, C. and KUFFLER, S. W. (1955*a*). Processes of excitation in the dendrites and in the soma of single isolated sensory nerve cells of the lobster and crayfish. *J. gen. Physiol.* **39**, 87–119.

EYZAGUIRRE, C. and KUFFLER, S. W. (1955*b*). Synaptic inhibition in an isolated nerve cell. *J. gen. Physiol.* **39**, 155–84.

FAIN, G. L. (1975). Quantum sensitivity of rods in the toad retina. *Science* **187**, 838–41.

FAIN, G. L., GOLD, G. H. and DOWLING, J. E. (1976). Receptor coupling in the toad retina. *Cold Spring Harb. Symp. quant. Biol.* **40**, 547–61.

FATT, P. and KATZ, B. (1951). An analysis of the end-plate potential recorded with an intracellular electrode. *J. Physiol., Lond.* **115**, 320–69.

FATT, P. and KATZ, B. (1952). Spontaneous subthreshold activity at motor nerve endings. *J. Physiol., Lond.* **117**, 109–28.

FAWCETT, D. W. and MCNUTT, N. S. (1969). The ultrastructure of the cat myocardium. I. Ventricular papillary muscle. *J. Cell Biol.* **42**, 1–45.

FECHNER, G. T. (1862). *Elemente der Psychophysik.* Leipzig: Breitkopf und Härtel.

FELDBURG, W. and GADDUM, J. H. (1934). The chemical transmitter at synapses in a sympathetic ganglion. *J. Physiol., Lond.* **81**, 305–19.

FENN, W. O. (1923). A quantitative comparison between the energy liberated and the work performed by the isolated sartorius of the frog. *J. Physiol., Lond.* **58**, 175–203.

FENN, W. O. (1924). The relation between the work performed and the energy liberated in muscular contraction. *J. Physiol., Lond.* **58**, 373–95.

FENN, W. O. and MARSH, B. S. (1935). Muscular force at different speeds of shortening. *J. Physiol., Lond.* **85**, 277–97.

FERTUCK, H. C. and SALPETER, M. N. (1974). Localization of acetylcholine receptor by ^{125}I-labelled α-bungarotoxin binding at mouse motor end-plates. *Proc. Natn Acad. Sci. U.S.A.* **71**, 1376–8.

FESSARD, A. (ed.) (1974). Electroreceptors and other specialized receptors in lower vertebrates. *Handbook of Sensory Physiology* (ed. H. Autrum *et al.*) **3**(3). Berlin: Springer.

FLETCHER, W. M. and HOPKINS, F. G. (1907). Lactic acid in amphibian muscle. *J. Physiol., Lond.* **35**, 247–309.

FLOCK, Å. (1964). Structure of the macula utriculi with special reference to directional interplay of sensory responses as revealed by morphological polarization. *J. Cell Biol.* **22**, 413–31.

FLOCK, Å. (1965). Transducing mechanisms in the lateral line canal organ receptors. *Cold Spr. Harb. Symp. quant. Biol.* **30**, 133–44.

FLOCK, Å. (1971). Sensory transduction in hair cells. In *Handbook of Sensory .Physiology* (ed. H. Autrum *et al.*) **1**, 396–441. Berlin: Springer.

FORD, L. E. and PODOLSKY, R. J. (1970). Regenerative calcium release within muscle cells. *Science* **167**, 58–9.

FRANK, G. B. (1960). Effects of changes in extracellular calcium concentration on the potassium-induced contracture of frog's skeletal muscle. *J. Physiol., Lond.* **151**, 518–38.

FRANK, K. and FUORTES, M. G. F. (1957). Presynaptic and post-synaptic inhibition of monosynaptic reflexes. *Fedn Proc. Fedn Am. Socs exp. Biol.* **16**, 39–40.

FRANKENHEUSER. B. and HODGKIN. A. L. (1957). The action of calcium on the electrical properties of squid axons. *J. Physiol., Lond.* **137**, 218–44.

FRANZINI-ARMSTRONG, C. (1970). Studies of the triad. I. Structure of the junction in frog twitch fibres. *J. Cell Biol.* **47**, 488–99.

FRANZINI-ARMSTRONG, C. (1975). Membrane particles and transmission at the triad. *Fedn Proc. Fedn Am. Socs exp. Biol.* **34**, 1382–9.

FRANZINI-ARMSTRONG, C. and PORTER, K. R. (1964). Sarcolemmal invaginations constituting the T system in fish muscle fibres. *J. Cell Biol.* **22**, 675–96.

FUORTES, M. G. F. (1959). The initiation of impulses in visual cells of *Limulus. J. Physiol., Lond.* **148**, 14–28.

FUORTES, M. G. F. and HODGKIN, A. L. (1964). Changes in time scale and sensitivity in the ommatidia of *Limulus. J. Physiol., Lond.* **172**, 239–63.

FUORTES, M. G. F. and MANTEGAZZINI, F. (1962). Interpretation of the repetitive firing of nerve cells. *J. gen. Physiol.* **45**, 1163–79.

FURSHPAN, E. J. and POTTER, D. D. (1959). Transmission at the giant synapses of the crayfish. *J. Physiol., Lond.* **145**, 289–325.

FURUKAWA, T. and FURSHPAN, E. J. (1963). Two inhibitory mechanisms in the Mauthner neurons of goldfish. *J. Neurophysiol.* **26**, 140–76.

FURUKAWA, T. and HANAWA, I. (1955). Effects of some common cations on electroretinogram of the toad. *Jap. J. Physiol.* **5**, 289–300.

GASSER, H. S. and HILL, A. V. (1924). The dynamics of muscular contraction. *Proc. R. Soc.* B **96**, 398–437.

GEREN, B. B. (1954). The formation from the Schwann cell surface of myelin in the peripheral nerves of chick embryos. *Exp. Cell Res.* **7**, 558.

GERGELY, J. (1966). Contractile proteins. *A. Rev. Biochem.* **35**, 691–722.

GERSCHENFELD, H. M. (1966). Chemical transmitters in invertebrate nervous systems. *Symp. Soc. exp. Biol.* **20**, 299–324.

GERSCHENFELD, H. M. and CHIARANDINI, D. J. (1965). Ionic mechanism associated with non-cholinergic synaptic inhibition in molluscan neurons. *J. Neurophysiol.* **28**, 710–23.

GILBERT, C., KRETSCHMAR, K. M., WILKIE, D. R. and WOLEDGE, R. C. (1971). Chemical change and energy output during muscular contraction. *J. Physiol., Lond.* **218**, 163–93.

GILBERT, D. L. (1971). Fixed surface charges. In *Biophysics and Physiology of Excitable membranes* (ed. W. J. Adelman, Jnr.), pp. 359–78. New York: Van Nostrand Reinhold.

GILULA, N. B. and EPSTEIN, M. L. (1976). Cell-to-cell communication, gap junctions and calcium. *Symp. Soc. exp. Biol.* **30**, 257–72.

GINSBORG, B. L. (1960). Some properties of avian skeletal muscle fibres with multiple neuromuscular junctions. *J. Physiol., Lond.* **154**, 581–98.

GOLDMAN, D. E. (1943). Potential, impedance, and rectification in membranes. *J. gen. Physiol.* **27**, 37–60.

GOLDSTEIN, E. B. (1968). Visual pigments and the early receptor potential of the isolated frog retina. *Vision Res.* **8**, 953–63.

GOLDSTEIN, E. B. (1969). Contribution of cones to the early receptor potential in the rhesus monkey. *Nature, Lond.* **222**, 1273–4.

GOLDSTEIN, E. B. and BERSON, E. L. (1969). Cone dominance of the human early receptor potential. *Nature, Lond.* **222**, 1272–3.

GONZALES-SERRATOS, H. (1971). Inward spread of activation in vertebrate muscle fibres. *J. Physiol., Lond.* **212**, 777–99.

GOODENOUGH, D. A. (1976). The structure and permeability of isolated hepatocyte gap junctions. *Cold Spring Harb. Symp. quant. Biol.* **40**, 37–43.

GÖPFERT, H. and SCHAEFER, H. (1938). Über den direkt und indirekt erregeten Aktionsstrom und die Funktion der motorischen Endplatte. *Pflügers Arch. ges Physiol.* **239**, 597–619.

GORDON, A. M., HUXLEY, A. F. and JULIAN, F. J. (1966*a*). Tension development in highly stretched vertebrate muscle fibres. *J. Physiol., Lond.* **184**, 143–69.

GORDON, A. M., HUXLEY, A. F. and JULIAN, F. J. (1966*b*). The variation in isometric tension with sarcomere length in vertebrate muscle fibres. *J. Physiol., Lond.* **184**, 170–92.

GORTER, E. and GRENDEL, F. (1925). On bimolecular layers of lipoids on the chromocytes of blood. *J. exp. Med.* **41**, 439–43.

GRANIT, R. (1933). The components of the retinal action potential and their relation to the discharge in the optic nerve. *J. Physiol., Lond.* **77**, 207–40.

GRANIT, R. (1947). *Sensory Mechanisms of the Retina.* London: Oxford University Press.

GRANIT, R. (1955). *Receptors and Sensory Perception.* New Haven: Yale University Press.

GRANIT, R. (1962). The visual pathway. In Davson (1962), vol. 2, 535–763.

GRAVES, R. (1958). What food the centaurs ate. In *Steps.* London: Cassell.

GRAY, E. G. (1957). The spindle and extrafusal innervation of a frog muscle. *Proc. R. Soc.* B **146**, 416–30.

GRAY, E. G. (1959). Axosomatic and axodendritic synapses of the cerebral cortex: an electron microscope study. *J. Anat.* **93**, 420–33.

GRAY, E. G. (1962). A morphological basis for presynaptic inhibition? *Nature, Lond.* **193**, 82–3.

GRAY, E. G. (1974). The synapse. In *The Cell in Medical Science*, vol. 2 (ed. F. Beck and J. B. Lloyd), pp. 385–416. London: Academic Press.

GRAY, E. G. and GUILLERY, R. W. (1966). Synaptic morphology in the normal and degenerating nervous system. *Int. Rev. Cytol.* **19**, 111–82.

GRAY, J. A. B. (1962). Coding in systems of primary receptor neurons. *Symp. Soc. exp. Biol.* **16**, 345–54.

GRAY, J. A. B. and MATTHEWS, P. B. C. (1951). A comparison of the adaptation of the Pacinian corpuscle with the accommodation of its own axon. *J. Physiol., Lond.* **114**, 454–64.

GRIFFIN, D. R. (1958). *Listening in the Dark.* New Haven: Yale University Press.

GRIFFIN, D. R. and GALAMBOS, R. (1941). The sensory basis of obstacle avoidance by flying bats. *J. exp. Zool.* **86**, 481–506.

GROEN, J. J., LOWENSTEIN, O. and VENDRIK, A. J. H. (1952). The mechanical analysis of the responses from the end-organs of the horizontal semicircular canal in the isolated elasmobranch labyrinth. *J. Physiol., Lond.* **117**, 329–46.

GRUNDFEST, H. (1957). Electrical inexcitability of synapses and some consequences in the central nervous system. *Physiol. Rev.* **37**, 337–61.

GRUNDFEST, H. (1966). Comparative electrobiology of excitable membranes. *Adv. comp. Physiol. Biochem.* **2**, 1–116.

HAGINS, W. A. (1972). The visual process: excitatory mechanisms in the primary receptor cells. *A. Rev. Biophys. Bioeng.* **1**, 131–58.

HAGINS, W. A., PENN, R. D. and YOSHIKAMI, S. (1970). Dark current and photocurrent in retinal rods. *Biophys. J.* **10**, 380–412.

HAGINS, W. A. and YOSHIKAMI, S. (1974). A role for Ca^{2+} in excitation of retinal rods and cones. *Exp. Eye Res.* **18**, 299–305.

HAGINS, W. A. and YOSHIKAMI, S. (1975). Ionic mechanisms in excitation of photoreceptors. *Ann. N.Y. Acad. Sci.* **264**, 314–23.

HAGIWARA, S. and MORITA, H. (1962). Electronic transmission between two nerve cells in leech ganglion. *J. Neurophysiol.* **25**, 721–31.

HAGIWARA, S. and MORITA, H. (1963). Coding mechanisms of electroreceptor fibres in some electric fish. *J. Neurophysiol.* **26**, 551–67.

HAGIWARA, S. and NAKA, K. (1964). The initiation of spike potential in barnacle muscle fibres under low intracellular Ca^{++}. *J. gen. Physiol.* **48**, 141–62.

HAGIWARA, S., SZABO, T. and ENGER, P. S. (1965). Electroreceptor mechanisms in a high-frequency weakly electric fish, *Sternarchus albifrons*. *J. Neurophysiol.* **28**, 784–99.

HAGIWARA, S. and TASAKI, I. (1958). Study of mechanism of impulse transmission across the giant synapse of the squid. *J. Physiol., Lond.* **143**, 114–37.

HAGIWARA, S., WATANABE, A. and SAITO, N. (1959). Potential changes in syncytial neurons of lobster cardiac ganglion. *J. Neurophysiol.* **22**, 554–72.

HAGOPIAN, M. (1966). Myofilament arrangement in femoral muscle of the cockroach, *Leucophaea maderae* Falemicius. *J. Cell Biol.* **28**, 545–62.

HAGOPIAN, M. and SPIRO, D. (1968). The filament lattice of cockroach thoracic muscle. *J. Cell Biol.* **36**, 433–42.

HALL, M. O. and BACHARACH, A. D. E. (1970). Linkage of retinal to opsin and absence of phospholipids in purified frog visual pigment$_{500}$. *Nature, Lond.* **225**, 637–8.

HANSON, F. E. and WOLBARSHT, M. L. (1962). Dendritic action potentials in insect chemoreceptors. *Amer. Zool.* **2**, 528.

HANSON, J. (1968). Recent X-ray diffraction studies of muscle. *Quart. Rev. Biophys.* **1**, 177–216.

HANSON, J. and HUXLEY, H. E. (1953). The structural basis of the cross-striations in muscle. *Nature, Lond.* **172**, 530–2.

HANSON, J. and HUXLEY, H. E. (1955). The structural basis of contraction in striated muscle. *Symp. Soc. exp. Biol.* **9**, 228–64.

HANSON, J. and LOWY, J. (1960). Structure and function of the contractile apparatus in the muscles of invertebrate animals. In *Structure and Function of Muscle* (ed. G. Bourne) vol. 1, pp. 263–365. New York: Academic Press.

HANSON, J. and LOWY, J. (1963). The structure of F-actin and of actin filaments isolated from muscle. *J. molec. Biol.* **6**, 46–60.

HANSON, J. and LOWY, J. (1964). The structure of actin filaments and the origin of the axial periodicity in the *I*-substance of vertebrate striated muscle. *Proc. R. Soc.* B **160**, 449–58.

HARTLINE, H. K. (1934). Intensity and duration in the excitation of single photo receptor units. *J. cell. comp. Physiol.* **5**, 229.

HARTLINE, H. K. (1938). The response of single optic nerve fibres of the vertebrate eye to illumination of the retina. *Am. J. Physiol.* **121**, 400–15.

HARTLINE, H. K. and RATLIFF, F. (1957). Inhibitory interaction of receptor units in the eye of *Limulus*. *J. gen. Physiol.* **40**, 357–76.

HARTLINE, H. K., RATLIFF, F. and MILLER, W. H. (1961). Inhibitory interaction in the retina and its significance in vision. In *Nervous Inhibition* (ed. E. Florey). Oxford: Pergamon Press.

HARTLINE, H. K., WAGNER, H. G. and RATLIFF, F. (1956). Inhibition in the eye of *Limulus*. *J. gen. Physiol.* **39**, 651–73.

HASSELBACH, W. (1964). Relaxing factor and the relaxation of muscle. *Progr. Biophys.* **14**, 167–222.

HASSELBACH, W. and MAKINOSE, M. (1963). Über den mechanismus des Calciumtransportes durch die Membranen des Sarkoplasmatichen Reticulums. *Biochem. Z.* **339**, 94–111.

HECHT, S. (1937). Rods, cones, and the chemical basis of vision. *Physiol. Rev.* **17**, 239–90.

HECHT, S. and SCHLAER, S. (1938). An adaptometer for measuring human dark adaptation. *J. opt. Soc. Amer.* **28**, 269–75.

HECHT, S., SCHLAER, S. and PIRENNE, M. (1942). Energy, quanta and vision. *J. gen. Physiol.* **25**, 819–40.

HELMHOLTZ, H. (1866). *Handbuch der Physiologischen Optik.* Leipzig.

HENSEL, H. and ZOTTERMAN, Y. (1951). Quantitative Beziehungen zwischen der Entladung einzelner Kältefasern und der Temperatur. *Acta physiol. scand.* **23**, 291–319.

HILL, A. V. (1936). The strength–duration relation for electric excitation of medullated nerve. *Proc. R. Soc.* B **119**, 440–53.

HILL, A. V. (1938). The heat of shortening and the dynamic constants of muscle. *Proc. R. Soc.* B **126**, 136–95.

HILL, A. V. (1948). On the time required for diffusion and its relation to processes in muscle. *Proc. R. Soc.* B **135**, 446–53.

HILL, A. V. (1949*a*). The abrupt transition from rest to activity in muscle. *Proc. R. Soc.* B **136**, 399–420.

HILL, A. V. (1949*b*). The onset of contraction. *Proc. R. Soc.* B **136**, 242–54.

HILL, A. V. (1950*a*). A challenge to biochemists. *Biochim. biophys. Acta* **4**, 4–11.

HILL, A. V. (1950*b*). The development of the active state of muscle during the latent period. *Proc. R. Soc.* B **137**, 320–9.

HILL, A. V. (1950*c*). The dimensions of animals and their muscular dynamics. *Sci. Prog., Lond.* **38**, 209–30.

HILL, A. V. (1953). The 'instantaneous' elasticity of active muscle. *Proc. R. Soc.* B **141**, 161–78.

HILL, A. V. (1960). *The Ethical Dilemma of Science and Other Writings.* New York: Rockefeller Institute Press.

HILL, A. V. (1964*a*). The effect of load on the heat of shortening of muscle. *Proc. R. Soc.* B **159**, 297–318.

HILL, A. V. (1964*b*). The effect of tension in prolonging the active state in a twitch. *Proc. R. Soc.* B **159**, 589–95.

HILL, A. V. (1965). *Trails and Trials in Physiology.* London: Edward Arnold.

HILL, A. V. and HARTREE, W. (1920). The four phases of heat production of muscle. *J. Physiol., Lond.* **54**, 84–128.

HILL, A. V. and HOWARTH, J. V. (1959). The reversal of chemical reactions in contracting muscle during an applied stretch. *Proc. R. Soc.* B **151**, 169–93.

HILL, D. K. (1953). The effect of stimulation on the diffraction of light by striated muscle. *J. Physiol., Lond.* **119**, 501–12.

HILL, R. B. and USHERWOOD, P. N. R. (1961). The action of 5-hydroxytryptamine and related compounds on neuromuscular transmission in the locust *Schistocerca gregaria*. *J. Physiol., Lond.* **157**, 393–401.

HILLE, B. (1970). Ionic channels in nerve membranes. *Progr. Biophys.* **21**, 1–32.

HILLE, B. (1971). Voltage clamp studies on myelinated nerve fibres. In *Biophysics and Physiology of Excitable Membranes* (ed. W. J. Adelman, Jnr.), pp. 230–46. New York: Van Nostrand Reinhold.

HILLE, B., WOODHULL, A. M. and SHAPIRO, B. I. (1975). Negative surface charge near sodium channels of nerve: divalent ions, monovalent ions, and pH. *Phil. Trans. R. Soc.* B **270**, 301–18.

HODGKIN, A. L. (1937). Evidence for electrical transmission in nerve. *J. Physiol., Lond.* **90**, 183–232.

HODGKIN, A. L. (1938). The subthreshold potentials in a crustacean nerve fibre. *Proc. R. Soc.* B **126**, 87.

HODGKIN, A. L. (1939). The relation between conduction velocity and the electrical resistance outside a nerve. *J. Physiol., Lond.* **94**, 560–70.

HODGKIN, A. L. (1948). The local electric changes associated with repetitive action in a non-medullated axon. *J. Physiol., Lond.* **107**, 165.

HODGKIN, A. L. (1954). A note on conduction velocity. *J. Physiol., Lond.* **125**, 221–4.

HODGKIN, A. L. (1958). Ionic movements and electrical activity in giant nerve fibres. *Proc. R. Soc.* B **148**, 1–37.

HODGKIN, A. L. (1964). *The Conduction of the Nervous Impulse.* Liverpool: University Press.

HODGKIN, A. L. (1975). The optimum density of sodium channels in an unmyelinated nerve. *Phil. Trans. R. Soc.* B **270**, 297–300.

HODGKIN, A. L. (1976). Chance and design in electrophysiology: an informal account of certain experiments on nerve carried out between 1934 and 1952. *J. Physiol., Lond.* **263**, 1–21.

HODGKIN, A. L. and HOROWICZ, P. (1959). The influence of potassium and chloride ions on the membrane potential of single muscle fibres. *J. Physiol., Lond.* **148**, 127–60.

HODGKIN, A. L. and HOROWICZ, P. (1960). Potassium contractures in single muscle fibres. *J. Physiol., Lond.* **153**, 386–403.

HODGKIN, A. L. and HUXLEY, A. F. (1939). Action potentials recorded from inside a nerve fibre. *Nature, Lond.* **140**, 710.

HODGKIN, A. L. and HUXLEY, A. F. (1945). Resting and action potentials in single nerve fibres. *J. Physiol., Lond.* **104,** 176.

HODGKIN, A. L. and HUXLEY, A. F. (1952*a*). Currents carried by sodium and potassium ions through the membrane of the giant axon of *Loligo. J. Physiol., Lond.* **116**, 449–72.

HODGKIN, A. L. and HUXLEY, A. F. (1952*b*). The components of membrane conductance in the giant axon of *Loligo. J. Physiol., Lond.* **116**, 473–96.

HODGKIN, A. L. and HUXLEY, A. F. (1952*c*). The dual effect of membrane potential on sodium conductance in the giant axon of *Loligo. J. Physiol., Lond.* **116**, 497–506.

HODGKIN, A. L. and HUXLEY, A. F. (1952*d*). A quantitative description of membrane current and its application to conduction and excitation in nerve. *J. Physiol., Lond.* **117**, 500–44.

HODGKIN, A. L. and HUXLEY, A. F. (1953). Movements of radioactive potassium and membrane current in a giant axon. *J. Physiol., Lond.* **121**, 403–14.

HODGKIN, A. L., HUXLEY, A. F. and KATZ, B. (1952). Measurement of current–voltage relations in the membrane of the giant axon of *Loligo. J. Physiol., Lond.* **116**, 424–48.

HODGKIN, A. L. and KATZ, B. (1949). The effect of sodium ions on the electrical activity of the giant axon of the squid. *J. Physiol.*, *Lond*. **108**, 37–77.

HODGKIN, A. L. and KEYNES, R. D. (1953). The mobility and diffusion coefficient of potassium in giant axons from *Sepia*, *J. Physiol.*, *Lond*. **119**, 513–28.

HODGKIN, A. L. and KEYNES, R. D. (1955). Active transport of cations in giant axons from *Sepia* and *Loligo*. *J. Physiol.*, *Lond*. **128**, 28–60.

HODGKIN, A. L. and KEYNES, R. D. (1957). Movements of labelled calcium in squid giant axons. *J. Physiol.*, *Lond*. **138**, 253–81.

HODGKIN, A. L. and RUSHTON, W. A. H. (1946). The electrical constants of a crustacean nerve fibre. *Proc. R. Soc.* B **133**, 444.

HODGSON, E. S., LETTVIN, J. Y. and ROEDER, K. D. (1955). Physiology of a primary chemoreceptor unit. *Science* **122**, 417–18.

HODGSON, E. S. and ROEDER, K. D. (1956). Electrophysiological studies of arthropod chemoreception. I. General properties of the labellar chemoreceptors of Diptera. *J. cell. comp. Physiol.* **48**, 51–76.

HOLMES, W. (1942). The giant myelinated nerve fibres of the prawn. *Phil. Trans. R. Soc.* B **231**, 293–311.

HOYLE, G. (1955*a*). The anatomy and innervation of locust skeletal muscle. *Proc. R. Soc.* B **143**, 281–92.

HOYLE, G. (1955*b*). Neuromuscular mechanisms of a locust skeletal muscle. *Proc. R. Soc.* B **143**, 343–67.

HOYLE, G. (1959). The neuromuscular mechanism of an insect spiracular muscle. *J. Insect Physiol.* **3**, 378–94.

HOYLE, G. and WIERSMA, C. A. G. (1958). Excitation at neuromuscular junctions in Crustacea. *J. Physiol.*, *Lond*. **143**, 403–25.

HUBBARD, J. I. (1963). Repetitive stimulation at the mammalian neuromuscular junction, and the mobilisation of transmitter. *J. Physiol.*, *Lond*. **169**, 641–62.

HUBBARD, J. I. and SCHMIDT, R. F. (1963). An electrophysiological investigation of mammalian motor nerve terminals. *J. Physiol.*, *Lond*. **166**, 145–67.

HUBBARD, R. (1959). The thermal stability of rhodopsin and opsin. *J. gen. Physiol.* **42**, 259–80.

HUBBARD, R. and KROPF, A. (1959). Molecular aspects of visual excitation. *Ann. N. Y. Acad. Sci.* **81**, 388–98.

HUBBARD, R. and WALD, G. (1952). *Cis–trans* isomers of vitamin A and retinene in the rhodopsin system. *J. gen. Physiol.* **36**, 269–315.

HUBBARD, S. J. (1958). A study of rapid mechanical events in a mechanoreceptor. *J. Physiol.*, *Lond*. **141**, 198–218.

HUBEL, D. H. and WIESEL, T. N. (1960). Receptive fields of optic nerve fibres in the spider monkey, *J. Physiol., Lond.* **154**, 572–80.

HUNT, C. C. (1954). Relation of function to diameter in afferent fibers of muscle nerves. *J. gen. Physiol.* **38**, 117–31.

HURSH, J. B. (1939). Conduction velocity and diameter of nerve fibres. *Am. J. Physiol.* **127**, 131–9.

HUTTER, O. F. (1961). Ion movements during vagus inhibition of the heart. In *Nervous Inhibition* (ed. E. Florey), pp. 114–23. Oxford: Pergamon Press.

HUTTER, O. F. and NOBLE, D. (1960). Rectifying properties of heart muscle. *Nature, Lond.* **188**, 495.

HUTTER, O. F. and TRAUTWEIN, W. (1956). Vagal and sympathetic effects on the pacemaker fibres in the sinus venosus of the heart. *J. gen. Physiol.* **39**, 715–33.

HUXLEY, A. F. (1957). Muscle structure and theories of contraction. *Progr. Biophys.* **7**, 255–318.

HUXLEY, A. F. (1959). Ion movements during nerve activity. *Ann. N. Y. Acad. Sci.* **81**, 221.

HUXLEY, A. F. (1974). Muscular contraction. *J. Physiol., Lond.* **243**, 1–43.

HUXLEY, A. F. and NIEDERGERKE, R. (1954). Structural changes in muscle during contraction. Interference microscopy of living muscle fibres. *Nature, Lond.* **173**, 971–3.

HUXLEY, A. F. and PEACHEY, L. D. (1961). The maximum length for contraction in vertebrate striated muscle. *J. Physiol., Lond.* **156**, 150–65.

HUXLEY, A. F. and SIMMONS, R. M. (1971). Proposed mechanism of force generation in striated muscle. *Nature, Lond.* **233**, 533–8.

HUXLEY, A. F. and SIMMONS, R. M. (1973). Mechanical transients and the origin of muscular force. *Cold Spring Harb. Symp. quant. Biol.* **37**, 669–80.

HUXLEY, A. F. and STÄMPFLI, R. (1949). Evidence for saltatory conduction in peripheral myelinated nerve fibres. *J. Physiol., Lond.* **108**, 315.

HUXLEY, A. F. and STÄMPFLI, R. (1951). Effect of potassium and sodium on resting and action potentials of single myelinated nerve fibres. *J. Physiol., Lond.* **122**, 496–508.

HUXLEY, A. F. and STRAUB, R. W. (1958). Local activation and interfibrillar structures in striated muscle. *J. Physiol., Lond.* **143**, 40–1 P.

HUXLEY, A. F. and TAYLOR, R. E. (1955). Function of Krause's membrane. *Nature, Lond.* **176**, 1068.

HUXLEY, A. F. and TAYLOR, R. E. (1958). Local activation of striated muscle fibres. *J. Physiol., Lond.* **144**, 426–41.

HUXLEY, H. E. (1953). X-ray analysis and the problem of muscle. *Proc. R. Soc.* B **141**, 59–66.

HUXLEY, H. E. (1957). The double array of filaments in cross-striated muscle. *J. biophys. biochem. Cytol.* **3**, 631–48.

HUXLEY, H. E. (1960). Muscle Cells. In *The Cell* (ed. J. Brachet and A. E. Mirsky) vol. 4, 365–481. New York: Academic Press.

HUXLEY, H. E. (1963). Electron microscope studies on the structure of natural and synthetic protein filaments from striated muscle. *J. molec. Biol.* **7**, 281–308.

HUXLEY, H. E. (1964). Evidence for continuity between the central elements of the triads and extracellular space in frog sartorius muscle. *Nature, Lond.* **202**, 1067–71.

HUXLEY, H. E. (1969). The mechanism of muscular contraction. *Science* **164**, 1356–66.

HUXLEY, H. E. (1971). The structural basis of muscular contraction. *Proc. R. Soc.* B **178**, 131–49.

HUXLEY, H. E. (1972). Molecular basis of contraction in cross-striated muscles. In *The Structure and Function of Muscle*, 2nd edn (ed. G. H. Bourne), vol. 1, 301–87. New York: Academic Press.

HUXLEY, H. E. (1973). Structural changes in the actin- and myosin-containing filaments during contraction. *Cold Spring Harb. Symp. quant. Biol.* **37**, 361–76.

HUXLEY, H. E. and BROWN, W. (1967). The low-angle X-ray diagram of vertebrate striated muscle and its behaviour during contraction and rigor. *J. molec. Biol.* **30**, 383–434.

HUXLEY, H. E., BROWN, W. and HOLMES, K. C. (1965). Constancy of axial spacings in frog sartorius muscle during contraction. *Nature, Lond.* **206**, 1358.

HUXLEY, H. E. and HANSON, J. (1954). Changes in the cross-striations of muscle during contraction and stretch and their structural interpretation. *Nature, Lond.* **173**, 973–6.

INFANTE, A. A. and DAVIES, R. E. (1962). Adenosine triphosphate breakdown during a single isotonic twitch of frog sartorius muscle. *Biochem. biophys. Res. Commun.* **9**, 410–15.

IVERSEN, L. L. (1963). Uptake of noradrenaline by isolated perfused rat heart. *Brit. J. Pharmacol.* **21**, 523–37.

IVERSEN, L. L. (1967). *The Uptake and Storage of Noradrenaline in Sympathetic Nerves.* London: Cambridge University Press.

IVERSEN, L. L. (1971). Role of transmitter uptake mechanisms in synaptic neurotransmission. *Brit. J. Pharmacol.* **41**, 571–91.

IVERSEN, L. L. and CALLINGHAM, B. A. (1971). Adrenergic Transmission. In *Fundamentals of Biochemical Pharmacology* (ed. Z. Bacq). Oxford: Pergamon Press.

JACK, J. J. B., NOBLE, D. and TSIEN, R. W. (1975). *Electric Current Flow in Excitable Cells*. London: Oxford University Press.

JENKINSON, D. H. (1973). Classification and properties of peripheral adrenergic receptors. *Brit. med. Bull.* **29**, 142–7.

JEWELL, B. R. (1959). The nature of the phasic and tonic responses of the anterior byssal retractor of *Mytilus. J. Physiol., Lond.* **149**, 154–77.

JEWELL, B. R. and RÜEGG, J. C. (1966). Oscillatory contraction of insect fibrillar muscle after glycerol extraction. *Proc. R. Soc.* B **164**, 428–59.

JEWELL, B. R. and WILKIE, D. R. (1958). An analysis of the mechanical components in frog striated muscle. *J. Physiol., Lond.* **143**, 515–40.

JEWELL, B. R. and WILKIE, D. R. (1960). The mechanical properties of relaxing muscle. *J. Physiol., Lond.* **152**, 30–47.

JÖBSIS, F. F. and O'CONNOR, M. J. (1966). Calcium release and reabsorption in the sartorius muscle of the toad. *Biochem. Biophys. Res. Comm.* **25**, 246–52.

JOHNSON, W. H. (1962). Tonic mechanisms in smooth muscles. *Physiol. Rev.* **42**, suppl. **5**, 113–43.

JOHNSTONE, B. M., TAYLOR, J. J. and BOYLE, A. J. (1970). Mechanics of guinea-pig cochlea. *J. acoust. Soc. Amer.* **47**, 504–9.

JULIAN, F. J. (1969). Activation in a skeletal muscle contraction model with a modification for insect fibrillar muscle. *Biophys. J.* **9**, 547–70.

JULIAN, F. J. (1971). The effect of calcium on the force–velocity relation of briefly glycerinated frog muscle fibres. *J. Physiol., Lond.* **218**, 117–45.

JULIAN, F. J. and SOLLINS, M. R. (1973). Regulation of force and speed of shortening in muscle contraction. *Cold Spring Harb. Symp. quant. Biol.* **37**, 635–46.

KALMIJN, A. J. (1966). Electro-perception in sharks and rays. *Nature, Lond.* **212**, 1232–3.

KALMIJN, A. J. (1971). The electric sense of sharks and rays. *J. exp. Biol.* **55**, 371–83.

KANEKO, A. (1970). Physiological and morphological identification of horizontal, bipolar and amacrine cells in goldfish retina. *J. Physiol., Lond.* **207**, 623–33.

KAO, C. Y. (1960). Postsynaptic electrogenesis in septate giant axons. II. Comparison of medial and lateral giant axons of crayfish. *J. Neurophysiol.* **23**, 618–35.

KARLIN, A., WEILL, C. A., McNAMEE, M. G. and VALDERRAMA, R. (1976). Facets of the structures of acetylcholine receptors from *Electrophorus* and *Torpedo*. *Cold Spring Harb. Symp. quant. Biol.* **40**, 203–210.

KATZ, B. (1937). Experimental evidence for a non-conducted response of nerve to subthreshold stimulation. *Proc. R. Soc.* B **124**, 244–76.

KATZ, B. (1949). Les constants électriques de la membrane du muscle. *Arch. Sci. physiol.* **3**, 285.

KATZ, B. (1950). Depolarization of sensory terminals and the initiation of impulses in the muscle spindle. *J. Physiol., Lond.* **111**, 261–82.

KATZ, B. (1962). The transmission of impulses from nerve to muscle, and the subcellular unit of synaptic action. *Proc. R. Soc.* B **155**, 455–79.

KATZ, B. (1966). *Nerve, Muscle and Synapse.* New York: McGraw-Hill.

KATZ, B. and MILEDI, R. (1963). A study of spontaneous miniature potentials in spinal motoneurones. *J. Physiol., Lond.* **168**, 389–422.

KATZ, B. and MILEDI, R. (1965*a*). Propagation of electrical activity in motor nerve terminals. *Proc. R. Soc.* B **161**, 453–82.

KATZ, B. and MILEDI, R. (1965*b*). The measurement of synaptic delay, and the time course of acetylcholine release at the neuromuscular junction. *Proc. R. Soc.* B **161**, 483–95.

KATZ, B. and MILEDI, R. (1967*a*). Tetrodotoxin and neuromuscular transmission. *Proc. R. Soc.* B **167**, 8–22.

KATZ, B. and MILEDI, R. (1967*b*). The release of acetylcholine from nerve endings by graded electric pulses. *Proc. R. Soc.* B **167**, 23–38.

KATZ, B. and MILEDI, R. (1967*c*). The timing of calcium action during neuromuscular transmission. *J. Physiol., Lond.* **189**, 535–44.

KATZ, R. and MILEDI, R. (1970). Membrane noise produced by acetylcholine. *Nature, Lond.* **226**, 962–3.

KATZ, R. and MILEDI, R. (1972). The statistical nature of the acetylcholine potential and its molecular components. *J. Physiol., Lond.* **224**, 665–700.

KATZ, B. and THESLEFF, S. (1957*a*). A study of the 'desensitization' produced by acetylcholine at the motor endplate. *J. Physiol., Lond.* **138**, 63–80.

KATZ, B. and THESLEFF, S. (1957*b*). On the factors which determine the amplitude of the 'miniature end-plate potential'. *J. Physiol., Lond.* **137**, 267–78.

KENDRICK-JONES, J., LEHMAN, W. and SZENT-GYÖRGYI, A. G. (1970). Regulation in molluscan muscles. *J. molec. Biol.* **54**, 313–26.

KERKUT, G. A. and COTTRELL, G. A. (1963). Acetylcholine and 5-hydroxytryptamine in the snail brain. *Comp. Biochem. Physiol.* **8**, 53–63.

KERKUT, G. A. and WALKER, R. J. (1962). The specific chemical sensitivity of *Helix* nerve cells. *Comp. Biochem. Physiol.* **7**, 277–88.

KEYNES, R. D. (1951). The ionic movements during nervous activity. *J. Physiol., Lond.* **114**, 119.

KEYNES, R. D. (1957). Electric organs. In *The Physiology of Fishes* (ed. M. E. Brown) vol. 2, pp. 323–43. New York: Academic Press.

KEYNES, R. D. (1963). Chloride in the squid giant axon. *J. Physiol., Lond.* **169**, 690–705.

KEYNES, R. D., BENNETT, M. V. L. and GRUNDFEST, H. (1961). Studies on the morphology and electrophysiology of electric organs. II. Electrophysiology of the electric organ of *Malapterurus electricus*. In *Bioelectrogenesis* (ed. C. Chagas and A. Paes de Carvalho), pp. 102–12. Amsterdam: Elsevier.

KEYNES, R. D., BEZANILLA, F., ROJAS, E. and TAYLOR, R. E. (1975). The rate of action of tetrodotoxin on sodium conductance in the squid giant axon. *Phil. Trans. R. Soc.* B **270**, 365–76.

KEYNES, R. D. and LEWIS, P. R. (1951). The sodium and potassium content of cephalopod nerve fibres. *J. Physiol., Lond.* **114**, 151–82.

KEYNES, R. D. and MARTINS-FERREIRA, H. (1953). Membrane potentials in the electroplates of the electric eel. *J. Physiol., Lond.* **119**, 315–51.

KEYNES, R. D. and ROJAS, E. (1974). Kinetics and steady-state properties of the charged system controlling sodium conductance in the squid giant axon. *J. Physiol., Lond.* **239**, 393–434.

KIANG, N. Y.-S., SACHS, M. B. and PEAKE, W. T. (1967). Shapes of tuning curves for single auditory nerve fibres. *J. acoust. Soc. Amer.* **42**, 1341.

KNAPPEIS, G. G. and CARLSEN, F. (1962). The ultrastructure of the Z disc in skeletal muscle. *J. Cell. Biol.* **13**, 323–35.

KOELLE, G. B. and FRIEDENWALD, J. S. (1949). A histochemical method for localising cholinesterase activity. *Proc. Soc. exp. Biol. Med.* **70**, 617–22.

KORDAS, M. (1969). The effect of membrane polarization on the time course of the end-plate current in frog sartorius muscle. *J. Physiol., Lond.* **204**, 493–502.

KORENBROT, J. I. and CONE, R. A. (1972). Dark ionic flux and effects of light in isolated rod outer segments. *J. gen. Physiol.* **60**, 20–45.

KRAVITZ, E. A., KUFFLER, S. W. and POTTER, D. D. (1963). Gamma-aminobutyric acid and other blocking compounds in Crustacea. III. Their relative concentrations in separated motor and inhibitory axons. *J. Neurophysiol.* **26**, 739–51.

KRETSCHMAR, K. M. and WILKIE, D. R. (1969). A new approach to freezing tissues rapidly. *J. Physiol., Lond.* **202**, 66–67P.

KRNJEVIĆ, K. (1974). Chemical nature of synaptic transmission in vertebrates. *Physiol. Rev.* **54**, 418–540.

KRNJEVIĆ, K. and MITCHELL, J. F. (1961). The release of acetylcholine in the isolated rat diaphragm. *J. Physiol., Lond.* **155**, 246–62.

KRNJEVIĆ, K. and PHILLIS, J. W. (1963*a*). Iontophoretic studies of neurones in the mammalian cerebral cortex. *J. Physiol., Lond.* **165**, 274–304.

KRNJEVIĆ, K. and PHILLIS, J. W. (1963*b*). Acetylcholine-sensitive cells in the cerebral cortex. *J. Physiol., Lond.* **166**, 296–327.

KUFFLER, S. W. (1946). The relation of electrical potential changes to contracture in skeletal muscle. *J. Neurophysiol.* **9**, 367–77.

KUFFLER, S. W. (1953). Discharge patterns and functional organization of mammalian retina. *J. Neurophysiol.* **16**, 37–68.

KUFFLER, S. W. (1967). Neuroglial cells: physiological properties and a potassium mediated effect of neuronal activity on the glial membrane potential. *Proc. Roy. Soc.* B **168**, 1–21.

KUFFLER, S. W. and EDWARDS, C. (1958). Mechanism of gamma-aminobutyric acid (GABA) action and its relation to synaptic inhibition. *J. Neurophysiol.* **21**, 589–610.

KUFFLER, S. W., HUNT, C. C. and QUILLIAM, J. P. (1951). Function of medullated small-nerve fibres in mammalian ventral roots: efferent muscle spindle innervation. *J. Neurophysiol.* **14**, 29–54.

KUFFLER, S. W. and NICHOLLS, J. G. (1966). The physiology of neuroglial cells. *Ergebn. Biol.* **57**, 1–90.

KUFFLER, S. W. and NICHOLLS, J. G. (1976). *From Neuron to Brain.* Sunderland, Mass: Sinauer Associates.

KUFFLER, S. W. and VAUGHAN WILLIAMS, E. M. (1953*a*). Small-nerve junctional potentials. The distribution of small motor nerves to frog skeletal muscle, and the membrane characteristics of the fibres they innervate. *J. Physiol., Lond.* **121**, 289–317.

KUFFLER, S. W. and VAUGHAN WILLIAMS, E. M. (1953*b*). Properties of the 'slow' skeletal muscle fibres of the frog. *J. Physiol., Lond.* **121**, 318–40.

KUFFLER, S. W. and YOSHIKAMI, D. (1975*a*). The distribution of acetylcholine sensitivity at the post-synaptic membrane of vertebrate skeletal twitch muscles: iontophoretic mapping in the micron range. *J. Physiol., Lond.* **244**, 703–30.

KUFFLER, S. W. and YOSHIKAMI, D. (1975*b*). The number of transmitter molecules in a quantum: an estimate from iontophoretic application of acetylcholine at the neuromuscular synapse. *J. Physiol., Lond.* **251**, 465–82.

KUNO, M. and RUDOMIN, P. (1966). The release of acetylcholine from the spinal cord of the cat by antidromic stimulation of motor nerves. *J. Physiol., Lond.* **187**, 177–93.

KUSHMERICK, M. J. (1977). Energy balance in muscle contraction: a biochemical approach. *Current Topics in Bioenergetics*, **7**, 1–37.

KUSHMERICK, M. J. and DAVIES, R. E. (1969). The chemical energetics of muscle contraction II. The chemistry, efficiency and power of maximally working sartorious muscles. *Proc. R. Soc.* B **174**, 315–53.

LANZAVECCHIA, G. (1968). Studi sulla muscolatura elicoidale e paramiosinica. II. Meccanismo di contrazione dei muscoli elicoidale. *Atti Accad. Naz. Lincei* **44**, 575–83.

LAPIQUE, L. (1907). *J. Physiol., Path. gén.* **9**, 622.

LAVERTY, R. (1973). The mechanisms of action of some hypertensive drugs. *Brit. med. Bull.* **29**, 152–7.

LEE, C. Y. (1970). Elapid neurotoxins and their mode of action. *Clinical Toxicol.* **3**, 457–72.

LEHMAN, W. (1976). Phylogenetic diversity of the proteins regulating muscular contraction. *Int. Rev. Cytol.* **44**, 55–92.

LEKSELL, L. (1945). The action potential and excitatory effects of the small ventral root fibres to skeletal muscle. *Acta physiol. scand. Suppl.* **31**, 1–84.

LESTER, H. A. (1977). The response to acetylcholine. *Sci. Amer.* **236**, (2), 106–18.

LEVICK, W. R. (1972). Receptive fields of retinal ganglion cells. In *Handbook of Sensory Physiology* (ed. H. Autrum *et al.*) **7**(2), 531–66. Berlin: Springer.

LEVIN, A. and WYMAN, J. (1927). The viscous elastic properties of muscle. *Proc. R. Soc.* B **101**, 218–43.

LEVINSON, S. R. and MEVES, H. (1975). The binding of tritiated tetrodotoxin to squid giant axons. *Phil Trans. R. Soc.* B **270**, 349–52.

LILEY, A. W. (1956). The effects of presynaptic polarization on the spontaneous activity at the mammalian neuromuscular junction. *J. Physiol., Lond.* **134**, 427–43.

LISSMANN, H. W. (1951). Continuous electrical signals from the tail of a fish, *Gymnarchus niloticus* Cuv. *Nature, Lond.* **167**, 201.

LISSMANN, H. W. (1958). On the function and evolution of electric organs in fish. *J. exp. Biol.* **35**, 156–91.

LISSMANN, H. W. and MACHIN, K. E. (1958). The mechanism of object location in *Gymnarchus niloticus* and similar fish. *J. exp. Biol.* **35**, 451–86.

LIVINGSTON, R. B. (1959). Central control of receptors and sensory transmission systems. In *Handbook of Physiology* section 1, vol. 1, pp. 741–60. Washington, D.C.: American Physiological Society.

LLOYD, D. P. C. (1949). Post-tetanic potentiation of response in monosynaptic reflex pathways of the spinal cord. *J. gen. Physiol.* **33**, 147–70.

LOEWENSTEIN, W. R. (1956). Modulation of cutaneous mechano-receptors by sympathetic stimulation. *J. Physiol., Lond.* **132**, 40–60.

LOEWENSTEIN, W. R. (1976). Permeable junctions. *Cold Spring Harb. Symp. quant. Biol.* **40**, 49–63.

LOEWI, O. (1921). Über humorale Übertragbarkeit der Herznerven-wirkung. *Pflügers Arch. ges. Physiol.* **189**, 239–42.

LONGENECKER, H. E., HURLBUT, W. P., MAURO, A. and CLARK, A. W. (1970). Effects of black widow spider venom on the frog neuromuscular junction. *Nature, Lond.* **225**, 701–5.

LOWENSTEIN, O. (1950). Labyrinth and equilibrium. *Symp. Soc. exp. Biol.* **4**, 60–82.

LOWENSTEIN, O., OSBORNE, H. P. and WERSÄLL, J. (1964). Structure and innervation of the sensory epithelia in the labyrinth of the thornback ray (*Raja clavata*). *Proc. R. Soc.* B **160**, 1–12.

LOWENSTEIN, O. and ROBERTS, T. D. M. (1950). The equilibrium function of the otolith organs of the thornback ray (*Raja clavata*). *J. Physiol., Lond.* **110**, 392–415.

LOWENSTEIN, O. and ROBERTS, T. D. M. (1951). The localization and analysis of the responses to vibration from the isolated elasmobranch labyrinth. A contribution to the problem of the evolution of hearing in vertebrates. *J. Physiol., Lond.* **114**, 471–89.

LOWENSTEIN, O. and SAND, A. (1936). The activity of the horizontal semicircular canal of the dogfish, *Scyllium canicula*. *J. exp. Biol.* **13**, 416–28.

LOWENSTEIN, O. and SAND, A. (1940*a*). The mechanism of the semicircular canal. A study of the responses of single-fibre preparations to angular accelerations and to rotation at constant speed. *Proc. R. Soc.* B **129**, 256–75.

LOWENSTEIN, O. and SAND, A. (1940*b*). The individual and integrated activity of the semicircular canals of the elasmobranch labyrinth. *J. Physiol., Lond.* **99**, 89–101.

LOWEY, S. and RISBY, D. (1971). Light chains from fast and slow muscle myosins. *Nature, Lond.* **234**, 81–5.

LOWY, J. and HANSON, J. (1962). Ultrastructure of invertebrate smooth muscle. *Physiol. Rev.* **42**, suppl. 5, 34–42.

LOWY, J. and MILLMAN, B. M. (1963). The contractile mechanism of the anterior byssus retractor muscle of *Mytilus edulis*. *Phil. Trans. R. Soc.* B **246**, 105–48.

LOWY, J., MILLMAN, B. M. and HANSON, J. (1964). Structure and function in smooth tonic muscles of lamellibranch molluscs. *Proc. Roy. Soc.* B **164**, 525–36.

LOWY, J., POULSON, F. R. and VIBERT, P. J. (1970). Myosin filaments in vertebrate smooth muscle. *Nature, Lond.* **225**, 1053–4.

LUDWIGH, E. and McCARTHY, E. F. (1938). Absorption of visible light by the refractive media of the human eye. *Archs Ophthal., N.Y.* **20**, 37–51.

LUNDBERG, A. and QUILISCH, H. (1953). On the effect of calcium on presynaptic potentiation and depression at the neuromuscular junction. *Acta physiol. scand. suppl.* **111**, 121–9.

LYMN, R. W. and TAYLOR, E. W. (1971). Mechanism of adenosine triphosphate hydrolysis by actomyosin. *Biochemistry* **10**, 4617–24.

MACHIN, K. E. (1962). Electric receptors. *Symp. Soc. exp. Biol.* **16**, 227–44.

MACHIN, K. E. and LISSMANN, H. W. (1960). The mode of operation of the electric receptors in *Gymnarchus niloticus. J. exp. Biol.* **37**, 801–11.

MACHIN, K. E. and PRINGLE, J. W. S. (1959). The physiology of insect fibrillar muscle. II. Mechanical properties of a beetle flight muscle. *Proc. R. Soc.* B **151**, 204–25.

MACHIN, K. E. and PRINGLE, J. W. S. (1960). The physiology of insect fibrillar muscle. III. The effect of sinusoidal changes of length on a beetle flight muscle. *Proc. R. Soc.* B **152**, 311–30.

MACHIN, K. E., PRINGLE, J. W. S. and TAMASIGE, M. (1962). The physiology of insect fibrillar muscle. IV. The effect of temperature on a beetle flight muscle. *Proc. R. Soc.* B **155**, 493–9.

MACLENNAN, D. H. and WONG, P. T. S. (1971). Isolation of a calcium-sequestering protein from sarcoplasmic reticulum. *Proc. Natn. Acad. Sci. U.S.A.* **68**, 1231–5.

MACNICHOL, E. F. and SVAETICHIN, G. (1958). Electric responses from isolated retinas of fishes. *Am. J. Ophthal.* **46**, 26–40.

MAGLEBY, K. L. and STEVENS, C. F. (1972*a*). The effect of voltage on the time course of end-plate currents. *J. Physiol., Lond.* **223**, 151–71.

MAGLEBY, K. L. and STEVENS, C. F. (1972*b*). A quantitative description of end-plate currents. *J. Physiol., Lond.* **223**, 173–97.

MANNHERZ, H. G. and GOODY, R. S. (1976). Proteins of contractile systems. *A. Rev. Biochem.* **45**, 427–65.

MARKS, W. B. (1965). Visual pigments of single goldfish cones, *J. Physiol., Lond.* **178**, 14–32.

MARKS, W. B., DOBELLE, W. H. and MACNICHOL, E. F. (1964). Visual pigments of single primate cones. *Science* **143**, 1181–3.

MARNAY, A. and NACHMANSOHN, D. (1938). Choline esterase in voluntary muscle. *J. Physiol., Lond.* **92**, 37–47.

MARSH, B. B. (1952). The effects of ATP on the fibre-volume of a muscle homogenate. *Biochim. biophys. Acta* **9**, 247–60.

MARTIN, A. R. (1955). A further study of the statistical composition of the end-plate potential. *J. Physiol., Lond.* **130**, 114–22.

MATTHEWS, B. H. C. (1933). Nerve endings in mammalian muscle. *J. Physiol., Lond.* **78**, 1–53.

MATTHEWS, P. B. C. (1962). The differentiation of two types of fusi-motor fibre by their effects on the dynamic response of muscle spindle primary endings. *Q. Jl exp. Physiol.* **47**, 324–33.

MATTHEWS, P. B. C. (1964). Muscle spindles and their motor control. *Physiol. Rev.* **44**, 219–88.

MATTHEWS, P. B. C. (1972). *Mammalian Muscle Receptors and their Central Actions.* London: Edward Arnold.

MATTHEWS, P. B. C. (1977). Muscle afferents and kinaesthesia. *Brit. med. Bull.* **33**, 137–42.

MATURANA, H. R., LETTVIN, J. Y., McCULLOCH, W. S. and PITTS, W. H. (1960). Anatomy and physiology of vision in the frog. *J. gen. Physiol.* **43**, suppl. 2, 129–75.

McNUTT, N. S. and WEINSTEIN, R. S. (1973). Membrane ultrastructure at mammalian intercellular junctions. *Progr. Biophys.* **26**, 45–101.

MILEDI, R. (1973). Transmitter release induced by injection of calcium ions into nerve terminals. *Proc. R. Soc.* B **183**, 421–5.

MILL, P. J. and KNAPP, M. F. (1970). The fine structure of obliquely striated body wall muscles in the earthworm, *Lumbricus terrestris* Linn. *J. Cell. Sci.* **7**, 233–61.

MILLER, R. F. and DOWLING, J. E. (1970). Intracellular responses of the Müller (glial) cells of the mudpuppy retina: their relation to the b-wave of the electroretinogram. *J. Neurophysiol.* **33**, 323–41.

MILLMAN, B. M., ELLIOTT, G. F. and LOWY, J. (1967). Axial period of actin filaments. X-ray diffraction studies. *Nature, Lond.* **213**, 356–8.

MOORE, P. B., HUXLEY H. E. and DEROSIER, D. J. (1970). Three-dimensional reconstruction of F-actin, thin filaments and decorated thin filaments. *J. molec. Biol.* **50**, 279–95.

MORIMOTO, K. and HARRINGTON, W. F. (1974). Substructure of the thick filament of vertebrate striated muscle. *J. Molec. Biol.* **83**, 83–97.

MORITA, H. and YAMASHITA, S. (1959). Generator potential of an insect chemoreceptor. *Science* **130**, 922.

MORTON, R. A. and PITT, G. A. J. (1955). Studies on rhodopsin 9. pH and the hydrolysis of indicator yellow. *Biochem. J.* **59**, 128–34.

MURRAY, R. W. (1962). The response of the ampullae of Lorenzini of elasmobranchs to electrical stimulation. *J. exp. Biol.* **39**, 119–28.

NAGAI, T., MAKINOSE, M. and HASSELBACH, W. (1960). Der physiologische Erschlaffungsfaktor und die Muskelgrana. *Biochim. Biophys. Acta* **54**, 338–44.

NARAHASHI, T. (1963). The properties of insect axons. *Adv. Insect Physiol.* **1**, 176–256.

NARAHASHI, T. (1974). Chemicals as tools in the study of excitable membranes. *Physiol. Rev.* **54**, 812–89.

NARAHASHI, T., MOORE, J. W. and SCOTT, W. R. (1964). Tetrodotoxin blockage of sodium conductance increase in lobster giant axons. *J. gen. Physiol.* **47**, 965–74.

NASTUK, W. L. (1953). The electrical activity of the muscle cell membrane at the neuro-muscular junction. *J. cell. comp. Physiol.* **42**, 249–72.

NASTUK, W. L. and HODGKIN, A. L. (1950). The electrical activity of single muscle fibres. *J. cell. comp. Physiol.* **35**, 39.

NEHER, E. and SAKMANN, B. (1976). Single-channel currents recorded from membrane of denervated frog muscle cells. *Nature, Lond.* **260**, 799–802.

NEVILLE, A. C. (1963). Motor unit distribution of the dorsal longitudinal flight muscles in locusts. *J. exp. Biol.* **40**, 123–36.

NIEDERGERKE, R. (1956). The potassium chloride contracture of the heart and its modification by calcium. *J. Physiol., Lond.* **134**, 584–99.

NISHI, S. and KOKETSU, K. (1960). Electrical properties and activities of single sympathetic neurons in frogs. *J. cell. comp. Physiol.* **55**, 15–30.

NOBLE, D. (1966). Applications of Hodgkin–Huxley equations to excitable tissues. *Physiol. Rev.* **46**, 1–50.

NOBLE, D. (1974). Cardiac action potentials and pacemaker activity. In *Recent Advances in Physiology*, 9th edition (ed. R. J. Linden), pp. 1–50. Edinburgh: Churchill Livingstone.

NOBLE, D. (1975). *The Initiation of the Heartbeat.* Oxford: University Press.

NOBLE, D. and STEIN, R. B. (1966). The threshold conditions for initiation of action potentials by excitable cells. *J. Physiol., Lond.* **187**, 129–62.

NONNER, W., ROJAS, E. and STÄMPFLI, R. (1975). Gating currents in the node of Ranvier: voltage and time dependence. *Phil. Trans. R. Soc.* B **270**, 483–92.

OFFER, G. (1974). The molecular basis of muscular contraction. In *Companion to Biochemistry* (ed. A. T. Bull, J. R. Lagnado, J. O. Thomas and K. F. Tipton), pp. 623–71. London: Longman.

OHTSUKI, I., MASAKI, T., NOMOMURA, T. and EBASHI, S. (1967). Periodic distribution of troponin along the thin filament. *J. Biochem., Tokyo* **61**, 817–19.

ORKAND, R. K., NICHOLLS, J. G. and KUFFLER, S. W. (1966). The effect of nerve impulses on the membrane potential of glial cells in the central nervous system of amphibia. *J. Neurophysiol.* **29**, 788–806.

ØSTERBERG, E. (1935). Topography of the layer of rods and cones in the human retina. *Acta ophthal.* suppl. **6**.

OTSUKA M., IVERSEN, L. L., HALL, Z. W. and KRAVITZ, E. A. (1966). Release of gamma-aminobutyric acid from inhibitory nerves of lobster. *Proc. Natn. Acad. Sci. U.S.A.* **56**, 1110–15.

OTTOSON, D. (1964). The effect of sodium deficiency on the response of the isolated muscle spindle. *J. Physiol., Lond.* **171**, 109–18.

PAGE, S. G. (1964). Filament lengths in resting and excited muscles. *Proc. R. Soc.* B **160**, 460–6.

PAGE, S. G. and HUXLEY, H. E. (1963). Filament lengths in striated muscle. *J. Cell Biol.* **19**, 369–90.

PAK, W. L. (1968). Rapid photoresponses in the retina and their relevance to vision research. *Photochem. Photobiol.* **8**, 495–503.

PALADE, G. E. and PALAY, S. L. (1954). Electron microscope observations of interneuronal and neuromuscular synapses. *Anat. Rec.* **118**, 335.

PARRY, D. A. D and SQUIRE, J. M. (1973). Structural role of tropomysin in muscle regulation: analysis of the X-ray diffraction patterns from relaxed and contracting muscles. *J. molec. Biol.* **75**, 33–55.

PATON, W. D. M. (1958). Central and synaptic transmission in the nervous system. *A. Rev. Physiol.* **20**, 431–70.

PEACHEY, L. D. (1961). Structure of the longitudinal body muscles of *Amphioxus. J. biophys. biochem. Cytol.* **10**, 159–76.

PEACHEY, L. D. (1965). The sarcoplasmic reticulum and transverse tubules of the frog's sartorius. *J. Cell Biol.* **25** (part 2), 209–32.

PENN, R. D. and HAGINS, W. A. (1969). Signal transmission along retinal rods and the origin of the electroretinographic *a*-wave. *Nature, Lond.* **223**, 201–5.

PENN, R. D. and HAGINS, W. A. (1972). Kinetics of the photocurrent of retinal rods. *Biophys. J.* **12**, 1073–94.

PENNYCUICK, C. J. (1964). Frog fast muscle. I. Mechanical power output in isotonic twitches. *J. exp. Biol.* **41**, 91–111.

PEPE, F. A. (1967*a*). The myosin filament. I. Structural organization from antibody staining observed in electron microscopy. *J. molec. Biol.* **27**, 203–25.

PEPE, F. A. (1967*b*). The myosin filament. II. Interaction between myosin and actin filaments observed using antibody staining in fluorescent and electron microscopy. *J. molec. Biol.* **27**, 227–36.

PERRY, S. V. (1975). The contractile and regulatory proteins of the myocardium. In *Contraction and Relaxation in the Myocardium* (ed. W. G. Nayler), pp. 29–77. London: Academic Press.

PIERCE, G. W. and GRIFFIN, D. R. (1938). Experimental determination of supersonic notes emitted by bats. *J. Mammal.* **19**, 454–5.

PIRENNE, M. H. (1943). Binocular and monocular threshold of vision. *Nature, Lond.* **152**, 698–9.

PIRENNE, M. H. (1944). Rods and cones, and Thomas Young's theory of colour vision. *Nature, Lond.* **154**, 741–2.

PIRENNE, M. H. (1962). Visual functions in man. Chaps 1–11 in vol. 2 of Davson (1962).

PODOLSKY, R. J. (1960). Kinetics of muscular contraction: the approach to the steady state. *Nature, Lond.* **188**, 666–8.

PODOLSKY, R. J. and NOLAN, A. C. (1971). Cross-bridge properties derived from physiological studies of frog muscle fibres. In *Contractility of Muscle Cells* (ed. R. J. Podolsky). New Jersey: Prentice-Hall.

PODOLSKY, R. J. and NOLAN, A. C. (1973). Muscle contraction transients, cross-bridge kinetics and the Fenn effect. *Cold Spring Harb. Symp. quant. Biol.* **37**, 661–80.

POINCELOT, R., MILLAR, P. G., KIMBEL, R. L. and ABRAHAMSON, E. W. (1969). Lipid to protein chromophore transfer in the photolysis of visual pigments. *Nature, Lond.* **221**, 256–67.

POLISSAR, M. (1952). Physical chemistry of contractile process in muscle. *Am. J. Physiol.* **168**, 766–811.

POLYAK, S. (1941). *The Retina.* Chicago: University Press.

PORTER, C. W. and BARNARD, E. A. (1975). The density of cholinergic receptors at the endplate postsynaptic membrane: ultrastructural studies in two mammalian species. *J. membrane Biol.* **20**, 31–49.

PORTER, K. R. and PALADE, G. E. (1957). Studies on the endoplasmic reticulum. III. Its form and distribution in striated muscle cells. *J. biophys. biochem. Cytol.* **3**, 269–300.

PORTZEHL, H. (1957). Die Bindung des Erschlaffungsfaktors von Marsh an die Muskelgrana. *Biochim. biophys. Acta* **26**, 373–7.

PORTZEHL, H., CALDWELL, P. C. and RÜEGG, J. C. (1964). The dependence of contraction and relaxation of muscle fibres from the crab *Maia squinado* on the internal concentration of free calcium ions. *Biochim. biophys. Acta* **79**, 581–99.

POST, R. L. and JOLLY, P. C. (1957). The linkage of sodium, potassium and ammonium active transport across the human erythrocyte membrane. *Biochim. biophys. Acta* **25**, 118–128.

PRINGLE, J. W. S. (1949). The excitation and contraction of the flight muscles of insects. *J. Physiol., Lond.* **108**, 226–32.

PRINGLE, J. W. S. (1954). The mechanism of the myogenic rhythm of certain insect striated muscles. *J. Physiol., Lond.* **124**, 269–91.

PRINGLE, J. W. S. (1957). *Insect Flight.* London: Cambridge University Press.

PRINGLE, J. W. S. (1960). Models of Muscle. *Symp. Soc. exp. Biol.* **14**, 41–68.

PRINGLE, J. W. S. (1967). The contractile mechanism of insect fibrillar muscle. *Progr. Biophys.* **17**, 1–60.

PUMPHREY, R. J. and YOUNG, J. Z. (1938). The rates of conduction of nerve fibres of various diameters of cephalopods. *J. exp. Biol.* **15**, 453–66.

QUILLIAM, T. A. and SATO, M. (1955). The distribution of myelin on nerve fibres from Pacinian corpuscles. *J. Physiol., Lond.* **129**, 167–76.

RAFTERY, M. A., VANDLEN, R. L., REED, K. L. and LEE, T. (1976). Characterization of *Torpedo californica* acetylcholine receptor: its subunit composition and ligand-binding properties. *Cold Spring Harb. Symp. quant. Biol.* **40**, 193–202.

RAMONYCAJAL, S. (1909). *Histologie du Système Nerveux de l'Homme et des Vertébres.* Paris: Maloine.

RASMUSSEN, G. L. (1953). Further observations on the efferent cochlear bundle. *J. comp. Neurol.* **99**, 61–74.

REEDY, M. K. (1964). *Proc. R. Soc.* B **160**, 458–60.

REEDY, M. K., HOLMES, K. C. and TREGEAR, R. T. (1965). Induced changes in orientation of the cross-bridges of glycerinated insect flight muscle. *Nature, Lond.* **207**, 1276–80.

REES, C. J. C. (1969). Chemoreceptor specificity associated with the choice of feeding site by the beetle *Chrysolina brunsvicensis* on its foodplant *Hypericum hirsutum. Entomologia exp. appl.* **12**, 565–83.

REICHARDT, W. (1962). Theoretical aspects of neural inhibition in the lateral eye of *Limulus. Proc. Int. Un. Physiol. Sci.* **3**, 65–84.

REUTER, H. (1973). Divalent cations as charge carriers in excitable membranes. *Progr. Biophys.* **26**, 1–43.

RIPLEY, S. H., BUSH, B. M. H. and ROBERTS, A. (1968). Crab muscle receptor which responds without impulses. *Nature, Lond.* **218**, 1170–1.

RITCHIE, J. M. (1954). The effect of nitrate on the active state of muscle. *J. Physiol., Lond.* **126**, 155–68.

RITCHIE, J. M. (1971). Electrogenic ion pumping in nervous tissue. In *Current topics in Bioenergetics*, **4**, 327–56. New York: Academic Press.

RITCHIE, J. M., ROGART, R. B. and STRICHARTZ, G. R. (1976). A new method for labelling saxitoxin and its binding to non-myelinated fibres of the rabbit vagus, lobster walking leg, and garfish olfactory nerves. *J. Physiol., Lond.* **261**, 477–94.

RITCHIE, J. M. and STRAUB, R. W. (1975). The movement of potassium ions during electrical activity, and the kinetics of the recovery process, in the non-myelinated fibres of the garfish olfactory nerve. *J. Physiol., Lond.* **249**, 327–48.

ROBERTSON, J. D. (1956). The ultrastructure of a reptilian myoneural junction. *J. biophys. biochem. Cytol.* **2**, 381–94.

ROBERTSON, J. D. (1960). The molecular structure and contact relationships of cell membranes. *Progr. Biophys.* **10**, 343.

ROEDER, K. D. (1951). Movements of the thorax and potential changes in the thoracic muscles of insects during flight. *Biol. Bull. mar. biol. Lab., Woods Hole* **100**, 95–106.

ROJAS, E. (1976). Gating mechanism for the activation of the sodium conductance in nerve membranes. *Cold Spring Harb. Symp. quant. Biol.* **40**, 305–20.

ROMANES, G. J. (1951). The motor cell columns of the lumbosacral spinal cord of the cat. *J. comp. Neurol.* **94**, 313–63.

RÜEGG, J. C. (1968). Contractile mechanisms of smooth muscle. *Symp. Soc. exp. Biol.* **22**, 45–66.

RÜEGG, J. C. and TREGEAR, R. T. (1966). Mechanical factors affecting the ATPase activity of glycerol-extracted insect fibrillar flight muscle. *Proc. R. Soc.* B **165**, 497–512.

RUSHTON, W. A. H. (1937). The initiation of the propagated disturbance. *Proc. R. Soc.* B **124**, 201–43.

RUSHTON, W. A. H. (1951). A theory of the effects of fibre size in medulated nerve. *J. Physiol., Lond.* **115**, 101–22.

RUSHTON, W. A. H. (1956). The difference spectrum and the photosensitivity of rhodopsin in the living human eye. *J. Physiol., Lond.* **134**, 11–29.

RUSHTON, W. A. H. (1961). Rhodopsin measurement and dark-adaptation in a subject deficient in cone vision. *J. Physiol., Lond.* **156**, 193–205.

RUSHTON, W. A. H. (1962). *Visual Pigments in Man.* Liverpool: University Press.

RUSHTON, W. A. H. (1963). A cone pigment in the protanope. *J. Physiol., Lond.* **168**, 345–59.

RUSHTON, W. A. H. (1965). Visual adaptation. *Proc. R. Soc.* B **162**, 20–46.

SAND, A. (1937). The mechanism of the lateral sense organs of fishes. *Proc. R. Soc.* B **123**, 472–95.

SANDOW, A. (1944). General properties of latency relaxation. *J. cell comp. Physiol.* **24**, 221–56.

SANDOW, A. (1961). Energetics of muscular contraction. In *Biophysics of Physiological and Pharmacological Actions* (ed. A. M. Shanes). Washington, D.C.: Am. Ass. Adv. Sci.

SANDOW, A. (1965). Excitation–contraction coupling in skeletal muscle. *Pharmacol. Rev.* **17**, 265–320.

SCHER, A. M. (1962). Excitation of the heart. In *Handbook of Physiology Section* 2, *Circulation* I, 287–322. Washington, D.C.: American Physiological Society.

SCHER, A. M. (1965). Mechanical events in the cardiac cycle. In *Physiology and Biophysics* (ed. T. C. Ruch and H. D. Patton). Philadelphia: Saunders.

SCHMIDT, W. J. (1938). Polarizations-optische Analyse eines Eiweiss-Lipoid-Systems erläutert am Aussenglied der Schzellen. *Kolloidzeitschrift* **85**, 137–48.

SCHMITT, F. O., DEV, P. and SMITH, B. H. (1976). Electronic processing of information by brain cells. *Science* **193**, 114–20.

SCHNEIDER, D. (1966). Chemical sense communication in insects. *Symp. Soc. exp. Biol.* **20**, 273–97.

SCHNEIDER, D., LACHER, V. and KAISSLING, K. E. (1964). Die Reaktionsweise und das Reaktionsspaktrum von Reichzellen bei Antherea pernyi (Lepidoptera, Saturniidae). *Z. vergl. Physiol.* **48**, 632–62.

SCHNEIDER, M. F. and CHANDLER, W. K. (1973). Voltage dependent charge movement in skeletal muscle: a possible step in excitation–contraction coupling. *Nature, Lond.* **242**, 244–6.

SCHOFFENIELS, E. (1959). Ion movements studied with single isolated electroplax. *Ann. N.Y. Acad. Sci.* **81**, 285–306.

SCHULTZE, M. (1866). Zur Anatomie und Physiologie der Retina. *Arch. mikrosk. Anat. EntwMech.* **2**, 175–286.

SELBY, C. C. and BEAR, R. S. (1956). The structure of the actin-rich filaments of muscles according to X-ray diffraction. *J. biophys. biochem. Cytol.* **2**, 71–85.

SHANES, A. M. (1958). Electrochemical aspects of physiological and pharmacological action in excitable cells. *Pharmac. Rev.* **10**, 59.

SHOENBERG, C. F. and HASELGROVE, J. C. (1974). Filaments and ribbons in vertebrate smooth muscle. *Nature, Lond.* **249**, 152–4.

SHOENBERG, C. F. and NEEDHAM, D. M. (1976). A study of the mechanism of contraction in vertebrate smooth muscle. *Biol. Rev.* **51**, 53–104.

SILINSKY, E. M. (1975). On the association between transmitter secretion and the release of adenine nucleotides from mammalian motor nerve terminals. *J. Physiol., Lond.* **247**, 145–62.

SILLMAN, A. J., ITO, H. and TOMITA, T. (1969). Studies on the mass receptor potential of the isolated frog retina. II. On the basis of the ionic mechanism. *Vision Res.* **9**, 1443–51.

SINGER, S. J. and NICOLSON, G. L. (1972). The fluid mosaic model of the structure of cell membranes. *Science* **175**, 720–31.

SKOGLUND, S. (1956). Anatomical and physiological studies on knee joint innervation in the cat. *Acta physiol. scand. Suppl.* **124**, 1–101.

SMITH, A. D. (1972). Subcellular localization of noradrenaline in sympathetic neurons. *Pharmacol. Rev.* **24**, 435–7.

SMITH, D. S., GUPTA, B. L. and SMITH, U. (1966). The organization and myofilament array of insect visceral muscles. *J. Cell Sci.* **1**, 49–57.

SOBIESZEK, A. and SMALL, J. V. (1976). Myosin-linked calcium regulation in vertebrate smooth muscle. *J. molec. Biol.* **101**, 75–92.

SQUIRE, J. M. (1974). Symmetry and three-dimensional arrangement of filaments in vertebrate striated muscle. *J. molec. Biol.* **90**, 153–60.

STÄMPFLI, R. (1954). Saltatory conduction in nerve. *Physiol. Rev.* **34**, 101–12.

STEINHAUSEN, W. (1931). Über den Nachweis der Bewegung der Cupula in der intakten Bogengangsampulle des Labyrinths bei der natürlichen rotatorischen und calorischen Reizung. *Pflüg. Arch. ges. Physiol.* **228**, 322–8.

STEINHAUSEN, W. (1933). Über die Beobachtung der Cupulla in den Bogengangsampullen des Labyrinths des lebenden Hechts. *Pflüg. Arch. ges. Physiol.* **232**, 500–12.

STEN-KNUDSEN, O. (1960). Is muscle contraction initiated by internal current flow? *J. Physiol., Lond.* **151**, 363–84.

STENSIÖ, E. A. (1927). *The Downtownian and Devonian vertebrates of Spitzbergen.* Oslo: Norske Videnskaps-Akademi.

STILES, W. S. (1939). The directional sensitivity of the retina and the spectral sensitivities of the rods and cones. *Proc. R. Soc.* B **127**, 64–105.

STRUMWASSER, F. (1962). Postsynaptic inhibition and excitation produced by different branches of a single neuron and the common transmitter involved. *Proc. Int. Un. Physiol. Socs.* **2**, 801.

SUMMERS, R. D., MORGAN, R. and REIMANN, S. P. (1943). The semicircular canals as a device for vectorial resolution. *Arch. Otolaryng.* **37**, 219–37.

SZABO, T. (1965). Sense organs of the lateral line system in some electric fish of the Gymnotidae, Mormyridae and Gymnarchidae. *J. Morph.* **117**, 229–50.

SZENT-GYÖRGYI, A. G. (1960). Proteins of the myofibril. In *Structure and Function of Muscle* (ed. G. H. Bourne) vol. 2, pp. 1–54. New York: Academic Press.

SZENT-GYÖRGYI, A. G., COHEN, C. and KENDRICK-JONES, J. (1971). Paramyosin and the filaments of molluscan 'catch' muscles. II. Native filaments: isolation and characterization. *J. molec. Biol.* **56**, 239–58.

TAKEUCHI, A. and TAKEUCHI, N. (1959). Active phase of frog's end-plate potential. *J. Neurophysiol.* **22**, 395–411.

TAKEUCHI, A. and TAKEUCHI, N. (1960). On the permeability of the end-plate membrane during the action of the transmitter. *J. Physiol., Lond.* **154**, 52–67.

TAKEUCHI, A. and TAKEUCHI, N. (1964). The effect on crayfish muscle of iontophoretically applied glutamate. *J. Physiol., Lond.* **170**, 296–317.

TAKEUCHI, N. (1963). Some properties of conductance changes at the end-plate membrane during the action of acetylcholine. *J. Physiol., Lond.* **167**, 128–40.

TASAKI, I. (1953). *Nervous Transmission.* Springfield: Thomas.

TASAKI, I., ISHI, K. and ITO, H. (1943). On the relation between the conduction-rate, the fiber-diameter and the internodal distance of the myelinated nerve fiber. *Jap. J. med. Sci., Biophys.* **9**, 189–99.

TASAKI, I. and SHIMAMURA, M. (1962). Further observations on resting and action potential of intracellularly perfused squid giant axon. *Proc. natn. Acad. Sci. U.S.A.* **48**, 1571–7.

TASAKI, I., SINGER, I. and TAKENAKA, T. (1965). Effects of internal and external ionic environment on excitability of squid giant axon. *J. gen. Physiol.* **48**, 1095–123.

TASAKI, I. and TAKEUCHI, T. (1942). Weitere Studien über den Aktionsstrom der markhaltigen Nervenfaser und über die elektrosaltorische Übertragung des Nervenimpulses. *Pflügers Arch. ges. Physiol.*, **245**, 274.

TAUC, L. (1958). Processus post-synaptique d'excitation et d'inhibition dans le some neuronique de l'Aplysie et de l'Escargot. *Archs. ital. Biol.* **96**, 78–110.

TAUC, L. (1959). Interaction non synaptique entre deux neurons adjacents du ganglion abdominal de l'Aplysie. *C. r. hebd. Séanc. Acad. Sci., Paris* **248**, 1857–9.

TAUC, L. (1962). Site of origin and propagation of spike in the giant neuron of *Aplysia. J. gen. Physiol.* **45**, 1077–97.

TAUC, L. and GERSCHENFELD, H. M. (1961). Cholinergic transmission mechanisms for both excitation and inhibition in molluscan central synapses. *Nature, Lond.* **192**, 366–7.

TAUC, L. and GERSCHENFELD, H. M. (1962). A cholinergic mechanism of inhibitory synaptic transmission in a molluscan nervous system. *J. Neurophysiol.* **25**, 236–62.

TAYLOR, E. W. (1973). Mechanism of actomyosin ATPase and the problem of muscle contraction. In *Current Topics in Bioenergetics*, **5**, 201–31.

TAYLOR, R. E. (1959). Effect of procaine on electrical properties of squid axon membrane. *Am. J. Physiol.* **196**, 1071–8.

TAYLOR, R. E. (1963). Cable Theory. In *Physical Techniques in Biological Research* (ed. W. L. Nastuk) vol. 5. New York: Academic Press.

THESLEFF, S. (1955). The mode of neuromuscular block caused by acetylcholine, nicotine, decamethonium and succinylcholine. *Acta physiol. scand.* **34**, 218–31.

THOMAS, R. C. (1969). Membrane current and intracellular sodium changes in a snail neurone during extrusion of injected sodium. *J. Physiol., Lond.* **201**, 495–514.

THOMAS, R. C. (1972). Intracellular sodium activity and the sodium pump in snail neurones. *J. Physiol., Lond.* **220**, 55–71.

THURM, U. (1965). An insect mechanoreceptor. I. Fine structure and adequate stimulus. *Cold Spring Harb. Symp. quant. Biol.* **30**, 75–82.

TOMITA, T. (1965). Electrophysiological study of the mechanisms subserving color coding in the fish retina. *Cold Spr. Harb. Symp. quant. Biol.* **30**, 559–66.

TOMITA, T. (1970). Electrical activity of vertebrate photoreceptors. *quart. Rev. Biophys.* **3**, 179–222.

TOMITA, T. (1972). Light-induced potential and resistance changes in vertebrate photoreceptors. In *Handbook of Sensory Physiology* (ed. H. Autrum *et al.*) **7**(2), 483–511. Berlin: Springer.

TOMITA, T., KANEKO, A., MURAKAMI, M. and PAUTTER, E. L. (1967). Spectral response curves of single cones in the carp. *Vision Res.* **7**, 519–31.

TOYODA, J., NOSAKI, H. and TOMITA, T. (1969). Light-induced resistance changes in single photoreceptors of *Necturus* and *Gekko*. *Vision Res.* **9**, 453–63.

TRANZER, J. P. and THOENEN, H. (1967). Electronmicroscopic localization of 5-hydroxydopamine (3, 4, 5-trihydroxyphenyl-ethylamine) a new 'false' sympathetic transmitter. *Experientia* **23**, 743–5.

TREGEAR, R. T. (1975). The biophysics of fibrillar flight muscle. In *Insect Muscle* (ed. P. N. R. Usherwood), pp. 357–403. London: Academic Press.

TREGEAR, R. T. and MILLER, A. (1969). Evidence of crossbridge movement during contraction of insect flight muscle. *Nature, Lond.* **222**, 1184–5.

USHERWOOD, P. N. R. (1967). Insect neuromuscular mechanisms. *Amer. Zoologist* **7**, 553–82.

USHERWOOD, P. N. R. and GRUNDFEST, H. (1965). Peripheral inhibition in skeletal muscle of insects. *J. Neurophysiol.* **28**, 497–518.

UCHIZONO, K. (1965). Characteristics of excitatory and inhibitory synapses in the central nervous system of the cat. *Nature, Lond.* **207**, 642–3.

USSING, H. H. (1949). The distinction by means of tracers between active transport and diffusion. *Acta physiol. scand.* **19**, 43–56.

VAN EGMOND, A. A. J., GROEN, J. J. and JONGKEES, L. B. W. (1949). The mechanics of the semicircular canal. *J. Physiol., Lond.* **110**, 1–17.

VAN HARREVELD, A., CROWELL, J. and MALHOTRA, S. K. (1965). A study of extracellular space in cortical nervous tissue by freeze substitution. *J. Cell Biol.* **25**, 117–37.

VIBERT, P. J., HASELMORE, J. C., LOWY, J. and POULSEN, E. R. (1972). Structural changes in actin-containing filaments of muscle. *Nature New Biol., Lond.* **236**, 182–3.

VON BÉKÉSY, G. (1960). *Experiments in Hearing.* New York: McGraw-Hill.

VON BÉKÉSY, G. (1962). The gap between the hearing of internal and external sounds. *Symp. Soc. exp. Biol.* **16**, 267–88.

VON EULER, U. S. (1955). *Noradrenaline.* Springfield: Charles C. Thomas.

VON FRISCH, K. (1936). Über den Gehörsinn der Fische. *Biol. Rev.* **11**, 210–46.

WAKABAYASHI, T., HUXLEY, H. E., AMOS, L. A. and KLUG, A. (1975). Three dimensional image reconstruction of actin-tropomyosin complex and actin-tropomyosin-troponin T-troponin I complex. *J. molec. Biol.* **93**, 477–97.

WALD, G. (1959). The photoreceptor process in vision. In *Handbook of Physiology*, section 1, vol. 1, pp. 671–92. Washington, D.C.: American Physiological Society.

WALD, G. (1965). In *Recent Progress in Photobiology* (ed. E. J. Bowen). Oxford: Blackwell.

WALD, G., BROWN, P. K. and GIBBONS, I. R. (1962). Visual excitation: a chemo-anatomical study. *Symp. Soc. exp. Biol.* **16**, 32–57.

WALLS, G. L. (1942). *The Vertebrate Eye and its Adaptive Radiation.* Michigan: Cranbrook Institute of Science.

WATANABE, A. and GRUNDFEST, H. (1961). Impulse propagation at the septal and commisural junctions of crayfish lateral giant axons. *J. gen. Physiol.* **45**, 267–308.

WAXMAN, S. G. and BENNETT, M. V. L. (1972). Relative conduction velocities of small myelinated and non-myelinated fibres in the central nervous system. *Nature New Biol., Lond.* **238**, 217–9.

WEBER, A. and HERZ, R. (1963). The binding of calcium to actomyosin systems in relation to their biological activity. *J. biol. Chem.* **238**, 599–605.

WEBER, E. H. (1846). Der Tastsinn und das Gemeingefühl. *Handwörtenbuch d. Physiologie* **3** no. 2, 481–588.

WEIDMANN, S. (1957). Resting and action potentials of cardiac muscle. *Ann. N.Y. Acad. Sci.* **65**, 693–9.

WEIS-FOGH, T. (1956). Tetanic force and shortening in locust flight muscle. *J. exp. Biol.* **33**, 668–84.

WERBLIN, F. S. and DOWLING, J. E. (1969). Organisation of the retina of the mudpuppy, *Necturus maculosus*. II. Intracellular recording. *J. Neurophysiol.* **32**, 339–55.

WERMAN, R. (1969). An electrophysiological approach to drug-receptor mechanisms. *Comp. Biochem. Physiol.* **30**, 997–1017.

WERMAN, R. and APRISON, M. H. (1968). Glycine: the search for a spinal cord inhibitory transmitter. In *Structure and Function of Inhibitory Neuronal Mechanisms* (ed. C. von Euler, S. Skoglund and U. Söderberg). Oxford: Pergamon.

WERSÄLL, J., FLOCK, Å. and LUNDQUIST, P.-G. (1965). Structural basis for directional sensitivity in cochlear and vestibular sensory receptors. *Cold Spring Harb. Symp. quant. Biol.* **30**, 115–32.

WHITBY, L. G., AXELROD, J. and WEIL-MALHERBE, H. (1961). The fate of H^3-norepinephrine in animals. *J. Pharmacol. exp. Ther.* **132**, 193–201.

WHITE, D. C. S. and THORSON, J. (1975). *The Kinetics of Muscle Contraction.* Pergamon Press: Oxford. (Originally published in *Progr. Biophys.* **27**).

WHITFIELD, I. C. (1967). *The Auditory Pathway.* London: Edward Arnold.

WHITTAKER, V. P., DOWDALL, M. J. and BOYNE, A. F. (1972). The storage and release of acetylcholine by cholinergic nerve terminals: recent results with non-mammalian preparations. *Biochem. Soc. Symp.* **36**, 49–68.

WHITTAKER, V. P. and GRAY, E. G. (1962). The synapse: biology and morphology. *Brit. med. Bull.* **18**, 223–8.

WHITTAKER, V. P., MICHAELSON, J. A. and KIRKLAND, R. S. (1964). The separation of synaptic vesicles from nerve-ending particles ('synaptosomes'). *Biochem. J.* **90**, 293.

WHITTAKER, V. P. and ZIMMERMANN, H. (1974). Biochemical studies on cholinergic synaptic vesicles. In *Synaptic Transmission and Neuronal Interaction* (ed. M. V. L. Bennett), pp. 217–38. New York: Raven Press.

WIESEL, T. N. (1960). Receptive fields of ganglion cells in the cat's retina. *J. Physiol., Lond.* **153**, 583–94.

WILKIE, D. R. (1954). Facts and theories about muscle. *Progr. Biophys.* **4**, 288–324.

WILKIE, D. R. (1968). Heat work and phosphorylcreatine breakdown in muscle. *J. Physiol., Lond.* **195**, 157–83.

WILKIE, D. R. (1976). Energy transformation in muscle. In *Molecular Basis of Motility* (ed. L. M. G. Heilmeyer Jr., Rüegg, J. C. and Wieland, Th.), pp. 69–80. Berlin: Springer-Verlag.

WILSON, H. R. (1966). *Diffraction of X-rays by Proteins, Nucleic Acids and Viruses*. London: Arnold.

WOLBARSHT, M. L. (1965). Receptor sites in insect chemoreceptors. *Cold Spring Harb. Symp. quant. Biol.* **30**, 281–8.

WOLBARSHT, M. L. and DETHIER, V. G. (1958). Electrical activity in the chemoreceptors of the blowfly. I. Responses to chemical and mechanical stimuli. *J. gen. Physiol.* **42**, 393–412.

WOLEDGE, R. C. (1961). The thermoelastic effect of change of tension in active muscle. *J. Physiol., Lond.* **155**, 187–208.

WOLEDGE, R. C. (1963). Heat production and energy liberation in the early part of a muscular contraction. *J. Physiol., Lond.* **166**, 211–24.

WOLEDGE, R. C. (1971). Heat production and chemical change in muscle. *Progr. Biophys.* **21**, 37–74.

WOLEDGE, R. C. (1973). In vitro calorimetric studies relating to the interpretation of muscle heat experiments. *Cold Spring Harb. Symp. quant. Biol.* **37**, 629–34.

WOOD, D. W. (1957). The effect of ions upon neuromuscular transmission in a herbivorous insect. *J. Physiol., Lond.* **138**, 119–39.

WOODBURY, J. W. and CRILL, W. E. (1961). On the problem of impulse conduction in the atrium. In *Nervous Inhibition* (ed. E. Florey), pp. 124–35. Oxford: Pergamon Press.

WORTHINGTON, C. R. (1959). Large axial spacings in striated muscle. *J. molec. Biol.* **1**, 398–401.

YOSHIKAMI, S. and HAGINS, W. A. (1973). Control of the dark current in vertebrate rods and cones. In *Biochemistry and Physiology of Visual Pigments* (ed. H. Langer), pp. 245–56. Berlin: Springer.

YOUNG, J. Z. (1936). The giant nerve fibres and epistellar body of cephalopods. *Q. Jl microsc. Sci.* **78**, 367.

YOUNG, S. (1973). *Electronics in the Life Sciences*. London: Macmillan.

YOUNG, T. (1802). On the theory of light and colours. *Phil. Trans. Roy Soc.* **92**, 12–48.

ZUCKERMAN, R. (1973). Ionic analysis of photoreceptor membrane currents. *J. Physiol., Lond.* **235**, 333–54.

Index

A-band, 219, 226, 233, 319–20
absorption spectrum, of visual pigment, 431–5
accessory structures, in sense organs, 350
accommodation: ionic basis of, 90; in nerves, 67–8; visual, 423
acetylcholine, 106–7; and action on end-plate membrane, 122–32; agonists of, 179; antagonists of, 179; and contracture, 251; and motor nerves, 110–12; noise, 128–32; release hypothesis of, 135–9; receptors, 124–5, 127–8, 131–2; and synaptic vesicles, 139–40; as transmitter, 177–8, 181–7
acetylcholinesterase, 107, 147–8; inhibitors of, 180–1
acoustico-lateralis system, 376–402; cochlea, 394–9; lateral line system of fish, 376–8; otolith organs, 392–4; receptor cells, 399–402; semicircular canals, 379–92
actin, 322; filament structure of, 230–3, 234–5; localisation of, 223–8; properties of, 218–19
action potentials, 2, 90–2; in axons, single, 37–42; compound, 69–70; of electroplaques, 337–41, 342, 344; of heart muscle, 307; in insect chemoreceptors, 413; ionic basis of, 78, 84–6; and local circuit theory, 53–7; in muscle fibres, 115–16, 117, 301; of retinal ganglion cells, 467; at sensory endings, 354–62; sodium theory of, 72–4; in spinal motoneurons, 156–9; and synaptic delay, 143–5
active increment, 214
active state, 294–7
active transport, of ions, 23–7
actomyosin, 218
adaptation: sensory, 353, 362–3; visual, 421; see also dark adaptation
adenosine triphosphate (ATP): active transport of ions and, 24–5, 27; and cross-bridge activity, 249–50; and muscle contraction, 214–19, 331–2; and

relaxing factor, 260; and smooth muscle, 196; in synaptic vesicles, 140
adrenaline (epinephrine), 106, 187–93, 374
aequorin, 94, 254–5
after-loaded contraction, 265–6, 278
after-potentials, 40, 41, 42
allethrin, 100
all-or-nothing law, 39, 70
α-actinin, 219
α-bungarotoxin, 124, 125
α-methyldopa, 193
α-methylnoradrenaline, 193
α-nerve fibres, 70, 373, 403
amacrine cells, retinal, 425, 426, 468, 471
amphetamine, 193
Amphioxus, muscle fibres of, 318–19
ampullary organs, 376n, 415–16, 418–20
anode break excitation, 68; ionic basis of, 90
Antherea pernyi, olfactory responses of, 414–15
anticholinesterases, 132, 135
antidromic impulses, 143 & n, 156, 157, 174
Aplysia (sea hare), synaptic transmission in, 169–70, 173, 186, 187
arecoline, 182
arthropods: inhibition in, 305–6; mechanoreceptors of, 350, 351; muscles of, 195, 303–7, 318; neuromuscular transmission, 196, 203
aspartic acid, 195
astrocytes, 205, 206
Astroscopus, electroplaques of, 341–2, 346
ATP *see* adenosine triphosphate
atropine (DL-hyoscyamine), 107, 111, 182, 183, 186
autoradiography, 124

Balanus (barnacle), muscle fibres of, 255
barbiturates, 194
basket cell, 200
bats, echolocation by, 373
β-actinin, 219

β-erythroidine, 180, 183, 184
bipolar cells, 425, 469
birds, muscle fibres of, 302
bleaching, of visual pigment, 440, 444, 446–7
Bombyx mori, sensilla of, 415
botulinum toxin, 179

cable theory *see* core-conductor theory
caesium fluoride, 101
calcium ions: activation of ATPase by, 219; active transport of, 26–7; and coupling in muscle, 252–60, 262–3, 318; and end–plate potential, 135; fluxes of, 92–5; and heart muscle, 309–11; and photocurrent of rods, 467; and synaptic transmission, 146–7.
Calliphora: chemoreceptors of, 411–11, 413; flight muscle of, 324
calsequestrin, 260
capacitance: of axon membrane, 45, 46, 47 & n, 50, 53; of myelin sheath, 59
carbachol, 179
Carcinus (shore crab), axons of, 42, 43, 44
cardiac muscle, 306–15
cat: electroretinogram of, 457; motoneurons of, 150–68; muscle spindles of, 404–10; receptors in foot pad, 362; tongue muscle, 110–11
catch muscles, of molluscs, 323, 333–5
catecholamines, 187–93; *see also* adrenaline
cathod ray oscilloscope, 10–12
cell membrane, structure of, 13–16
Cephalaspis, semi-circular canals of, 388, 389
channels, ionic, 72, 103–5, 198
chelating agents: and muscles, 253, 254; and photocurrent of rods, 466
chemoreceptors, 349, 350; of insects, 410–15; mammalian, 351
chlorolabe, 448
chloralose, 194
chloride ions: active transport of, 26, 27; and arthropod muscle IPSP, 305; and end-plate potential, 119, 121; and motoneuron IPSP, 161, 162; and resting potential, 32–4
chlorpromazine, 193
choline, 179
cholinergic neurons, 178
chronaxie, 65
Chrysolina brunsvicensis, chemoreceptors of, 414
cicada, muscle fibres of, 324, 331
cocaine, and effect on saltatory conduction, 60

cochlea, 394–9
coding, sensory, 252–4, 362–72, 417–18
collagenase, 124, 133
colour vision, 445, 474–5
compound action potentials, 69–70
concentration cell, 16–18
conductance: during action potential, 82–5, 90–2, during end-plate potential, 126; equations for, 86–9; ionic, 71–2; *see also* potassium and sodium conductance
conduction velocity: in different fibre groups, 70; factors affecting, 54–6; in giant fibres of earthworm, 39, 40; in myelinated fibres, 64; and temperature effects, 90
cones: 424, 425, 427, 449, 470, 472; role in vision, 428–30; visual pigments of, 445–7
constant field theory, 29–30, 97
contact receptors, 349
contractile component, 277
contraction, of muscle: biochemistry of, 214–19, 297–300; dynamics of, 272–80; and heat production, 267–72, 274; isometric, 211–14, 264; isotonic, 265–7; skeletal, 317–18; theories and models of, 283–91; twitch, analysis of, 292–7
contracture, 251–2, 253
core-conductor theory, 47–53, 114
Corti, organ of, 394–5, 402
crab, muscle of, 257, 362
crayfish: giant fibres of, 172, 173–5, 202–3; muscle, 195, 307; stretch receptors of, 357, 358
creatine phosphate, 298–300
cross-bridge, 225, 227–8, 231, 320; biochemical events of, 249–50; and filament overlap, 238, 241–2, 290–1; movement of, 245–9; 245–9; in smooth muscle, 323–4
cupula, 376, 377, 379, 381, 382, 383, 385
curare, 112, 113, 178, 180, 182, 186; and effect on end-plate potential, 117, 118, 120, 132, 140
cuttlefish *see Sepia*
cyanide, and effects on active transport of sodium, 25
cyclic AMP, 189–90
cysteic acid, 195

dark adaptation, 421, 429–30, 444, 473–4
DDT (dichlorodiphenyltrichloroethane), and effects on nervous conduction, 100
decamethonium, 179
dendrites, 199, 202
density spectra, 431, 433, 436, 437
depression, neuromuscular, 140, 141, 142

desensitization, 127–8, 141, 179
DFP *see* diisopropylphosphorofluoridate
dichloroisoprenaline, 193
difference spectrum, 434, 436
3:4-dihydroxyphenylalanine (Dopa), 188
diisopropylphosphorofluoridate (DFP), 181
2:4-dinitrophenol (DNP), and effects on active transport, 24–5, 26, 101
direct inhibitory pathway, 159–60
distance receptors, 349
5,5′-dithiobis-(2-nitrobenzoate) (DTNB), 217
Donnan equilibrium system, 20–3, 32
Dopa (3:4-dihydroxyphenylalanine), 188
dopamine, 188, 189
dorsal root potential, 168
Dosidicus, sodium conductance in giant axon of, 90–1; *see also* squid
drugs: and adrenergic synapses, 192–3; and nervous conduction, 99–101; and synaptic transmission, 177–96
DTNB, 217, 300
duplicity theory, of vision, 428–30

early receptor potential, 457–9
earthworm, giant axons of, 37–40, 172
echolocation, of bats, 373
edrophonium, 180, 181
Eigenmannia, electroreceptors of, 418
elastic body theory, of muscle, 272
elastic elements, of muscle, 279, 289–91
electrical transmission at synapses, 170–6
electric eel *see Electrophorus electricus*
electric organs, 125, 173, 204, 336–47; functions and evolution of, 346–7; synaptic vesicles of, 140
electrocardiogram (ECG), 311–12
electrocyte *see* electroplaque
electrodes, recording, 7–8, 9, 184
electron microscopy, techniques of, 220–1
Electrophorus electricus (electric eel), 125, 336–41, 346
electrophysiological methods, 7–12
electroplaques, 336–46
electroreceptors, 349, 350, 415–20
electroretinogram, 456–7
electrotonic potential, 44, 47–53
end-plate, 108, 109, 110
end-plate membrane, action of acetylcholine in, 122–32
end-plate potential (EPP), 112–22; ionic basis of, 116–22; quantal nature of, 135–9; and synaptic delay, 143–5
end-plate potentials, miniature, 132–5, 138, 139
energy release, by muscle, 269, 273

ephedrine, 193
epinephrine *see* adrenaline
equilibrium receptors, 349; of vertebrates, 376–94
erythrolabe, 448
eserine (physostigmine), 107, 110, 112, 180–1, 183, 184, 186
excitation-contraction coupling, 250–63, 318–19; and calcium ions, 252–6; T system, 256–60
excitatory postsynaptic potentials (EPSP): in *Aplysia* neurons, 169–70; in crustacean muscle fibres, 305–6, 307; and initiation of action potentials, 156–9; and interactions with IPSP, 163–5; of spinal moto-neurons, 151–6; at squid giant synapse, 168–9; of sympathetic ganglion cells, 155; *see also* end-plate potential
exteroceptors, 349, 372
extrafusal muscle fibres, 373
extrinsic potentials, 54, 55
eye, vertebrate, 421–75; absolute threshold, 450–6; retinal electrical activity, 456–73; structure of, 422–7; visual pigments, 430–50

facilitation, neuromuscular, 140, 141, 142
Faraday's constant, 17, 77
FDNB (1-fluoro-2,4-dinitrobenzene), 215, 216
Fenn effect, 273, 283
ferritin, 259
fibrillar muscles, of insects, 324–33
Fidicina, tymbal muscles of, 333
fish: electric organs of, 336–47; electroreceptors of, 415–20; lateral line system, 364–5, 376–8; Mauthner fibres, 176; muscle fibres of, 302; semi-circular canals of, 382
fixation and flash technique, for visual thresholds, 428–30
fixed charges, on axon membrane, 92–5
fleas, analogy of the, 129–30
flight muscles, 305, 321, 324–33
fluid-mosaic model, of the cell membrane, 14–16, 25–6
1-fluoro-2,4-dinitrobenzene (FDNB), 215, 216
fluxes, ionic, 18: potassium, 74–5, 77, 97; sodium, 74, 76–7, 96
focal recording technique, 143
force-velocity curve, 266, 276–9, 279–80, 282
fovea, 425, 428, 444, 445
Forcipomyia (midge), flight muscles of, 324
free-loaded contraction, 265

frequency code, 353, 361; *see also* coding, sensory
fusimotor fibres *see* γ nerve fibres
fusion frequency, 214

GABA (γ-aminobutyric acid), 195
γ-aminobutyric acid (GABA), 195, 305
γ nerve fibres, 405–6, 409–10, 70, 373, 402
ganglion cells, retinal, 425, 467–9, 471–2
ganglion cells, sympathetic, 155
gap junctions (nexuses), 171, 172; in cardiac muscle, 313; in smooth muscle, 315
gating currents, and sodium densities, 101–5
generator potential, 354–6, 358–9
giant axons: and active transport, 23–6; of crayfish *see* crayfish; of cuttlefish *see* *Sepia*; drug effects on, 99–101; of earthworm, 37–40, 172; gating currents, 102–5; and membrane potentials, 30–1; perfusion experiments on, 95–9; septal synapses in, 172; of squid *see* squid
glial cells, 205–10
glutamic acid, 195
glycerinated muscle fibres, 218, 327
glycine, 194
Goldman equation, 30
Golgi tendon organs, 368–9, 402–3
Gymnarchus: electric organs of, 347; electroreceptors of, 416–18
Gymnotus carapo, electroplaques of, 343–4, 347

H zone, 219, 226–7
heart muscle, vertebrate, 107, 306–15
heat of shortening, 272, 273, 274; effect of load on, 280
heat production, of muscle, 267–72; and creatine phosphate, 297–300
hemicholinium-3, 178
Hill's analysis and equation, 273–79; limitations of, 280–3
Homarus, core-conductor theory and axons of, 52–3
horizontal cells, retinal, 425
Huxley's theory (1957) of muscle contraction, 283–6
Huxley and Simmon's theory (1971) of muscle contraction, 287–91
5-hydroxydopamine (serotonin), 191, 196

I-band, 219, 226–7, 233, 235
impedance, 45; of axon membrane, 46–7
increment threshold, 364

inhibition: in arthropod muscles, 305–6; of crustacean receptors, 195, 374; of heart muscle, 314; lateral, 371–2; pharmacology of, 193–4; postsynaptic, 159–63; presynaptic, 165–8; in spinal motoneurons, 159–68
inhibitory postsynaptic potential (IPSP): in *Aplysia* neurons, 170; in arthropod muscle fibres, 305; and interactions with EPSP, 163–5; in spinal motoneurons, 160–5; effect of strychnine on, 193
insects: chemoreception in, 410–15; fibrillar muscles of, 324–33; neuro-muscular transmission in, 195, 196
integration, neuronal, 197
intercalated discs, 313
interoreceptors, 349
intrafusal muscle fibres, 373, 404–5
indoacetic acid, 215
ionophoresis, of acetylcholine, 122–3, 133
iproniazid, 193
isometric contractions, 211–14, 264, 270–1, 280–2
*iso*prenaline, 189, 193
isotonic contraction, 265–7
isotonic release, 266–7, 168; transient changes following, 282–3
isotopes, radioactive, 74–6, 96
IS spike, 156–9

junction potential *see* postsynaptic potential

L-zone (pseudo H-zone), 228
lactic acid, 214, 215
latency relaxation, 293
latent addition, 66–7
latent period, of muscle, 293
lateral inhibition, 371–2
lateral line system, of fish, 364–5, 376–8; electroreceptors, 418–20
leech: glial cells, 206, 207; muscle of, 110–11; neurons of, 172–3
length-tension relation, of muscle, 213–14, 240, 242
Lethocerus cordofanus, flight muscle of, 248, 331–2, 333
Limulus, eye of, 358, 370, 371
lobster: inhibitory fibres of, 195; pacemaker cells of, 173
local circuit theory, 53–7
local response, of axon, 44
Loligo, giant axon of, 25, 56
Lota vulgaris, labyrinth of, 400–1
LSD (Lysergic acid diethylamide), 196

M-line, 219

magnesium ions, neuromuscular block by, 135, 136, 141, 142, 143, 146
Maia (spider crab), muscle fibres of, 254
Malapterurus electricus, 337; electroplaques of, 344–5, 346
Mauthner fibres, of fish, 176, 194
mechanoreceptors, 349, 350, 365, 374
membrane system, in muscle, 256–60
membrane theory, of nervous conduction, 71–105; fixed charges and calcium ions, 92–5; gating currents, 101–3; sodium and potassium conductance, 72–92, 95–9
meromyosins, 216–17, 249
metanephrine, 193
microglia, 205
microphonic potential, 398
microspectrophotometry, 448–9
miotene, 180, 181
modalities, sensory, 349
molluscs: catch muscle of, 323, 333–5; muscles of, 319, 321; synapses of, 168–70, 172; transmitters in, 187, 196
monosynaptic spinal reflex, 141, 149, 151, 156
moths, chemoreception in, 414–15
motoneurons, mammalian spinal, 149–68; inhibition in, 159–68; synaptic excitation in, 149–59
mud-puppy *see Necturus*
Müller fibres, 425
multiple channel sensory systems, 365–71
multiterminal innervation, 301, 302
murexide, 256
murexine, 179
muscarine, 182
muscle: Donnan equilibrium system, 20–3; comparative physiology of, 301–35; heat production of, 267–72; mechanical properties of, 324–35; membrane potential and, 28–9; *see also* contraction, of muscle
muscle fibres, 19; acetylcholine sensitivity and, 131–2; extrafusal, 373; intrafusal, 373; myofibril structure, 219–35
muscle receptor organs *see* stretch receptors
muscle spindles, 352–4, 359, 362, 374, 402–10
myasthenia gravis, 178
myelin, 14, 59
myelinated nerves, saltatory conduction in, 58–64; gating currents of, 105; membrane potentials of, 92
myofibril, structure of, 219–35; striation pattern of, 223–8; ultra-structure of, 228–35

myofilaments, 320
myogenic rhythm: of heart muscle, 306; of insect fibrillar muscle, 324; of vertebrate smooth muscle, 315–16
myosin, 263, 322; filament structure of, 228–31, 234, 249, 290; localisation of, 223–8; properties of, 216–18
Mytilis, catch muscle of, 333–5
Myxine, semi-circular canal of, 388

Narcine, electroplaques of, 341–2
Necturus (mud-puppy), glial cells of, 206–8, 469
neostigmine (prostigmine), 180, 181, 184
Nernst equation, 17, 20; for potassium and sodium ions, 72
nervous conduction, ionic theory of, 71–105
neurin, 147
neurogenic rhythm, of heart muscle, 306–7
neuromasts, 376, 378, 379
neuromuscular junction, vertebrate, 106–48; structure of, 108–10, 316
neuromuscular transmission, 106–48; action of drugs on, 178–81
neurons, 2; electrical interconnections between, 172–3; functional divisions of, 198–202; inout–output relations of, 202–4
nexuses (gap junctions), 171, 172
nicotine, 179, 181, 183, 184
nociceptive stimuli, 349
nodes of Ranvier, 58, 59, 60
noise, acetylcholine, 128–32
noradrenaline, 178, 188–91, 374
nuclear bag fibres, 405
nuclear chain fibres, 405
Nyquist plot, 327–31, 333
nystagmus, 380–1

octopamine, 193
Ohm's law, 45, 61, 83
olfaction, in insects, 414–15
oligodendrocytes, 205, 206
orthodromic impulses, 143n, 157, 158
Oryctes (rhinoceros beetle), flight muscle of, 325–7
otolith organs, 392–4
ouabain, 25–6, 35, 36, 100, 465–6
overshoot, of action potential, 41

pacemaker cells, 204, 313
pacemaker potential, 204, 308, 310
Pacinian corpuscle, 349–50, 359, 363
Pacini's law, 340
pain, 349, 402
Panulirus (rock lobster), muscles of, 305
papain, 217

paramyosin, 321, 323, 335
perfusion experiments, on giant axons, 95–9, 103
peripheral nervous system, 3–4
phenoxybenzamine, 193
phentolamine, 193
Phormia, chemoreceptors of, 411–11, 413
photocurrent, of rods, 461–7
photopic vision, 428 *see also* cones
photoreceptor potential, 459–61
photoreceptors, 349, 350, 363, 424–5, 469–73
physostigmine *see* eserine
picrotoxin, 194, 195
pilocarpine, 182
Podolsky and Nolan's model, for muscle contraction, 287
polyneuronal innervation, 304
postsynaptic potential, 143, 144; of anthropod muscle fibres, 305; of electroplaques, 341; of vertebrate smooth muscle, 316–17; *see also* end-plate potential, excitatory postsynaptic potential, inhibitory postsynaptic potential
post-tetanic potentiation, 141–2
potassium conductance: of axon membrane, 71–2, 78, 81, 82, 85, 87; direct measurement of, 90–2; and drug effects on, 99–100; equations of, 86–8, 89–90; during end-plate potential, 116–22; during motoneuron inhibitory postsynaptic potential, 161–3; of heart muscle, 308, 310–11
potassium contracture, 319
potassium ions: 19, 22, 30; and contracture, 251–2, 253, 254; fluxes of, 74–5, 77, 97; and glial cells, 207–9; and resting membrane potential, 28–31
power density curve *see* spectral density curve
prawns, axons of, 59
presynaptic events, producing transmitter release, 142–7
presynaptic inhibition, 165–8, 305–6
presynaptic membrane potential, 133–5
primary endings, 405, 406, 407–10
procaine, 100
pronase, 100
propanolol, 192, 193
proprioceptors, 349
prostaglandin E, 191
prostigmine, 132
pseudo H-zone, 228
Purkinje cells of cerebellum, 201
Purkinje fibres of heart, 307–8, 309, 311
Purkinje shift, 428
pyramidal cell, 201

quanta, of light, 357; absorption of retina, 450–6
quantal release hypothesis of transmitter substance, 135–9
quick release, 265, 278

Raia, 337; electroplaques of, 341–2; otolith organs of, 393; semi-circular canals, 389
range fractionation, 368
receptor cell, 351–2
receptor potential, 356–7; in insect chemoreceptors, 412–14; in vertebrate eye, 457–9
receptors, adrenergic, 189–90, 193
receptors, sensory, 348–75; central control of, 373–5; coding of sensory information, 352–4, 362–72; and multiple channel systems, 365–71; and initiation of nerve impulses, 354–62
rectification, 34; synaptic, 175
reflex, 4
refractory period, 69
relaxing factor, 260
Renshaw cells, 183–5, 201, 204
reserpine, 193
resistance: of axon membrane, 45, 47 & n, 50, 53; of myelin sheath, 59
responses, electrically and non-electrically excited, 197–8
resting potential, 28–36, 98; and chloride ions, 32–4; of electroplaques, 336–7; ionic basis of, 95; of glial cells, 206–9; of motoneurons, 151; potassium electrode hypothesis, 28–31
reticular theory, 2–3
retina: electrical activity of, 456–73; structure of, 423–7
retinaldehyde, 439–9, 440
retinal densitometry, 442–8, 455
reversal potential, of end-plate current, 120–1
rheobase, 65
rhinocerus beetle *see Oryctes*
rhodopsin (visual purple), 435–41, 472, 473–4; chemistry of, 438–41; organisation of, in rods, 449–50
rods, 427, 470, 472; photocurrent in, 461–7; quanta for excitation of, 450–6; role in vision, 428–30; structure of, 424; visual pigment of, *see* rhodopsin

S-potentials, 470
Sachs, organ of, 339–40
saltatory conduction, in myelinated nerves, 58–64

sarcomere, 219, 236, 319; length of, and isomeric tension, 238–44
sarcoplasm, 254
sarcoplasmic reticulum, 258, 260–1, 318
sarcosomes, 219
sartorius muscle, of frog, 113, 117, 128, 132, 211; active state of, 295–7; and excitation-contraction coupling process, 251, 257; heat production of, 271, 297–300; myofibril structure of, 219, 224; responses of, 198; and synaptic delay, 143, 147
saxitoxin, and action on nervous conduction, 99, 103–4
Schwann cells, 14, 58–9; at neuromuscular junction, 108, 109, 110
scotopic vision, 428, 435; *see also* rods
SD spike, 156–9
sea hare *see Aplysia*
secondary endings, 405, 406, 407–10
semicircular canals, 379–92
sense organs *see* receptors, sensory
sensilla, insect chemoreceptive, 413–14, 415
sensory information, coding of, 352–4, 362–72
Sepia (cuttlefish), giant axons of, 23, 24, 25, 30; and conduction velocity, 56; and effect of dinitrophenol, 100–1; ionic movements during activity of, 74–7
series elastic component, 277–8, 279, 289–91
serotonin (5-hydroxydopamine), 191, 196
shortening velocity, transient changes in, 282
skeletal muscles, 301–3; of arthropods, 303–6, 318; contraction of, 317–18; *see also* striated muscle
sliding filament theory, 235–49; and smooth muscle, 323–4
smooth muscle: invertebrate, 323; and sliding filament theory, 323–4; vertebrate, 315-17, 322
snakes, facial pit receptors of, 365
sodium channel densities, and gating currents, 101–5
sodium conductance: of axon membrane, 71–2, 78, 80–1, 85, 87, 98; by direct measurement, 90–2; and drug effects on, 99–101, 104; during end-plate potential, 116–22; equations of, 88, 89–90; of heart muscle, 308, 310–11; in muscle action potential, 115; in receptors, 358–9, 361
sodium dodecyl sulphate, 217
sodium ions: active transport of, 19, 23–6; and end-plate potential, 119, 120–1;

fluxes of, during nervous activity, 74, 76–7, 96; and photo-current of rods, 465–6; and receptor/generator potential, 358–9; and resting potential, 30; *see also* sodium pump, sodium theory
sodium pump, electrogenic nature of, 34–6
sodium theory, of the action potential, 72–4
soma, of nerve cell, 199, 202
space constant, 49, 51–2
spatial summation, 153–4
spectral density curve (power density spectrum), 129, 130
spectrum, visible, 421–2
spider crab *see Maia*
spike *see* action potential
spinal cord, presynaptic inhibition in, 166–8
spontaneous activity, 204, 364–5
squid, giant axon of, 19, 46–7; action potential of, 39–41; calcium ion fluxes in, 94–5; drug effects on nervous conduction, 99–100; gating current in, 102–5; perfusion experiments, 95–9; and sodium theory of action potential, 73–4; and voltage clamp experiments, 78–92
squid, synapses of, 168–9
Sternachus: electric organ of, 345–6; electroreceptors of, 418–19
stenin, 147
Sternopygus, electroreceptors of, 418
stimulus artefact, 38–9
strength-duration relation, 64–6
stretch receptors, 195, 354–8; *see also* muscle spindles
striated muscles, 319–22; *see also* muscle
striation pattern, of muscle, 219–28
strychnine, 193, 194
subthreshold potentials, 42–5
succinylcholine, 179–80, 184
summation pools of retina, 456
summation, temporal and spatial, 203
sympathin, 188
synapses, 202; adrenergic, 192–3; chemically-transmitting, 149, 150; electrically transmitting, 170–6; molluscan, 168–70
synaptic cleft, 110, 145, 149, 171
synaptic delay, 143–5, 146
synaptic transmission, 106–8; and role of calcium ions, 146–7
synaptosomes, 139, 147

T system (transverse tubular system), and sarcoplasmic reticulum, 256–60, 261, 318
taste, 351, 370; *see also* chemoreceptors

TEA *see* tetraethylammonium ions
temporal summation, 154
TEPP (tetraethyl pyrophosphate), 181
tetanus, 213, 271
tetanus toxin, 193–4
tetraethylammonium (TEA) ions, and effect on action potential, 100
tetraethyl pyrophosphate (TEPP), 181
tetrodotoxin, 94–5, 96, 259; and action on nervous conduction, 99, 101, 103, 145–6
thermoreceptors, 349, 350
threshold: increment, 364; sensory, 363–4; threshold membrane potential, 44–5; threshold stimulus intensity, 39, 60; visual, 450–6
tight junction (zonal occludens), 171
time constant, 49, 51, 52
tonic muscle fibres, 301–3
Torpedo (electric ray), 140, 337; electroplaques of, 341–2, 346
touch receptors, 362
transducer mechanism, 357
transmitter-receiver systems, 372–3
transmitter substance, 106, 177; release of, 142–7; *see also* acetylcholine
triads, 258
trichromatic basis, of colour vision, 475
tropomyosin, 218–19, 233, 261–3
troponin, 218–19, 261–3
trypsin, 216
tryptamine, 304
D-tubocurarine, 180, 184
twitch: active state, 294–7; analysis of, 292–7; isometric, 213
twitch muscle fibres, 203, 251, 301–3

tyramine, 193
tyrosine, 188

unmyelinated nerve fibres, 59, 64; sodium channel densities of, 105
urethane, and effect on saltatory conduction, 60

varicosities, 190
vector modulus plot, 328–31
Venus (clam), muscle filaments of, 231
veratridine, 100, 251
vesicles: adrenergic, 190; synaptic, 139–40, 149, 150, 165
vesiculin, 140
visceroreceptors, 349
visco-elastic theory, of muscle, 273
visual pigments, 421, 430–50; *see also* rhodopsin
visual purple *sse* rhodopsin
voltage clamp experiments, 35, 77–92, 161; of end-plate, 115, 117–20; of heart muscle, 308–10; of squid axon, 78–92, 96–7

Weber–Fechner relation, 364
Wheatstone bridge circuit, 45–6

X-ray diffraction, 221–3; of contracting muscle, 244–9; of resting muscle, 228–34

yohimbrine, 193

Z-line, 219, 226, 235
zonula occludens (tight junction), 171